PRACTICAL ANALOG DESIGN TECHNIQUES

1. SINGLE-SUPPLY AMPLIFIERS
2. HIGH SPEED OP AMPS
3. HIGH RESOLUTION SIGNAL CONDITIONING ADCs
4. HIGH SPEED SAMPLING ADCs
5. UNDERSAMPLING APPLICATIONS
6. MULTICHANNEL APPLICATIONS
7. OVERVOLTAGE EFFECTS ON ANALOG ICs
8. DISTORTION MEASUREMENTS
9. HARDWARE DESIGN TECHNIQUES

INDEX

ANALOG DEVICES TECHNICAL REFERENCE BOOKS

PUBLISHED BY PRENTICE HALL

    Analog-Digital Conversion Handbook
    Digital Signal Processing Applications Using the ADSP-2100 Family
       (Volume 1:1992, Volume 2:1994)
    Digital Signal Processing in VLSI
    DSP Laboratory Experiments Using the ADSP-2101
    ADSP-2100 Family User's Manual

PUBLISHED BY ANALOG DEVICES

    Practical Analog Design Techniques
    Linear Design Seminar
    ADSP-21000 Family Applications Handbook
    System Applications Guide
    Applications Reference Manual
    Amplifier Applications Guide
    Mixed Signal Design Seminar Notes
    High-Speed Design Seminar Notes
    Nonlinear Circuits Handbook
    Transducer Interfacing Handbook
    Synchro & Resolver Conversion
    THE BEST OF *Analog Dialogue*, 1967-1991

PRACTICAL ANALOG DESIGN TECHNIQUES

ACKNOWLEDGMENTS

Thanks are due the many technical staff members of Analog Devices in Engineering and Marketing who provided invaluable inputs during this project. Particular credit is due the individual authors whose names appear at the beginning of their material.

In addition to authoring much of the material, Walter G. (Walt) Jung acted as primary technical reviewer. His comments and revisions have added much to the accuracy and clarity of the text and diagrams.

Linda Grimes Brandon of Brandon's WordService prepared the new illustrations and typeset the text. Ernie Lehtonen of the Analog Devices' art department supplied many camera-ready drawings. Judith Douville compiled the index, and printing was done by R. R. Donnelley and Sons, Inc.

Walt Kester
1995

Copyright © 1995 by Analog Devices, Inc.
Printed in the United States of America

All rights reserved. This book, or parts thereof, must not be reproduced in any form without permission of the copyright owner.

Information furnished by Analog Devices, Inc., is believed to be accurate and reliable. However, no responsibility is assumed by Analog Devices, Inc., for its use.

Analog Devices, Inc., makes no representation that the interconnections of its circuits as described herein will not infringe on existing or future patent rights, nor do the descriptions contained herein imply the granting of licenses to make, use, or sell equipment constructed in accordance therewith.

Specifications are subject to change without notice.

ISBN 0-916550-16-8

PRACTICAL ANALOG DESIGN TECHNIQUES

SECTION 1
SINGLE-SUPPLY AMPLIFIERS

- Rail-to-Rail Input Stages
- Rail-to-Rail Output Stages
- Single-Supply Instrumentation Amplifiers

SECTION 2
HIGH SPEED OP AMPS

- Driving Capacitive Loads
- Cable Driving
- Single-Supply Considerations
- Applications Circuits

SECTION 3
HIGH RESOLUTION SIGNAL CONDITIONING ADCs

- Sigma-Delta ADCs
- High Resolution, Low Frequency Measurement ADCs

SECTION 4
HIGH SPEED SAMPLING ADCs

- ADC Dynamic Considerations
- Selecting the Drive Amplifier Based on ADC Dynamic Performance
- Driving Flash Converters
- Driving the AD9050 Single-Supply ADC
- Driving ADCs with Switched Capacitor Inputs
- Gain Setting and Level Shifting
- External Reference Voltage Generation
- ADC Input Protection and Clamping
- Applications for Clamping Amplifiers
- Noise Considerations in High Speed Sampling

SECTION 5
UNDERSAMPLING APPLICATIONS

- Fundamentals of Undersampling
- Increasing ADC SFDR and ENOB using External SHAs
- Use of Dither Signals to Increase ADC Dynamic Range
- Effect of ADC Linearity and Resolution on SFDR and Noise in Digital Spectral Analysis Applications
- Future Trends in Undersampling ADCs

SECTION 6
MULTICHANNEL APPLICATIONS

- Data Acquisition System Considerations
- Multiplexing
- Filtering Considerations for Data Acquisition Systems
- SHA and ADC Settling Time Requirements in Multiplexed Applications
- Complete Data Acquisition Systems on a Chip
- Multiplexing into Sigma-Delta ADCs
- Simultaneous Sampling Systems
- Data Distribution Systems using Multiple DACs

SECTION 7
OVERVOLTAGE EFFECTS ON ANALOG ICs

- Amplifier Input Stage Overvoltage
- Amplifier Output Voltage Phase Reversal
- Understanding and Protecting Integrated Circuits from Electrostatic Discharge (ESD)

SECTION 8
DISTORTION MEASUREMENTS

- High Speed Op Amp Distortion
- High Frequency Two-Tone Generation
- Using Spectrum Analyzers in High Frequency Low Distortion Measurements
- Measuring ADC Distortion using FFTs
- FFT Testing
- Troubleshooting the FFT Output
- Analyzing the FFT Output

SECTION 9
HARDWARE DESIGN TECHNIQUES

- Prototyping Analog Circuits
- Evaluation Boards
- Noise Reduction and Filtering for Switching Power Supplies
- Low Dropout References and Regulators
- EMI/RFI Considerations
- Sensors and Cable Shielding

INDEX

PRACTICAL ANALOG DESIGN TECHNIQUES

SINGLE-SUPPLY AMPLIFIERS — 1

HIGH SPEED OP AMPS — 2

HIGH RESOLUTION SIGNAL CONDITIONING ADCs — 3

HIGH SPEED SAMPLING ADCs — 4

UNDERSAMPLING APPLICATIONS — 5

MULTICHANNEL APPLICATIONS — 6

OVERVOLTAGE EFFECTS ON ANALOG ICs — 7

DISTORTION MEASUREMENTS — 8

HARDWARE DESIGN TECHNIQUES — 9

INDEX — I

SECTION 1

SINGLE-SUPPLY AMPLIFIERS

- Rail-to-Rail Input Stages
- Rail-to-Rail Output Stages
- Single-Supply Instrumentation Amplifiers

Single-Supply Amplifiers

SECTION 1
SINGLE-SUPPLY AMPLIFIERS
Adolfo Garcia

Over the last several years, single-supply operation has become an increasingly important requirement as systems get smaller, cheaper, and more portable. Portable systems rely on batteries, and total circuit power consumption is an important and often dominant design issue, and in some instances, more important than cost. This makes low-voltage/low supply current operation critical; at the same time, however, accuracy and precision requirements have forced IC manufacturers to meet the challenge of "doing more with less" in their amplifier designs.

SINGLE-SUPPLY AMPLIFIERS

- Single Supply Offers:
    - Lower Power
    - Battery Operated Portable Equipment
    - Simplifies Power Supply Requirements

- But Watch Out for:
    - Signal-swings limited, therefore more sensitive to errors caused by offset voltage, bias current, finite open-loop gain, noise, etc.
    - More likely to have noisy power supply because of sharing with digital circuits
    - DC coupled, multi-stage single-supply circuits can get very tricky!
    - Rail-to-rail op amps needed to maximize signal swings

Figure 1.1

In a single-supply application, the most immediate effect on the performance of an amplifier is the reduced input and output signal range. As a result of these lower input and output signal excursions, amplifier circuits become more sensitive to internal and external error sources. Precision amplifier offset voltages on the order of 0.1mV are less than a 0.04 LSB error source in a 12-bit, 10V full-scale system. In a single-supply system, however, a "rail-to-rail" precision amplifier with an offset voltage of 1mV represents a 0.8LSB error in a 5V FS system, and 1.6LSB error in a 2.5V FS system.

Furthermore, amplifier bias currents, now flowing in larger source resistances to keep current drain from the battery low, can generate offset errors equal to or greater than the amplifier's own offset voltage.

Gain accuracy in some low voltage single-supply devices is also reduced, so device selection needs careful consideration. Many amplifiers having open-loop gains in the millions typically operate on dual supplies: for example, the OP07 family types. However, many single-supply/rail-to-rail amplifiers for precision applications typically have open-loop gains between 25,000 and 30,000 under light loading (>10kΩ). Selected devices, like the OPX13 family, do have high open-loop gains (i.e., >1V/μV).

Many trade-offs are possible in the design of a single-supply amplifier: speed versus power, noise versus power, precision versus speed and power, etc. Even if the noise floor remains constant (highly unlikely), the signal-to-noise ratio will drop as the signal amplitude decreases.

Besides these limitations, many other design considerations that are otherwise minor issues in dual-supply amplifiers become important. For example, signal-to-noise (SNR) performance degrades as a result of reduced signal swing. "Ground reference" is no longer a simple choice, as one reference voltage may work for some devices, but not others. System noise increases as operating supply current drops, and bandwidth decreases. Achieving adequate bandwidth and required precision with a somewhat limited selection of amplifiers presents significant system design challenges in single-supply, low-power applications.

Most circuit designers take "ground" reference for granted. Many analog circuits scale their input and output ranges about a ground reference. In dual-supply applications, a reference that splits the supplies (0V) is very convenient, as there is equal supply headroom in each direction, and 0V is generally the voltage on the low impedance ground plane.

In single-supply/rail-to-rail circuits, however, the ground reference can be chosen anywhere within the supply range of the circuit, since there is no standard to follow. The choice of ground reference depends on the type of signals processed and the amplifier characteristics. For example, choosing the negative rail as the ground reference may optimize the dynamic range of an op amp whose output is designed to swing to 0V. On the other hand, the signal may require level shifting in order to be compatible with the input of other devices (such as ADCs) that are not designed to operate at 0V input.

Early single-supply "zero-in, zero-out" amplifiers were designed on bipolar processes which optimized the performance of the NPN transistors. The PNP transistors were either lateral or substrate PNPs with much poorer performance than the NPNs. Fully complementary processes are now required for the new-breed of single-supply/rail-to-rail operational amplifiers. These new amplifier designs do not use lateral or substrate PNP transistors within the signal path, but incorporate parallel NPN and PNP input stages to accommodate input signal swings from ground to the positive supply rail. Furthermore, rail-to-rail output stages are designed with bipolar NPN and PNP common-emitter, or N-channel/P-channel common-source amplifiers

SINGLE-SUPPLY AMPLIFIERS

"RAIL-TO-RAIL" AMPLIFIERS

- What exactly is "rail-to-rail"

- Does the input common mode range (for guaranteed CMRR) include: 0V, +Vs, both, or neither?

- Output Voltage Swing (how close to the rails can you get under load?)

- Where is "ground"?

- Complementary bipolar processes make rail-to-rail inputs and outputs feasible (within some fundamental physical limitations)

- Implications for precision single-supply instrumentation amps

Figure 1.2

whose collector-emitter saturation voltage or drain-source channel on-resistance determine output signal swing with the load current.

The characteristics of a single-supply amplifier input stage (common-mode rejection, input offset voltage and its temperature coefficient, and noise) are critical in precision, low-voltage applications. Rail-to-rail input operational amplifiers must resolve small signals, whether their inputs are at ground, or at the amplifier's positive supply. Amplifiers having a minimum of 60dB common-mode rejection over the entire input common-mode voltage range from 0V to the positive supply (V_{POS}) are good candidates. It is not necessary that amplifiers maintain common-mode rejection for signals beyond the supply voltages: what is required is that they do not self-destruct for momentary overvoltage conditions. Furthermore, amplifiers that have offset voltages less than 1mV and offset voltage drifts less than 2μV/°C are also very good candidates for precision applications. Since *input* signal dynamic range and SNR are equally if not more important than *output* dynamic range and SNR, precision single-supply/rail-to-rail operational amplifiers should have noise levels referred-to-input (RTI) less than 5μVp-p in the 0.1Hz to 10Hz band.

Since the need for rail-to-rail amplifier output stages is driven by the need to maintain wide dynamic range in low-supply voltage applications, a single-supply/rail-to-rail amplifier should have output voltage swings which are within at least 100mV of either supply rail (under a nominal load). The output voltage swing is very dependent on output stage topology and load current, but the voltage swing of a good output stage should maintain its rated swing for loads down to 10kΩ. The smaller the V_{OL} and the larger the V_{OH}, the better. System parameters, such as "zero-scale" or "full-scale" output voltage, should be determined by an amplifier's V_{OL} (for zero-scale) and V_{OH} (for full-scale).

Since the majority of single-supply data acquisition systems require at least 12- to 14-bit performance, amplifiers which exhibit an open-loop gain greater than 30,000 for all loading conditions are good choices in precision applications.

SINGLE-SUPPLY/RAIL-TO-RAIL OP AMP INPUT STAGES

With the increasing emphasis on low-voltage, low-power, and single-supply operation, there is some demand for op amps whose input common-mode range includes *both* supply rails. Such a feature is undoubtedly useful in some applications, but engineers should recognize that there are relatively few applications where it is absolutely essential. These should be carefully distinguished from the many applications where common-mode range *close* to the supplies or one that includes one of the supplies is necessary, but input rail-rail operation is not.

In many single-supply applications, it is required that the input go to only one of the supply rails (usually ground). Amplifiers which will handle zero-volt inputs are relatively easily designed using either PNP transistors (see OP90 and the OPX93 in Figure 1.3) or N-channel JFETs (see AD820 family in Figure 1.4). P-channel JFETs can be used where inputs must include the positive supply rail (but not the negative rail) as shown in Figure 1.4 for the OP282/OP482.

In the FET-input stages of Figure 1.4, the possibility exists for phase reversal as input signals approach and exceed the amplifier's linear input common-mode voltage ranges. As described in Section 7, internal amplifier stages saturate, forcing subsequent stages into cutoff. Depending on the structure of the input stage, phase reversal forces the output voltage to one of the supply rails. For n-channel JFET-input stages, the output voltage goes to the negative output rail during phase reversal. For p-channel JFET-input stages, the output is forced to the positive output rail. New FET-input amplifiers, like the AD820 family of amplifiers, incorporate design improvements that prevent output voltage phase reversal for signals within the rated supply voltage range. Their input stage and second gain stage even offer protection against output voltage phase reversal for input signals 200mV *more* positive than the positive supply voltage.

OP90 AND OPX93 INPUT STAGE ALLOWS INPUT TO GO TO THE NEGATIVE RAIL

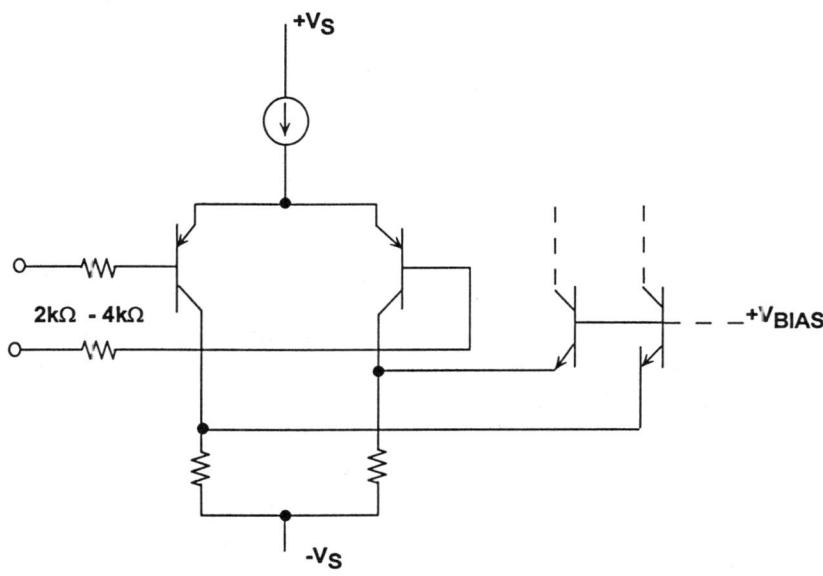

Figure 1.3

AD820/AD822/AD824 INPUT INCLUDES NEGATIVE RAIL, OP-282/OP-482 INCLUDES POSITIVE RAIL

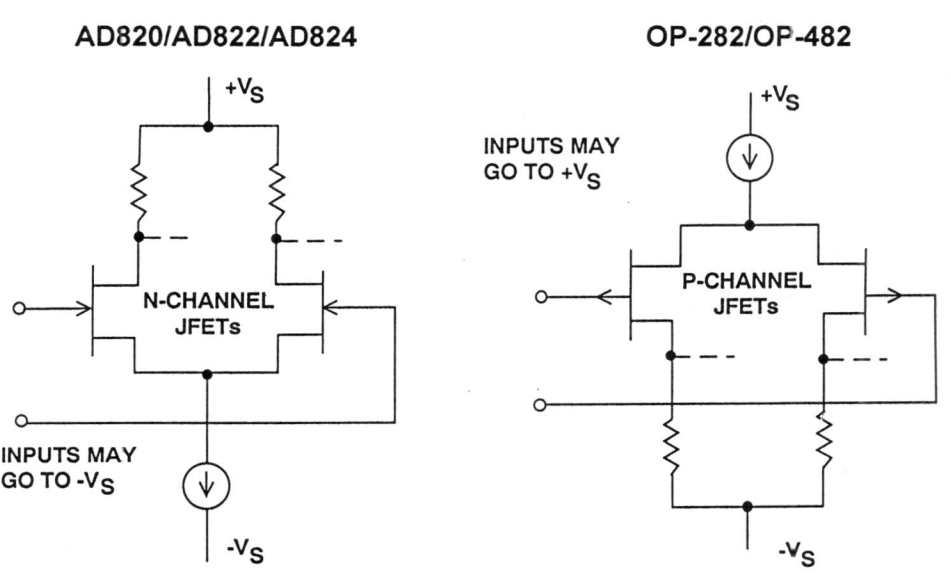

Figure 1.4

SINGLE-SUPPLY AMPLIFIERS

As shown in Figure 1.5, true rail-to-rail input stages require two long-tailed pairs, one of NPN bipolar transistors (or N-channel FETs), the other of PNP transistors (or p-channel FETs). These two pairs exhibit *different* offsets and bias currents, so when the applied input common-mode voltage changes, the amplifier input offset voltage and input bias current does also. In fact, when both current sources (I1 and I2) remain active throughout the entire input common-mode range, amplifier input offset voltage is the *average* offset voltage of the NPN pair and the PNP pair. In those designs where the current sources are alternatively switched off at some point along the input common-mode voltage, amplifier input offset voltage is dominated by the PNP pair offset voltage for signals near the negative supply, and by the NPN pair offset voltage for signals near the positive supply.

Amplifier input bias current, a function of transistor current gain, is also a function of the applied input common-mode voltage. The result is relatively poor common-mode rejection (CMR), and a changing common-mode input impedance over the common-mode input voltage range, compared to familiar dual supply precision devices like the OP07 or OP97. These specifications should be considered carefully when choosing a rail-rail input op amp, especially for a non-inverting configuration. Input offset voltage, input bias current, and even CMR may be quite good over *part* of the common-mode range, but much worse in the region where operation shifts between the NPN and PNP devices.

RAIL-TO-RAIL INPUT STAGE TOPOLOGY

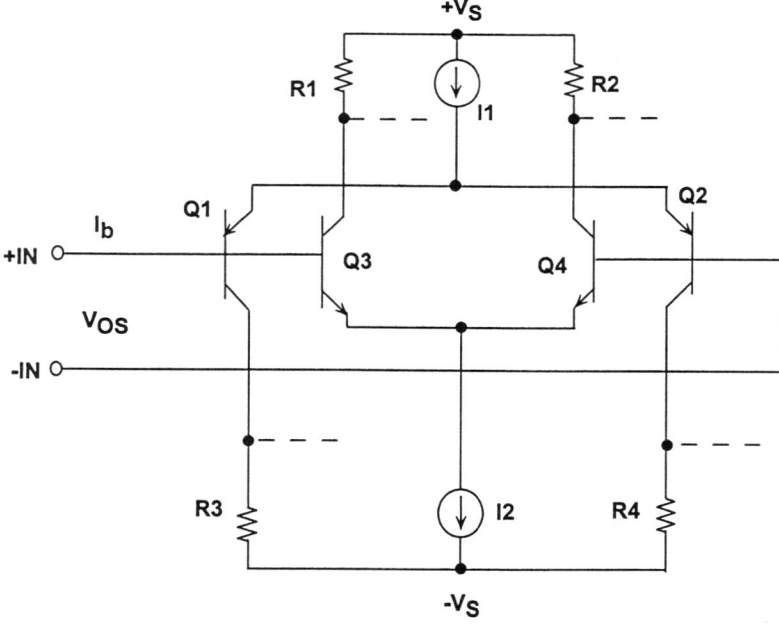

Figure 1.5

SINGLE-SUPPLY AMPLIFIERS

Many rail-to-rail amplifier input stage designs switch operation from one differential pair to the other differential pair somewhere along the input common-mode voltage range. Devices like the OPX91 family and the OP279 have a common-mode crossover threshold at approximately 1V below the positive supply. In these devices, the PNP differential input stage remains active; as a result, amplifier input offset voltage, input bias current, CMR, input noise voltage/current are all determined by the characteristics of the PNP differential pair. At the crossover threshold, however, amplifier input offset voltage becomes the average offset voltage of the NPN/PNP pairs and can change rapidly. Also, amplifier bias currents, dominated by the PNP differential pair over most of the input common-mode range, change polarity and magnitude at the crossover threshold when the NPN differential pair becomes active. As a result, source impedance levels should be balanced when using such devices, as mentioned before, to minimize input bias current offsets and distortion.

An advantage to this type of rail-to-rail input stage design is that input stage transconductance can be made constant throughout the entire input common-mode voltage range, and the amplifier slews symmetrically for all applied signals.

Operational amplifiers, like the OP284/OP484, utilize a rail-to-rail input stage design where both PNP and NPN transistor pairs are active throughout the entire input common-mode voltage range, and there is no common-mode crossover threshold. Amplifier input offset voltage is the average offset voltage of the NPN and the PNP stages. Amplifier input offset voltage exhibits a smooth transition throughout the entire input common-mode voltage range because of careful laser-trimming of resistors in the input stage. In the same manner, through careful input stage current balancing and input transistor design, amplifier input bias currents also exhibit a smooth transition throughout the entire common-mode input voltage range. The exception occurs at the extremes of the input common-mode range, where amplifier offset voltages and bias currents increase sharply due to the slight forward-biasing of parasitic p-n junctions. This occurs for input voltages within approximately 1V of either supply rail.

When *both* differential pairs are active throughout the entire input common-mode range, amplifier transient response is faster through the middle of the common-mode range by as much as a factor of 2 for bipolar input stages and by a factor of √2 for FET input stages. Input stage transconductance determines the slew rate and the unity-gain crossover frequency of the amplifier, hence response time degrades slightly at the extremes of the input common-mode range when either the PNP stage (signals approaching V_{POS}) or the NPN stage (signals approaching GND) are forced into cutoff. The thresholds at which the transconductance changes occur approximately within 1V of either supply rail, and the behavior is similar to that of the input bias currents.

Applications which initially appear to require true rail-rail inputs should be carefully evaluated, and the amplifier chosen to ensure that its input offset voltage, input bias current, common-mode rejection, and noise (voltage and current) are suitable. A true rail-to-rail input amplifier should not generally be used if an input range which includes only one rail is satisfactory.

SINGLE-SUPPLY/RAIL-TO-RAIL OP AMP OUTPUT STAGES

The earliest IC op amp output stages were NPN emitter followers with NPN current sources or resistive pull-downs, as shown in Figure 1.6. Naturally, the slew rates were greater for positive-going than for negative-going signals. While all modern op amps have push-pull output stages of some sort, many are still asymmetrical, and have a greater slew rate in one direction than the other. This asymmetry, which generally results from the use of IC processes with better NPN than PNP transistors, may also result in the ability of the output to approach one supply more closely than the other.

In many applications, the output is required to swing only to one rail, usually the negative rail (i.e., ground in single-supply systems). A pulldown resistor to the negative rail will allow the output to approach that rail (provided the load impedance is high enough, or is also grounded to that rail), but only slowly. Using an FET current source instead of a resistor can speed things up, but this adds complexity.

OP AMP OUTPUT STAGES USING COMPLEMENTARY DEVICES ALLOW PUSH-PULL DRIVE

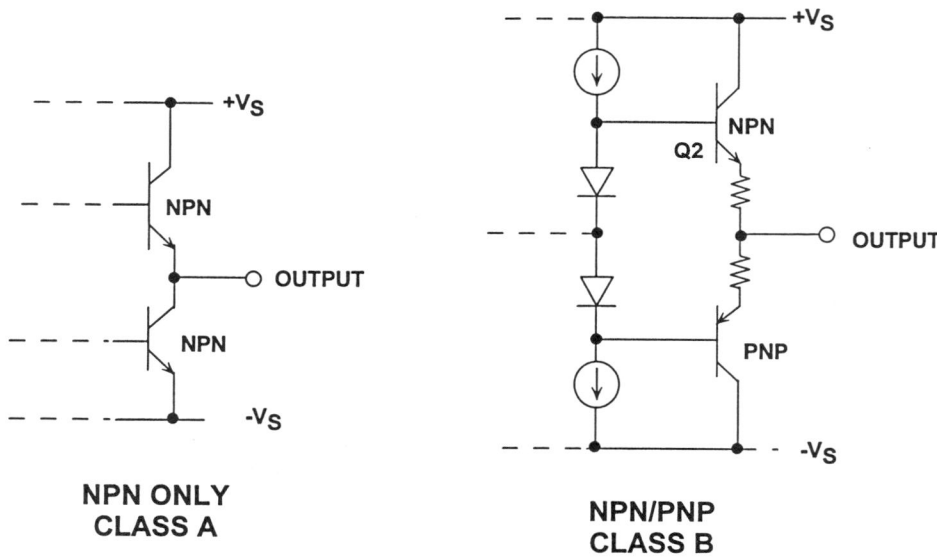

Figure 1.6

An IC process with relatively well-matched (AC and DC) PNP and NPN transistors allows both the output voltage swing and slew rate to be reasonably well matched. However, an output stage using BJTs cannot swing completely to the rails, but only to within the transistor saturation voltage (V_{CESAT}) of the rails (see Figure 1.7). For small amounts of load current (less

SINGLE-SUPPLY AMPLIFIERS

than 100µA), the saturation voltage may be as low as 5 to 10mV, but for higher load currents, the saturation voltage can increase to several hundred mV (for example, 500mV at 50mA).

On the other hand, an output stage constructed of CMOS FETs can provide true rail-to-rail performance, but only under no-load conditions. If the output must source or sink current, the output swing is reduced by the voltage dropped across the FETs internal "on" resistance (typically, 100Ω).

RAIL-TO-RAIL OUTPUT STAGE SWING IS LIMITED BY V_{cesat}, R_{on}, AND LOAD CURRENT

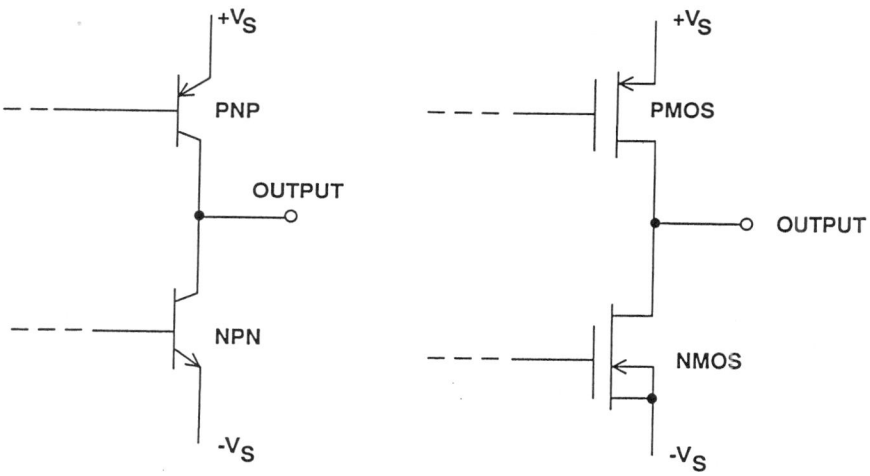

SWINGS TO RAILS LIMITED BY SATURATION VOLTAGE

SWINGS TO RAILS LIMITED BY FET "ON" RESISTANCE (~100Ω)

Figure 1.7

In summary, the following points should be considered when selecting amplifiers for single-supply/rail-to-rail applications:

First, input offset voltage and input bias currents can be a function of the applied input common-mode voltage (for true rail-to-rail input op amps). Circuits using this class of amplifiers should be designed to minimize resulting errors.

An inverting amplifier configuration with a false ground reference at the non-inverting input prevents these errors by holding the input common-mode voltage constant. If the inverting amplifier configuration cannot be used, then amplifiers like the OP284/OP484 which do not exhibit any common-mode crossover thresholds should be used.

SINGLE-SUPPLY AMPLIFIERS

Second, since input bias currents are not always small and can exhibit different polarities, source impedance levels should be carefully matched to minimize additional input bias current-induced offset voltages and increased distortion. Again, consider using amplifiers that exhibit a smooth input bias current transition throughout the applied input common-mode voltage.

Third, rail-to-rail amplifier output stages exhibit load-dependent gain which affects amplifier open-loop gain, and hence closed-loop gain accuracy. Amplifiers with open-loop gains greater than 30,000 for resistive loads less than 10kΩ are good choices in precision applications. For applications not requiring full rail-rail swings, device families like the OPX13 and OPX93 offer DC gains of 0.2V/μV or more.

Lastly, no matter what claims are made, rail-to-rail output voltage swings are functions of the amplifier's output stage devices and load current. The saturation voltage (V_{CESAT}), saturation resistance (R_{SAT}), and load current all affect the amplifier output voltage swing.

These considerations, as well as those regarding rail-to-rail precision, have implications in many circuits, namely instrumentation amplifiers, which will be covered in the next sections.

THE TWO OP AMP INSTRUMENTATION AMPLIFIER TOPOLOGY

There are several circuit topologies for instrumentation amplifier circuits suitable for single-supply applications. The *two op amp* configuration is often used in cost- and space-sensitive applications, where tight matching of input offset voltage, input bias currents, and open-loop gain is important. Also, when compared to other topologies, the two op amp instrumentation amplifier circuit offers the lowest power consumption and low total drift for moderate-gain (G=10) applications. Obviously, it also has the merit of using a single dual op amp IC.

Figure 1.8 shows the topology of a two op amp instrumentation circuit which uses a 5th gain-setting resistor, R_G. This additional gain-setting resistor is optional, and should be used in those applications where a fine gain trim is required. Its effect will be included in this analysis.

Circuit resistor values for this topology can be determined from Equations 1.1 through 1.3, where R1 = R4. To maintain low power consumption in single-supply applications, values for R should be no less than 10kΩ:

$$R1 = R4 = R \qquad \text{Eq. 1.1}$$

$$R2 = R3 = \frac{R}{0.9G - 1} \qquad \text{Eq. 1.2}$$

$$R_G = \frac{2R}{0.06G} \qquad \text{Eq. 1.3}$$

where G equals the desired circuit gain. Note that in those applications where fine gain trimming is not required, Eq. 1.2 reduces to:

$$R2 = R3 = \frac{R}{G - 1} \qquad \text{Eq. 1.4}$$

SINGLE-SUPPLY AMPLIFIERS

A nodal analysis of the topology will illustrate the behavior of the circuit's nodal voltages and the amplifier output currents as functions of the applied common-mode input voltage (V_{CM}), the applied differential (signal) voltage (V_{IN}), and the output reference voltage (V_{REF}). These expressions are summarized in Equations 1.5 through 1.8, Eq. 1.12, and in Eq. 1.13 for positive, input differential voltages. Due to the structure of the topology, expressions for voltages and currents are similar in form and magnitude for negative, input differential voltages.

From the figure, expressions for the four nodal voltages A, B, C, and V_{OUT} as well as the output stage currents of A1 (I_{OA1}) and A2 (I_{OA2}) have been developed. Note that the direction of the amplifier output currents, I_{OA1} and I_{OA2}, is defined to be *into* the amplifier's output stage. For example, if the nodal analysis shows that I_{OA1} and I_{OA2} are positive entities, their direction is *into* the device; thus, their output stages are *sinking* current. If the analysis shows that they are negative quantities, their direction is opposite to that shown; therefore, their output stages are *sourcing* current.

Resistors R_{P1} and R_{P2} at the inputs to the circuit are optional input current limiting resistors used to protect the amplifier input stages against input overvoltage. Although any reasonable value can be used, these resistors should be less than 1kΩ to prevent the unwanted effects of additional resistor noise and bias current-generated offset voltages. For protection against a specific level of overvoltage, the interested reader should consult the section on overvoltage effects on integrated circuits, found in Section 7 of this book.

THE TWO OP AMP INSTRUMENTATION AMPLIFIER TOPOLOGY IN SINGLE-SUPPLY APPLICATIONS

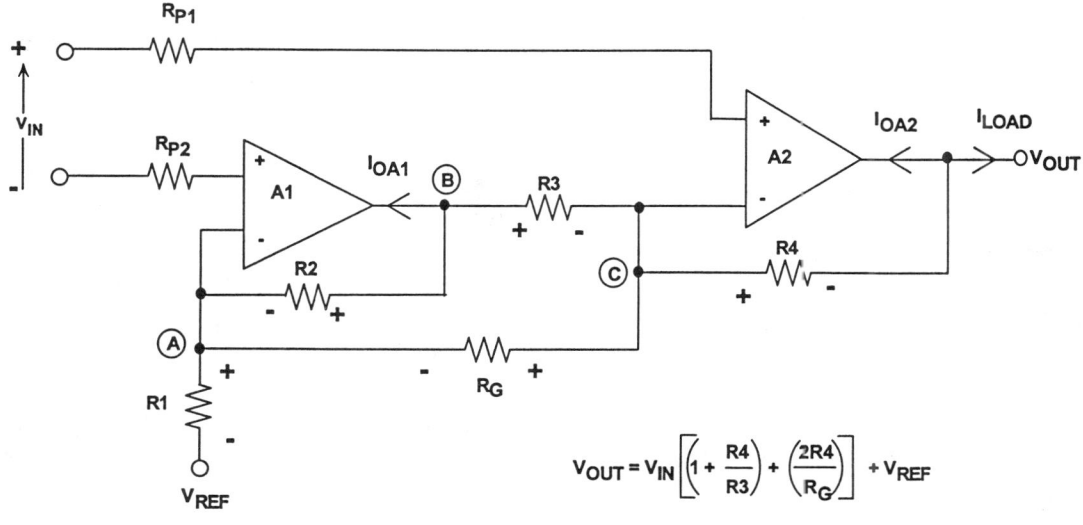

$$V_{OUT} = V_{IN}\left[\left(1 + \frac{R4}{R3}\right) + \left(\frac{2R4}{R_G}\right)\right] + V_{REF}$$

WHERE R1 = R4 & R2 = R3

Figure 1.8

SINGLE-SUPPLY AMPLIFIERS

Using half-circuit concepts and the principle of superposition, the input signal voltage, V_{IN-}, on the non-inverting input of A1 is set to zero. Since the input signal, V_{IN+}, is applied to the non-inverting terminal of A2, an expression for the nodal voltage at the inverting terminal of A1 is given by Eq. 1.5:

$$V_A = V_{CM} \qquad \text{Eq. 1.5}$$

An expression for the output voltage of A1 (node B) shows that it is dependent on all three externally applied voltages (V_{IN}, V_{CM}, and V_{REF}), and is illustrated in Eq. 1.6:

$$V_B = (-V_{IN+})\left(\frac{R2}{R_G}\right) + V_{CM}\left(1 + \frac{R2}{R1}\right) - V_{REF}\left(\frac{R2}{R1}\right) \qquad \text{Eq. 1.6}$$

Since the input signal, V_{IN+}, as well as the applied input common-mode voltage, V_{CM}, is applied to the non-inverting terminal of A2, then the expression for the voltage at A2's inverting input (node C) is given by:

$$V_C = V_{CM} + V_{IN+} \qquad \text{Eq. 1.7}$$

For the case where R1 = R4 and R2 = R3, combining the results in Eq. 1.5, 1.6, and 1.7 yields the familiar expression for the circuit's output voltage:

$$V_{OUT} = (V_{IN+})\left(1 + \frac{R4}{R3} + \frac{2R4}{R_G}\right) + V_{REF} \qquad \text{Eq. 1.8}$$

At this point, it is worth noting the behavior of the circuit's nodal voltages based on the applied external voltages. From Eq. 1.5 and Eq. 1.7, the common-mode component of the current through R_G is equal to zero, whereas the full differential input voltage appears across it. Furthermore, Eq. 1.6 has shows that A1 *amplifies* the applied common-mode input voltage by a factor of (1 + R2/R1). In low-gain applications, the ratio of R2 to R_G can be as small as 1:1 (for circuit gains ≥ 2). Therefore, Equation 1.6 sets the upper bound on the input common-mode voltage in low-gain applications. If the output of A1 is allowed to saturate at high input common-mode voltages, then it will not have enough "headroom" to amplify the input signal, as shown in Eq. 1.6. Therefore, in order for A1 to amplify accurately input signal voltages for any circuit gain > 1 (circuit gains equal to 1 are not permitted in this topology) requires that an upper bound on the total applied input voltage (common-mode plus differential-mode voltages) be determined to prevent amplifier output voltage saturation. This upper bound can be determined by the desired circuit gain, G, and the amplifier's minimum output high voltage:

$$V_{IN(TOTAL)} < V_{OH(MIN)}\left(\frac{0.9G - 1}{0.9G}\right) - V_{IN+} \qquad \text{Eq. 1.9}$$

SINGLE-SUPPLY AMPLIFIERS

In a similar fashion, a lower bound on the total applied input voltage is also determined by circuit gain and the amplifier's maximum output low voltage:

$$V_{IN(TOTAL)} > V_{OL(MAX)}\left(\frac{0.9G-1}{0.9G}\right) + V_{IN+} \qquad \text{Eq. 1.10}$$

For example, if a rail-to-rail operational amplifier exhibited a $V_{OL(MAX)}$ equal to 10mV and a $V_{OH(MIN)}$ equal to 4.95V, and if the application required a circuit gain of 10 to produce a 1V full-scale output, then the total input voltage range would be bounded by:

$$0.109 \text{ V} < V_{IN(TOTAL)} < 4.3 \text{ V}$$

Therefore, the range over which the circuit will handle input voltages without amplifier output voltage saturation is given by:

$$V_{OL(MAX)}\left(\frac{0.9G-1}{0.9G}\right) + V_{IN+} < V_{IN(TOTAL)} < V_{OH(MIN)}\left(\frac{0.9G-1}{0.9G}\right) - V_{IN+}$$
$$\text{Eq. 1.11}$$

In low-gain instrumentation circuits, the usable input voltage range is limited and asymmetric about the supply mid-point voltage. To complete the nodal analysis of the two op amp instrumentation circuit, expressions for operational amplifier output stage currents are shown in Equations 1.12 and 1.13:

$$I_{OA1} = (V_{IN+})\left(\frac{2}{R_G} + \frac{1}{R3}\right) + (V_{REF} - V_{CM})\left(\frac{2}{R1}\right) \qquad \text{Eq. 1.12}$$

$$I_{OA2} = (-V_{IN+})\left(\frac{2}{R_G} + \frac{1}{R3}\right) + (V_{CM} - V_{REF})\left(\frac{1}{R4}\right) \qquad \text{Eq. 1.13}$$

Equation 1.12 illustrates that A1's output stage must be able to sink current as a function of the applied differential input voltage and the output reference voltage. On the other hand, A1's output stage is required to source current over the entire input voltage range. *In the single-supply case where the circuit is required to sense small differential signals near ground, Eq. 1.6 and Eq. 1.12 both illustrate that A1's output stage is required to sink current while trying to maintain a more negative output voltage than its own negative supply. A1 is thus forced into saturation.*

As shown in Eq. 1.13, A2's output stage sources current for positive differential input voltages with no differential or common-mode voltage constraints placed upon its output by Eq. 1.8. *Note, however, that as a function of the applied common-mode voltage, A2 is required to sink current. Unfortunately, in the absence of an input signal, Eq. 1.13 shows that A2's output stage may be forced into saturation, trying to sink current while maintaining its output voltage at V_{OL}.*

Single-Supply Amplifiers

To circumvent the circuit topological and amplifier output voltage limitations, a reference voltage should be used to bias the output of the circuit (A2's output) *in the middle of its output voltage swing*, and not at exactly one-half the supply voltage:

$$V_{REF} = \frac{V_{OH(MIN)} + V_{OL(MAX)}}{2} \qquad \text{Eq. 1.14}$$

The output reference voltage allows the output stages of A1 and A2 to sink or source current without any output voltage constraints. So long as Eq. 1.11 is used to define to total input voltage range, then amplifier behavior for differential- and common-mode operation is linear. To maximize output signal dynamic range and output SNR, the gain of the instrumentation amplifier circuit should be set according to Eq. 1.15:

$$\text{Circuit Gain} = \frac{V_{OH(MIN)} - V_{OL(MAX)}}{2 \cdot V_{IN(MAX)}} \qquad \text{Eq. 1.15}$$

Under these operating conditions, the differential output voltage of the instrumentation amplifier circuit is now measured relative to V_{REF} and not to GND. Thus, negative full-scale input signals produce output voltages near A2's V_{OL}, and positive full-scale signals produce output voltages near A2's V_{OH}. Therefore, the circuit's input common-mode range and output dynamic range are optimized in terms of the desired circuit gain and amplifier output voltage characteristics.

For minimal impact of amplifier output load currents on V_{OH} and V_{OL}, circuit resistor values should be greater than 10kΩ in most single-supply applications. Thus, Equations 1.11, 1.14, and 1.15 can all be used to design accurate and repeatable two op amp instrumentation amplifier circuits with single-supply/rail-to-rail operational amplifiers.

One fundamental limitation of the two operational amplifier instrumentation circuit is that since the two amplifiers are operating at different closed-loop gains (and thus at different bandwidths), there will be generally poor AC common-mode rejection without the use of an AC CMR trim capacitor. For optimal AC CMR performance, a trimming capacitor should be connected between the inverting terminal of A1 to ground.

A Two Op Amp, FET-Input Instrumentation Amplifier

Figure 1.9 illustrates a two op amp instrumentation amplifier using the AD822, a dual JFET-input, rail-to-rail output operational amplifier. The output offset voltage is set by V_{REF}.

Dual operational amplifiers, like the AD822, make these types of instrumentation amplifiers both cost- and power-efficient. In fact, when operating on a single, +3 V supply, total circuit power consumption is less than 3.5mW. The AD822's 2pA bias currents minimize offset errors caused by unbalanced source impedances.

Circuit performance is enhanced dramatically by the use of a matched resistor network. A thin-film resistor array sets the circuit gain to either 10 or 100 through a DPDT (double-pole, double-throw) switch. The array's resistors are laser-trimmed for a ratio match of 0.01%, and exhibit a maximum differential temperature coefficient of 5ppm/°C. Note that in this application circuit, the fifth gain-setting resistor is not used. The use of this gain trim resistor would introduce serious gain and linearity errors due to the resistance of the double-pole, double-throw switches.

A performance summary and transient response of this instrumentation amplifier is shown in Figure 1.10. Note that the small-signal bandwidth of the circuit is independent of supply voltage, and that the rail-to-rail output pulse response is well-behaved. For greater bandwidth at the expense of higher supply current, the functionally similar AD823 can also be used.

A SINGLE-SUPPLY, PROGRAMMABLE, FET-INPUT INSTRUMENTATION AMPLIFIER

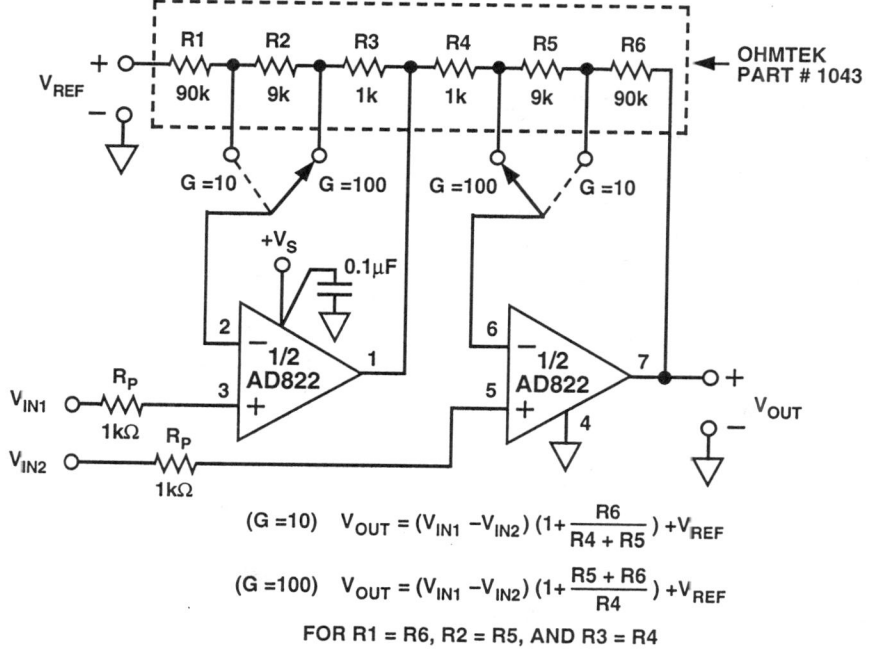

$(G = 10) \quad V_{OUT} = (V_{IN1} - V_{IN2})(1 + \frac{R6}{R4 + R5}) + V_{REF}$

$(G = 100) \quad V_{OUT} = (V_{IN1} - V_{IN2})(1 + \frac{R5 + R6}{R4}) + V_{REF}$

FOR R1 = R6, R2 = R5, AND R3 = R4

Figure 1.9

PERFORMANCE SUMMARY OF AD822 IN-AMP

Parameters	V_S = 3V, 0V	V_S = ± 5V
CMRR	74dB	80dB
Common-Mode Voltage Range,		
G = 10	+0.51 to +1.75V	–4.49 to +3.75V
G = 100	+0.06 to +1.98V	–4.90 to +3.98V
3dB BW, G=10	180kHz	180kHz
3dB BW, G=100	18kHz	18kHz
$t_{settling}$		
2V Step (V_S = 0V, 3V)	2 µs	
5V Step (V_S = ±5V)		5 µs
Noise @ 1kHz, G=10	270 nV/√Hz	270 nV/√Hz
(RTO) G=100	2.2 µV/√Hz	2.2 µV/√Hz
Supply Current	1.10mA	1.15mA

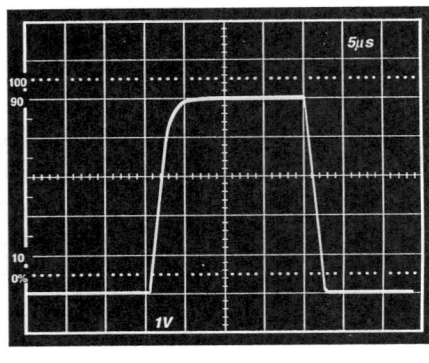

Input = 500mV, G=10, Vs = +5V, 0V
Vertical Scale: 1V/div
Horizontal Scale: 5 µs/div

Figure 1.10

THE THREE OP AMP INSTRUMENTATION AMPLIFIER TOPOLOGY

For the highest precision and performance, the *three op amp* instrumentation amplifier topology is optimum for bridge and other offset transducer applications where high accuracy and low nonlinearity are required. This is at the expense of additional power consumption over the two op amp instrumentation circuit (3 amplifiers versus 2 amplifiers). Furthermore, like the two op amp configuration, the input amplifiers can use one dual op amp for tight matching of input offset voltage matching, input bias current, and open-loop gain. Or, a single quad operational amplifier can be used for the whole circuit, including a reference voltage buffer, if required.

Single-supply/rail-to-rail amplifiers can be used in this topology, like that shown for two op amp designs, if the output characteristics of the single-supply/rail-to-rail amplifiers are understood. As shown in Figure 1.11, a generalized, comprehensive analysis of the structure will illustrate the behavior of the nodal voltages and amplifier output currents as functions of the applied common-mode input voltage (V_{CM}), the applied differential (signal) voltage (V_{IN}), and the output reference voltage (V_{REF}). As shown in Eq. 1.16 through 1.27, the nodal analysis was carried out for positive-input differential voltages; because of the symmetry in the circuit, the expressions for the nodal voltages and amplifier output currents carried out for negative-input differential voltages are identical.

THE UBIQUITOUS 3 OP AMP INSTRUMENTATION AMPLIFIER IN SINGLE-SUPPLY APPLICATIONS

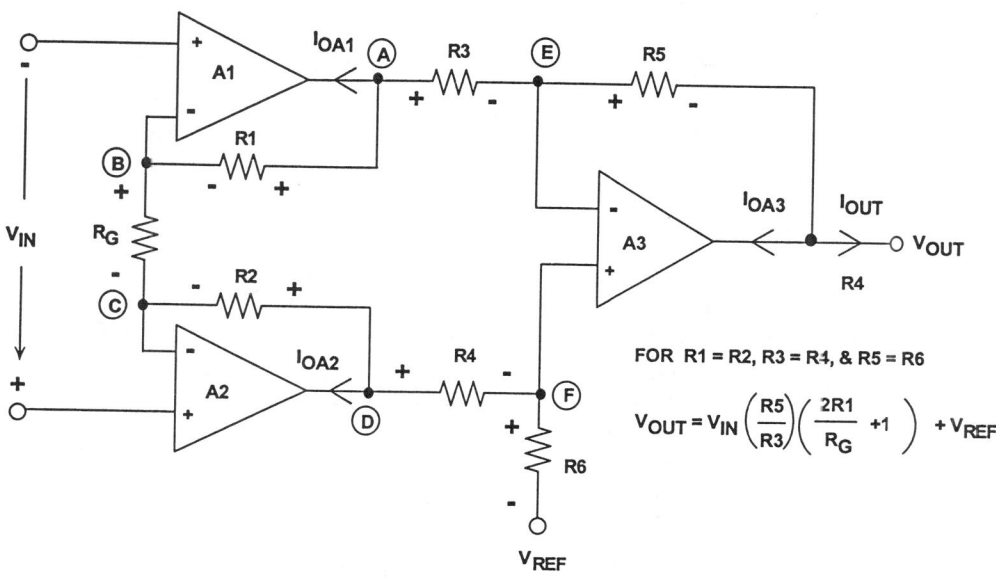

Figure 1.11

Using half-circuit concepts and the principle of superposition, the signal voltage applied to the non-inverting terminal of A1 is set to zero. Since the input signal is applied to the non-inverting terminal of A2, then an expression for the output voltage of amplifier A1 (node A) for positive, differential input signals is given by Eq. 1.16:

$$V_A = (-V_{IN+})\left(\frac{R1}{R_G}\right) + V_{CM} \qquad \text{Eq. 1.16}$$

Since the voltage at the inverting input of A1 must equal the voltage at its non-inverting terminal, then an expression for the voltage at amplifier A1's inverting terminal (node B) is given by Eq. 1.17:

$$V_B = V_{CM} \qquad \text{Eq. 1.17}$$

In a similar manner, the voltage at A2's inverting terminal must equal the voltage on A2's non-inverting terminal:

$$V_C = V_{IN+} + V_{CM} \qquad \text{Eq. 1.18}$$

The expression for the output voltage of A2 (node D) shows that it is dependent upon both the input signal and the applied input common-mode voltage:

SINGLE-SUPPLY AMPLIFIERS

$$V_D = (V_{IN+})\left(1+\frac{R2}{R_G}\right)+V_{CM} \qquad \text{Eq. 1.19}$$

At this point, it is worth noting the behavior of the nodal voltages of the input amplifiers as functions of the applied differential input voltage and the input common-mode voltage. From Eqs. 1.17 and 1.18, the common-mode component of the current through the gain setting resistor, R_G, is zero — the input stages simply buffer the applied input common-mode voltage. In other words, *the input stage common-mode gain is unity*.

On the other hand, the full differential input voltage appears across R_G. In fact, Eq. 1.16 shows that A1 multiplies and inverts the input differential voltage by a factor of $(-R1/R_G)$, while Eq. 1.19 shows that A2 multiplies the input voltage by a factor of $(1+R1/R_G)$. For the case where the output subtractor stage is configured for a gain of 1, all the differential gain is set in the input stage. Therefore, the ratio of R1 to R_G (or R2 to R_G) could be as small as 1:1 or as large as 5000:1. Therefore, to avoid input amplifier output voltage saturation requires an upper and a lower bound be placed on the total input voltage (defined to be common-mode plus differential-mode voltages). These bounds are set by the gain of the instrumentation amplifier and the output high and low voltage limits of the amplifier. The lower bound on the total applied input voltage is given by Eq. 1.20:

$$V_{IN(TOTAL)} > V_{OL(MAX)} + \left(\frac{G-1}{2}\right)(V_{IN+}) \qquad \text{Eq. 1.20}$$

An upper bound on the total input voltage can be determined in a similar fashion and is also dependent on the circuit gain and the amplifier's minimum output high voltage:

$$V_{IN(TOTAL)} < V_{OH(MIN)} - \left(\frac{G+1}{2}\right)(V_{IN+}) \qquad \text{Eq. 1.21}$$

For example, if a rail-to-rail operational amplifier exhibited a $V_{OL(MAX)}$ equal to 10mV and a $V_{OH(MIN)}$ equal to 4.95V, and if the application required a circuit gain of 10 for a 1V full-scale output, then the total input voltage range would be bounded by:

$$0.46\ V < V_{IN(TOTAL)} < 4.4\ V$$

Therefore, for the three op amp instrumentation circuit, the total applied input voltage range expressed in terms of circuit gain and amplifier output voltage limits is given by:

$$V_{OL(MAX)} + \left(\frac{G-1}{2}\right)(V_{IN+}) < V_{IN(TOTAL)} < V_{OH(MIN)} - \left(\frac{G+1}{2}\right)(V_{IN+})$$

$$\text{Eq. 1.22}$$

Since the non-inverting input of the subtractor amplifier A3 determines the voltage on its inverting terminal, an expression for the voltages at Nodes E and F is given by Eq. 1.23:

$$V_E = V_F = (V_{IN+})\left(\frac{R6}{R4+R6}\right)\left(1+\frac{R2}{R_G}\right) + V_{CM}\left(\frac{R6}{R4+R6}\right) + V_{REF}\left(\frac{R4}{R4+R6}\right)$$
Eq. 1.23

For the case where R3, R4, R5, and R6 are all equal to R (typically the case for instrumentation amplifier gains ≥ 1), then these nodal voltages will set up at one-half the applied output voltage reference (V_{REF}) and at one-half the applied input common-mode voltage (V_{CM}). Furthermore, the component due to the amplified differential input signal is also attenuated by a factor of two. Finally, Eq. 1.24 shows an expression for the circuit's output voltage in its familiar form for R4 = R3 and R6 = R5:

$$V_{OUT} = (V_{IN+})\left(\frac{R5}{R3}\right)\left(1+\frac{2R1}{R_G}\right) + V_{REF}$$
Eq. 1.24

From Eq. 1.24, the circuit output voltage is only a function of the amplified input differential voltage and the output reference voltage. Provided that R4 = R3 and R6 = R5, the component of the output voltage due to the applied input common-mode voltage is completely suppressed. The only remaining error voltage is that due to the finite CMR of A3 and the ratio match of R3 to R5 and R4 to R6. Also, in the absence of either an input signal or an output reference voltage, A3's output voltage is equal to zero; in a single-supply application where rail-to-rail output amplifiers are used, it is equal to V_{OL}.

To complete the analysis of this instrumentation circuit, expressions for operational amplifier output stage currents have been developed and are shown in Eqs. 1.25 through 1.27:

$$I_{OA1} = \left(\frac{V_{IN+}}{R3}\right)\left[\left(\frac{R1}{R_G}\right)\left(1+\frac{R3}{R1}\right) + \left(1+\frac{R2}{R_G}\right)\left(\frac{R4}{R4+R6}\right)\right] + \frac{V_{REF} - V_{CM}}{R3}\left(\frac{R4}{R4+R6}\right)$$
Eq. 1.25

$$I_{OA2} = (-V_{IN+})\left\{\left(1+\frac{R2}{R_G}\right)\left[\frac{1}{R2}+\frac{1}{R4}-\frac{1}{R4}\left(\frac{R6}{R4+R6}\right)\right] - \left(\frac{1}{R2}\right)\right\} + \frac{V_{REF} - V_{CM}}{R4+R6}$$
Eq. 1.26

$$I_{OA3} = \frac{-(V_{IN+})}{R3+R5}\left(\frac{R1}{R_G} + \frac{R5}{R3} + \frac{2 \cdot R1 \cdot R5}{R3 \cdot R_G}\right) + \frac{V_{CM} - V_{REF}}{R3+R5} \qquad \text{Eq. 1.27}$$

Recall in the analysis of the two-amplifier instrumentation circuit that amplifier output stage currents were defined to be positive, if current flow is *into* the device, the amplifier is *sinking* current. Conversely, if the nodal analysis shows that output currents are negative quantities, then current flow is *out of* the amplifier, and the amplifier is *sourcing* current.

Equation 1.25 illustrates that A1's output stage must be able to sink current as a function of the applied differential input voltage and the output reference voltage. On the other hand, A1's output stage is required to source current throughout the applied common-mode voltage. In the single-supply case where the circuit is required to sense small differential signals near ground, Eq. 1.16 and Eq. 1.25 both illustrate that A1's output stage is required to sink current while trying to maintain a more negative output voltage than its own negative supply. A1 cannot sustain this operating point, and thus is forced into output saturation.

As shown in Eq. 1.26, A2's output stage sources current for positive input signal voltages with no differential nor common-mode voltage constraints placed upon its output by Eq. 1.19. A3's output stage is also required to source current around its feedback resistor as a function of the positive input differential voltage. Note, however, that as a function of the applied common-mode voltage, it is required to sink current. *Unfortunately Eq. 1.24 showed that in the absence of an input signal, A3's output stage can be forced into saturation, trying to sink current while maintaining its output voltage at A3's V_{OL}.*

To circumvent circuit topological and amplifier output voltage limitations, the results shown in Eq. 1.14 and Eq. 1.15 for the two op amp instrumentation circuit apply equally well here. The output reference voltage is chosen in the middle of A1 and A2's output voltage swing:

$$V_{REF} = \frac{V_{OH(MIN)} + V_{OL(MAX)}}{2} \qquad \text{Eq. 1.14}$$

Similarly, output signal dynamic range and output SNR are maximized if the gain of the instrumentation circuit is set according to Eq. 1.15:

$$\text{Circuit Gain} = \frac{V_{OH(MIN)} - V_{OL(MAX)}}{2 \cdot V_{IN(MAX)}} \qquad \text{Eq. 1.15}$$

Under these operating conditions, the differential output voltage of the instrumentation amplifier circuit is now measured relative to V_{REF} and not to GND. Thus, negative full-scale input signals yield output voltages near A3's V_{OL}, and positive full-scale signals produce output voltages near A3's V_{OH}. Thus, circuit input common-mode range and output dynamic range are optimized in terms of the desired circuit gain and amplifier output voltage characteristics.

For minimal impact on V_{OH} and V_{OL} due to amplifier output load currents, circuit resistor values should be greater than 10kΩ in single-supply applications. Thus, Equations 1.22, 1.14, and 1.15 can all be used to design accurate and repeatable three op amp instrumentation amplifier circuits with single-supply/rail-to-rail operational amplifiers.

A Composite, Single-supply Instrumentation Amplifier [3]

As it has been shown throughout this chapter, operation of high performance linear circuits from a single, low-voltage supply (5V or less) is a common requirement. While there are many precision single supply operational amplifiers (some rail-rail), such as the OP213, the OP291, and the OP284, and some good single-supply instrumentation amplifiers, such as the AMP04 and the AD626 (both covered later), the highest performance instrumentation amplifiers are still specified for dual-supply operation.

One way to achieve both high precision and single-supply operation takes advantage of the fact that several popular transducers (e.g. strain gauges) provide an output signal centered around the (approximate) mid-point of the supply voltage (or the reference voltage), where the inputs of the signal conditioning amplifier need not operate near "ground" or the positive supply voltage.

Under these conditions, a dual-supply instrumentation amplifier referenced to the supply mid-point followed by a "rail-to-rail" operational amplifier gain stage provides very high DC precision. Figure 1.12 illustrates one such high-performance instrumentation amplifier operating on a single, +5V supply. This circuit uses an AD620 low-cost precision instrumentation amplifier for the input stage, and an AD822 JFET-input dual rail-to-rail output operational amplifier for the output stage.

In this circuit, R1 and R2 form a voltage divider which splits the supply voltage in half to +2.5V, with fine adjustment provided by a trimming potentiometer, P1. This voltage is applied to the input of an AD822 which buffers it and provides a low-impedance source needed to drive the AD620's output reference port. The AD620's REFERENCE input has a 10kΩ input resistance and an input signal current of up to 200µA. The other half of the AD822 is connected as a gain-of-3 inverter, so that it can output ±2.5V, "rail-to-rail," with only ±0.83V required of the AD620. This output voltage level of the AD620 is well within the AD620's capability, thus ensuring high linearity for the "dual-supply" front end. *Note that the final output voltage must be measured with respect to the +2.5V reference, and not to GND.*

A PRECISION SINGLE-SUPPLY INSTRUMENTATION AMPLIFIER WITH RAIL-TO-RAIL OUTPUT

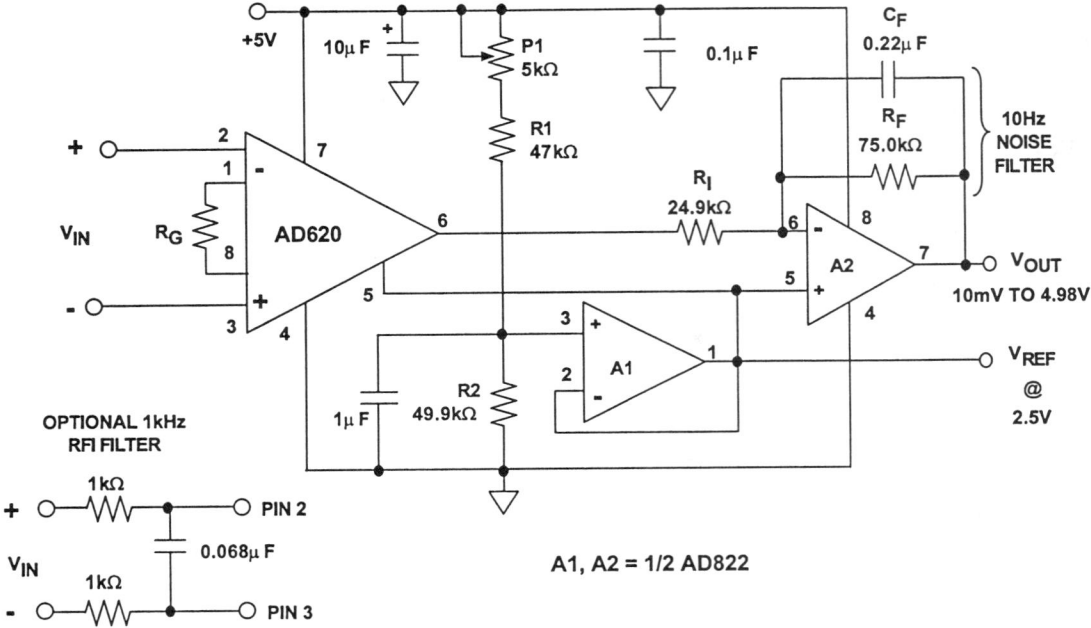

Figure 1.12

The general gain expression for this composite instrumentation amplifier is the product of the AD620 and the inverting amplifier gains:

$$\text{GAIN} = \left(\frac{49.4\text{k}\Omega}{R_G} + 1\right)\left(\frac{R_F}{R_I}\right) \quad \text{Eq. 1.28}$$

For this example, an overall gain of 10 is realized with $R_G = 21.5\text{k}\Omega$ (closest standard value). The table (Figure 1.13) summarizes various R_G/gain values.

In this application, the total input voltage applied to the inputs of the AD620 can be up to +3.5V with no loss in precision. For example, at an overall circuit gain of 10, the common-mode input voltage range spans 2.25V to 3.25V, allowing room for the ±0.25V full-scale differential input voltage required to drive the output ±2.5V about V_{REF}.

The inverting configuration was chosen for the output buffer to facilitate system output offset voltage adjustment by summing currents into the buffer's feedback summing node. These offset currents can be provided by an external DAC, or from a resistor connected to a reference voltage.

The AD822 rail-to-rail output stage exhibits a very clean transient response (not shown) and a small-signal bandwidth over 100kHz for gain configurations up to 300. Figure 1.13 summarizes the performance of this composite instrumentation amplifier. To reduce the effects of unwanted noise pickup, a capacitor is recommended across A2's feedback resistance limit the circuit bandwidth to the frequencies of interest. Also, to prevent the effects of input-stage rectification, an optional 1kHz filter is recommended at the inputs of the AD620.

PERFORMANCE SUMMARY OF THE +5V SINGLE-SUPPLY AD620/AD822 COMPOSITE INSTRUMENTATION AMP WITH RAIL-TO-RAIL OUTPUTS

CIRCUIT GAIN	R_G (Ω)	Vos, RTI (µV)	TCVos, RTI (µV/°C)	Nonlinearity*	Bandwidth (kHz)
10	21.5k	1000	4.5	<0.005%	600
30	5.49k	430	1.5	<0.005%	600
100	1.53k	215	1.4	<0.005%	300
300	499	150	1.1	<0.005%	120
1000	149	150	1.0	<0.005%	30

*Nonlinearity measured over the output voltage range: $0.1V < V_{OUT} < 4.90V$

■ Input bias current (2nA, max) must have DC return path to power supply

■ Total V_{IN} (CM + DM) = $V_{IN,CM} \pm \dfrac{V_{IN,DM}}{2}$

Figure 1.13

Low-side and High-side Signal Conditioning

As previous discussions have shown, single-supply and rail-to-rail operational amplifiers in two and three op amp instrumentation amplifier circuits impose certain limits on the usable input common-mode and output voltage ranges of the circuit. There are, however, many single-supply applications where low- and high-side signal conditioning is required. For these applications, novel circuit design techniques allow sensing of very small differential signals at GND or at V_{POS}. Two such devices, the AMP04 and AD626, have been designed specifically for these applications.

As illustrated in Figure 1.14, the AMP04, a single-supply instrumentation amplifier, uses a inverting-mode output gain architecture, where an external resistor, R_G (connected between the AMP04's Pins 1 and 8), is used as the input resistor to A4, and an internal 100kΩ thin-film resistor, R1, serves as the output amplifier's feedback resistance. Unity-gain input buffers A1 and A2 both serve two functions: they present a high impedance to the source, and provide a DC level shift to the applied common-mode input voltage of one V_{BE} for amplifiers A3 and A4. As a result, their output stages can operate very close the negative supply without saturating.

The input buffers are designed with PNP transistors that allow the applied common-mode voltage range to extend to 0V. In fact, the usable input common-mode voltage range of the AMP04 actually extends 0.25V *below* the negative supply (although not guaranteed, applied input voltages to any integrated circuit should always remain within its total supply voltage range). On the other hand, since the input buffers are PNP stages, the input common-mode voltage range does not include the AMP04's positive supply voltage. When the inputs are driven within 1V of the positive rail, the PNP input transistors are forced into cutoff; and, as a result, input offset voltages and bias currents increase, and CMR degrades.

SINGLE-SUPPLY INSTRUMENTATION AMP HANDLES ZERO-VOLTS INPUT AND ZERO-VOLTS OUTPUT (AMP04)

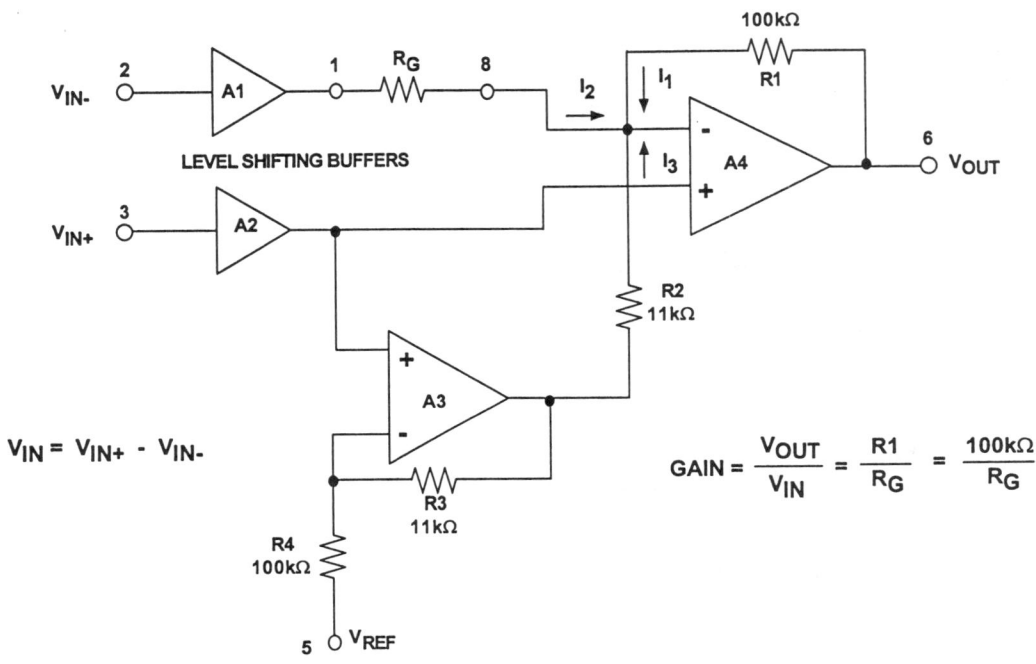

Figure 1.14

A pulsed-bridge transducer-driver/amplifier illustrates the utility of this low-power, single-supply instrumentation amplifier circuit as shown in Figure 1.15. Commonly available 350Ω strain-gauge bridges are difficult to apply in low-voltage, low-power systems for a number of reasons, including the requirements for high bridge drive currents and high sensitivity. For low-speed measurements, power limitations can be overcome by operating the bridge in a pulsed-power mode, reading the amplified output on a low-speed, low-duty-cycle basis.

SINGLE-SUPPLY AMPLIFIERS

A LOW POWER, PULSED LOAD CELL BRIDGE AMPLIFIER

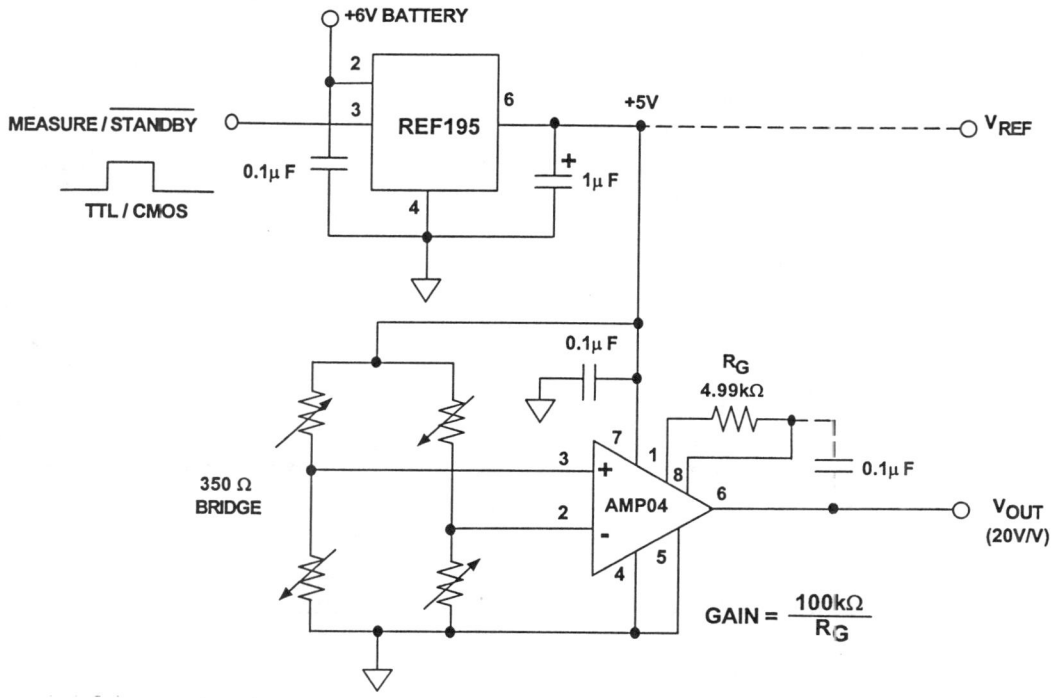

Figure 1.15

In this circuit, an externally generated 800µs TTL/CMOS pulse is applied to the SHUTDOWN input to the REF195, a +5V precision voltage reference. The REF195's shutdown feature is used to switch between a normal +5V DC output if left open (or at logic HIGH), and a low-power-down standby state (5µA maximum current drain) with the shutdown pin held low. The switched 5V output from the REF195 drives the bridge and supplies power to the AMP04. The AMP04 is programmed for a gain of 20 by the 4.99kΩ resistor, which should be a stable film type (TCR = 50ppm/°C or better) in close physical proximity to the amplifier. Dynamic performance of the circuit is excellent, because the AMP04's output settles to within 0.5mV of its final value in about 230µs (not shown).

This approach allows fast measurement speed with a minimum standby power. Generally speaking, with all active circuitry essentially being switched by the measurement pulse, the average current drain of this circuit is determined by its duty cycle. On-state current drain is about 15mA from the 6V battery during the measurement interval (90mW peak power). Therefore, an 800µs measurement strobe once per second will dissipate an average of 72µW, to which is added the 30µW standby power of the REF195. In any event, overall operation is enhanced by the REF195's low-dropout regulation characteristics. The REF195 can operate with supply voltages as low as +5.4 V and still maintain +5V output operation.

Single-Supply Amplifiers

If low-frequency filtering is desired, an optional capacitor can be connected between pins 6 and 8 of the AMP04. However, a much longer strobe pulse must be used so that the filter can settle to the circuit's required accuracy. For example, if a 0.1µF capacitor is used for noise filtering, then the R-C time constant formed with the AMP04's internal 100kΩ resistor is 10ms. Therefore, for a 10-bit settling criterion, 6.9 time constants, or 70ms, should be allowed. Obviously, this will place greater demands upon system power, so trade-offs may be necessary in the amount of filtering used.

Of course, the amplified bridge output appears only during the a measurement interval, and is valid after 220µs unless filtering is used. During this time, a sampled-input ADC (analog-to-digital converter) reads V_{OUT}, eliminating the need for a dedicated sample-and-hold circuit to retain the output voltage. If 10-bit measurements are sufficient, the 5V bridge drive can also be assumed to be constant (for 10-bit accuracy), because the REF195 exhibits a ±1mV (±0.02 %) output voltage tolerance. For more accurate measurements, a ratiometric reading of the bridge status can be obtained by reading the bridge drive (V_{REF}) as well as V_{OUT}.

On the other hand, single-supply instrumentation amplifiers, like the AD626, shown in Figure 1.16, exhibit an input stage architecture that allows the sensing of small differential input signals, not only at its positive supply, but beyond it as well. The AD626 is a differential amplifier consisting of a precision balanced attenuator, a very low-drift preamplifier (A1), and an output buffer amplifier (A2). It has been designed so that small differential signals can be accurately amplified and filtered in the presence of large common-mode voltages, without the use of any other external active or passive components.

The simplified equivalent circuit in Figure 1.16 illustrates the main elements of the AD626. The signal inputs at Pins 1 and 8 are first applied to the dual resistive attenuators R1 through R4, whose purpose is to reduce the peak common-mode voltage at the inputs of A1. This allows the applied differential voltage to be accurately amplified in the presence of large common-mode voltages six times greater than that which can be tolerated by the actual input to A1. As a result, input common-mode rejection extends to $6 \times (V_S - 1V)$. The overall common-mode error is minimized by precise laser trimming of R3 and R4, thus giving the AD626 a common-mode rejection ratio (CMRR) of at least 10,000:1 (80dB).

To minimize the effect of spurious RF signals at the inputs due to rectification at the inputs to A1, small filter capacitors C1 and C2, internal to the AD626, limit the input bandwidth to 1MHz.

The output of A1 is connected to the input of A2 via a 100kΩ resistor (R12) to allow the low-pass filtering to the signals of interest. To use this feature, a capacitor is connected between Pin 4 and the circuit's common. Equation 1.29 can be used to determine the value of the capacitor, based on the corner frequency of this low-pass filter:

$$C_{LP} = \frac{1}{2\pi \cdot (100 \text{ k}\Omega) \cdot f_{LP}} \qquad \text{Eq. 1.29}$$

where f_{LP} = the desired corner frequency of the low-pass filter, in Hz.

SINGLE-SUPPLY AMPLIFIERS

AD626 SCHEMATIC ILLUSTRATES INPUT PROTECTION AND SCALING RESISTORS AND ALLOWS INPUT COMMON MODE VOLTAGE UP TO 6 × (V_S – 1V)

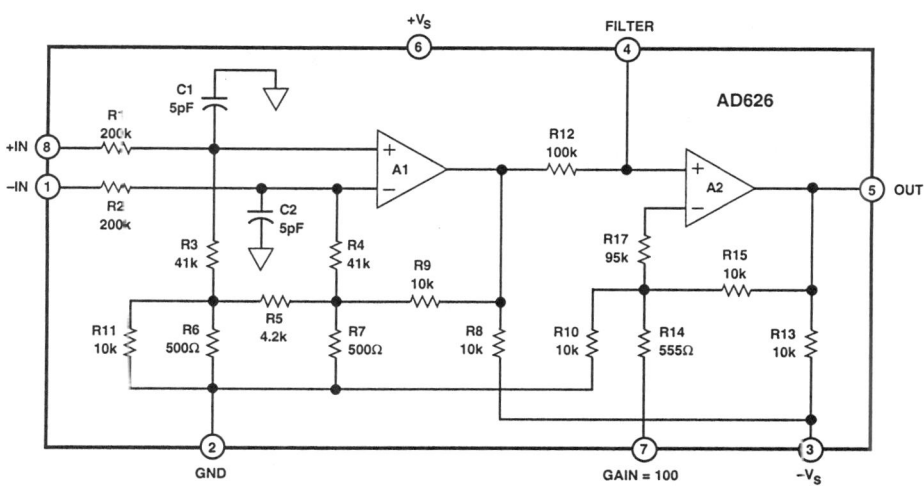

Figure 1.16

The 200kΩ input impedance of the AD626 requires that the source resistance driving this amplifier should be less than 1kΩ to minimize gain error. Also, any mismatch between the total source resistance of either input will affect gain accuracy and common-mode rejection. For example, when operating at a gain of 10, an 80Ω mismatch in the source resistance between the inputs will degrade circuit CMR to 68dB.

Output amplifier, A2, operates at a gain of 2 or 20, thus setting the overall, precalibrated gain of the AD626 (with no external components) at 10 or 100. The gain is set by the feedback network around amplifier A2.

The output of A2 uses an internal 10kΩ resistor to –V_S to "pull down" its output. In single-supply applications where –V_S equals GND, A2's output can drive a 10kΩ ground-referenced load to at least +4.7V. The minimum nominal "zero" output voltage of the AD626 is 30mV.

If pin 7 is left unconnected, the gain of the AD626 is 10. By connecting pin 7 to GND, the AD626's gain can be set to 100. To adjust the gain of the AD626 for gains between 10 and 100, a variable resistance network can be used between pin 7 and GND. This variable resistance network includes a fixed resistor with a rheostat-connected potentiometer in series. The interested reader should consult the AD626 data sheet for complete details for adjusting the gain of the AD626. For these applications, a ±20% adjustment range in the gain is required. This is due to the on-chip resistors absolute tolerance of 20% (these resistors, however, are ratio-matched to within 0.1%).

SINGLE-SUPPLY AMPLIFIERS

An example of the AD626 high-side sensing capabilities, Figure 1.17 illustrates a typical current sensor interface amplifier. The signal current is sensed across the current shunt, R_S. For reasons mentioned earlier, the value of the current shunt should be less than 1Ω and should be selected so that the average differential voltage across this resistor is typically 100mV. To generate a full-scale output voltage of +4V, the AD626 is configured in a gain of 40. To accommodate the tolerance in the current shunt, the variable gain-setting resistor network shown in the circuit has an adjustment range of ±20%. Note that sufficient headroom exists in the gain trim to allow at least a 10% overrange (+4.4V).

AD626 HIGH-SIDE CURRENT MONITOR INTERFACE

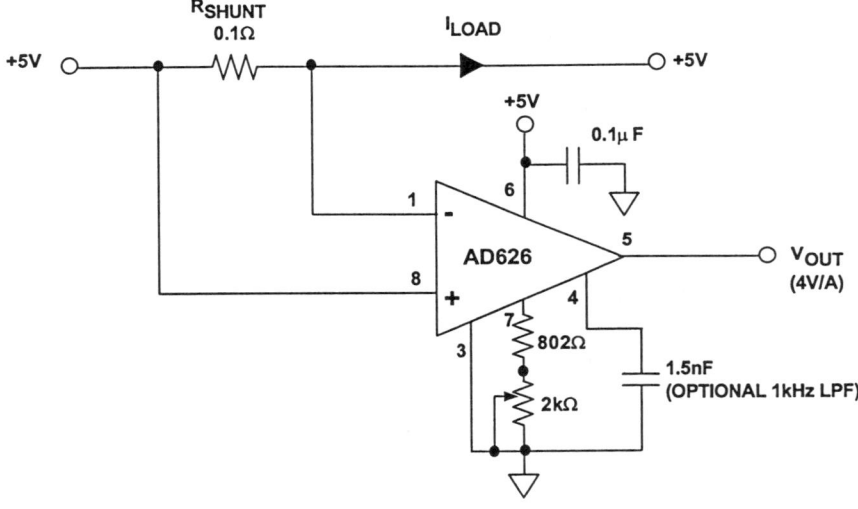

Figure 1.17

Instrumentation Amplifier Input-Stage Rectification

A well-known phenomenon in analog integrated circuits is RF rectification, particularly in instrumentation amplifiers and operational amplifiers. While amplifying very small signals, these devices can rectify unwanted high-frequency, out-of-band signals. The results are DC errors at the output in addition to the wanted sensor signal. Unwanted out-of-band signals enter sensitive circuits through the circuit's conductors which provide a direct path for interference to couple into a circuit. These conductors pick up noise through capacitive, inductive, or radiation coupling. Regardless of the type of interference, the unwanted signal is a voltage which appears in series with the inputs.

All instrumentation and operational amplifier input stages are either emitter-coupled (BJT) or source-coupled (FET) differential pairs with resistive or current-source loading. Depending on the quiescent current level in the devices and the frequency of the interference, these differential pairs can behave as high-frequency detectors. As it has been shown in [1], this detection process produces spectral components at the harmonics of the interference as well at DC. It is the DC component that shifts internal bias levels of the input stages causing errors, which can lead to system inaccuracies. For a complete treatment of this issue, including analytical and empirical results, the interested reader should consult Reference [1].

Since it is required to prevent unwanted signals and noise from entering the input stages, input filtering techniques are used for these types of devices. As illustrated in Reference [1], this technique uses an equivalent approach suggested for operational amplifiers. As shown in Figure 1.18, low-pass filters are used in series with the differential inputs to prevent unwanted noise from reaching the inputs. Here, capacitors, C_{X1}, C_{X2}, and C_{X3}, connected across the inputs of the instrumentation amplifier, form common-mode (C_{X1} and C_{X2}) and differential-mode (C_{X3}) low-pass filters with the two resistors, R_X. Time constants R_X-C_{X1} and R_X-C_{X2} should be well-matched (1% or better), because imbalances in these impedances can generate a differential error voltage which will be amplified.

On the other hand, an additional benefit of using a differentially-connected capacitor is that it can reduce common-mode capacitive imbalance. This differential connection helps to preserve high-frequency AC common-mode rejection. Since series resistors are required to form the low-pass filter, errors due to poor layout (CMR imbalance), component tolerance of R_X (input bias current-induced offset voltage) and resistor thermal noise must be considered in the design process. In applications where the sensor is an RTD or a resistive strain gauge, R_X can be omitted, provided the sensor is close to the amplifier.

EXTERNAL COMMON-MODE AND DIFFERENTIAL-MODE INPUT FILTERS PREVENT RFI RECTIFICATION IN INSTRUMENTATION AMPLIFIER CIRCUITS

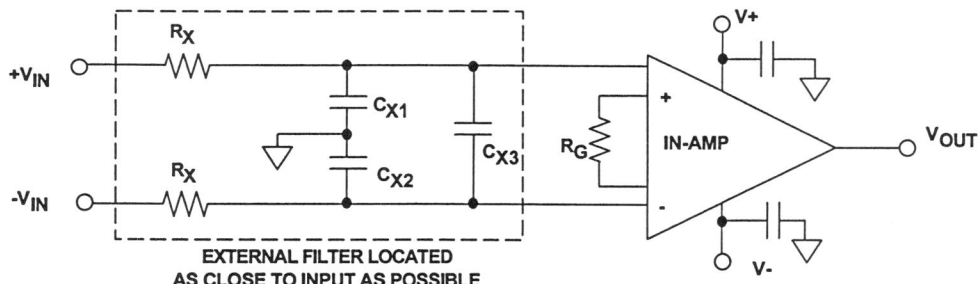

- C_{X1}, C_{X2} forms common-mode RC filter with R_X; $\tau_{CM}(LPF) = R_X C_{X1}$
- C_{X3} forms differential RC with R_X; $\tau_{diff}(LPF) = 2R_X C_{X3}$
- C_{X3} reduces common-mode capacitance imbalance → AC CMRR
- Choose $\tau_{diff}(LPF) \gg \tau_{cm}(LPF)$
- Evaluate R_X errors -- CMR, Bias Current, Noise
- R_X may be omitted if transducer is resistive and close to amplifier

Figure 1.18

REFERENCES

1. **Systems Application Guide**, Chapter 1, pg. 21-55, Analog Devices, Incorporated, Norwood, MA, 1993.

2. **Linear Design Seminar**, Section 1, pp. 19-22, Analog Devices, Incorporated, Norwood, MA, 1995.

3. Lew Counts, Product Line Director, Advanced Linear Products, Analog Devices, Incorporated, personal communication, 1995.

4. Walt Jung, and James Wong, *Op amp selection minimizes impact of single-supply design*, **EDN**, May 27, 1993, pp. 119-124.

5. E. Jacobsen, and J. Baum, *Home-brewed circuits tailor sensor outputs to specialized needs*, **EDN**, January 5, 1995, pp. 75-82.

6. Walt Jung, Corporate Staff Applications Engineer, Analog Devices, Incorporated, personal communication, January 27, 1995.

7. Walt Jung, *Analog-Signal-Processing Concepts Get More Efficient*, **Electronic Design Analog Applications Issue**, June 24, 1993, pp. 12-27.

8. C. Kitchin and L. Counts, **Instrumentation Amplifier Application Guide**, Analog Devices, Incorporated, Norwood, MA, 1991.

Single-Supply Amplifiers

SECTION 2

HIGH SPEED OP AMPS

- Driving Capacitive Loads
- Cable Driving
- Single-Supply Considerations
- Application Circuits

HIGH SPEED OP AMPS

SECTION 2

HIGH SPEED OP AMPS
Walt Jung and Walt Kester

Modern system design increasingly makes use of high speed ICs as circuit building blocks. With bandwidths going up and up, demands are placed on the designer for faster and more power efficient circuits. The default high speed amplifier has changed over the years, with high speed complementary bipolar (CB) process ICs such as the AD846 and AD847 in use just about ten years at this writing. During this time, the general utility/availability of these and other ICs have raised the "high speed" common performance denominator to 50MHz. The most recent extended frequency complementary bipolar (XFCB) process high speed devices such as the AD8001/AD8002, the AD9631/9632 and the AD8036/AD8037 now extend the operating range into the UHF region.

Of course, a traditional performance barrier has been speed, or perhaps more accurately, *painless* speed. While fast IC amplifiers have been around for some time, until more recently they simply haven't been the easiest to use. As an example, devices with substantial speed increases over 741/301A era types, namely the 318-family, did so at the expense of relatively poor settling and capacitive loading characteristics. Modern CB process parts like the AD84X series provide far greater speed, faster settling, and do so at low user cost. Still, the application of high performance fast amplifiers is never entirely a cookbook process, so designers still need to be wary of many inter-related key issues. This includes not just the amplifier selection, but also control of parasitics and other potentially performance-limiting details in the surrounding circuit.

It is worth underscoring that reasons for the "speed revolution" lie not just in affordability of the new high speed ICs, but is also rooted in their *ease of use*. Compared to earlier high speed ICs, CB process devices are generally more stable with capacitive loads (with higher phase margins in general), have lower DC errors, consume less power for a given speed, and are all around more "user friendly". Taking this a step further, XFCB family devices, which extend the utility of the op amp to literally hundreds of MHz, are understandably less straightforward in terms of their application (as is any amplifier operating over such a range). Thus, getting the most from these modern devices definitely stresses the "total environment" aspects of design.

Another major ease of use feature found in today's linear ICs is a much wider range of supply voltage characterization. While the older ±15V standard is still much in use, there is a trend towards including more performance data at popular lower voltages, such as ±5V, or +5V only, single supply operation. The most recent devices using the lower voltage XFCB process use supply voltages of either ±5V, or simply +5V only. The trend towards lower supply voltages is unmistakable, with a goal of squeezing the highest performance from a given voltage/power circuit environment. These "ease of use" design aspects with current ICs are illustrated in this chapter, along with parasitic issues, optimizing performance over supply ranges, and low distortion stages in a variety of applications.

Driving Capacitive Loads

From system and signal fidelity points of view, transmission line coupling between stages is best, and is described in some detail in the next section. However, complete transmission line system design may not always be possible or practical. In addition, various other parasitic issues need careful consideration in high performance designs. One such problem parasitic is amplifier load capacitance, which potentially comes into play for all wide bandwidth situations which do not use transmission line signal coupling.

A general design rule for wideband linear drivers is that capacitive loading (cap loading) effects should *always* be considered. This is because PC board capacitance can build up quickly, especially for wide and long signal runs over ground planes insulated by thin, higher K dielectric. For example, a 0.025" PC trace using a G-10 dielectric of 0.03" over a ground plane will run about 22pF/foot (Reference 1). Even relatively small load capacitance (i.e., 100 pF) can be troublesome, since while not causing outright oscillation, it can still stretch amplifier settling time to greater than desirable levels for a given accuracy.

The effects of cap loading on high speed amplifier outputs are not simply detrimental, they are actually an anathema to high quality signals. However, before-the-fact designer knowledge still allows high circuit performance, by employing various tricks of the trade to combat the capacitive loading. If it is not driven via a transmission line, remote signal circuitry should be checked for capacitive loading very carefully, and characterized as best possible. Drivers which face poorly defined load capacitance should be bullet-proofed accordingly with an appropriate design technique from the options list below.

Short of a true matched transmission line system, a number of ways exist to drive a load which is capacitive in nature while maintaining amplifier stability.

Custom capacitive load (cap load) compensation, includes two possible options, namely a); overcompensation, and b); an intentionally forced-high loop noise gain allowing crossover in a stable region. Both of these steps can be effective in special situations, as they reduce the amplifier's effective closed loop bandwidth, so as to restore stability in the presence of cap loading.

Overcompensation of the amplifier, when possible, reduces amplifier bandwidth so that the additional load capacitance no longer represents a danger to phase margin. As a practical matter however, amplifier compensation nodes to allow this are available on few high speed amplifiers. One such useful example is the AD829, compensated by a single capacitor at pin 5. In general, almost any amplifier using external compensation can always be over compensated to reduce bandwidth. This will restore stability against cap loads, by lowering the amplifier's unity gain frequency.

HIGH SPEED OP AMPS

CAPACITIVE LOADING ON OP AMP GENERALLY REDUCES PHASE MARGIN AND MAY CAUSE INSTABILITY, BUT INCREASING THE NOISE GAIN OF THE CIRCUIT IMPROVES STABILITY

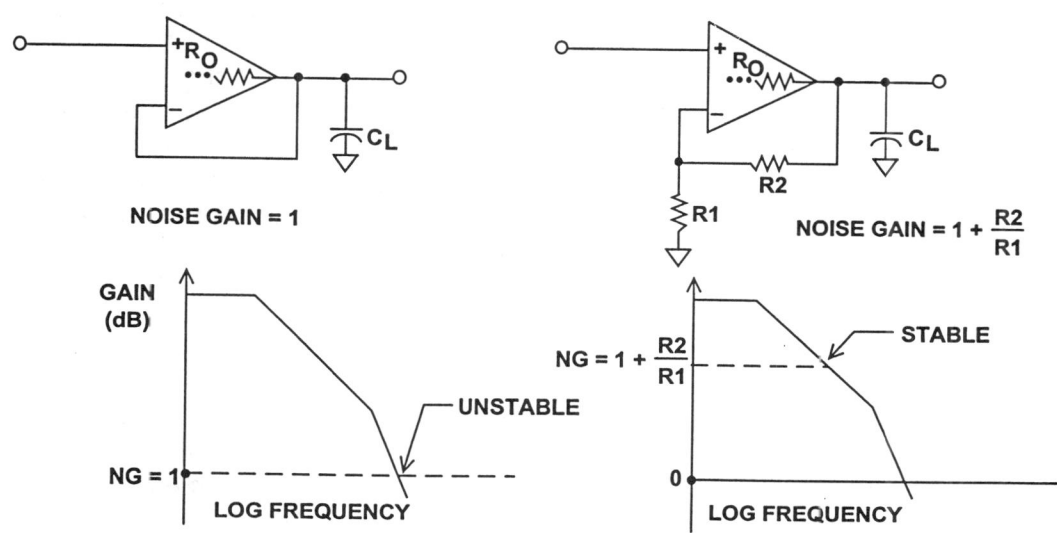

Figure 2.1

RAISING NOISE GAIN (DC OR AC) FOR FOLLOWER OR INVERTER STABILITY

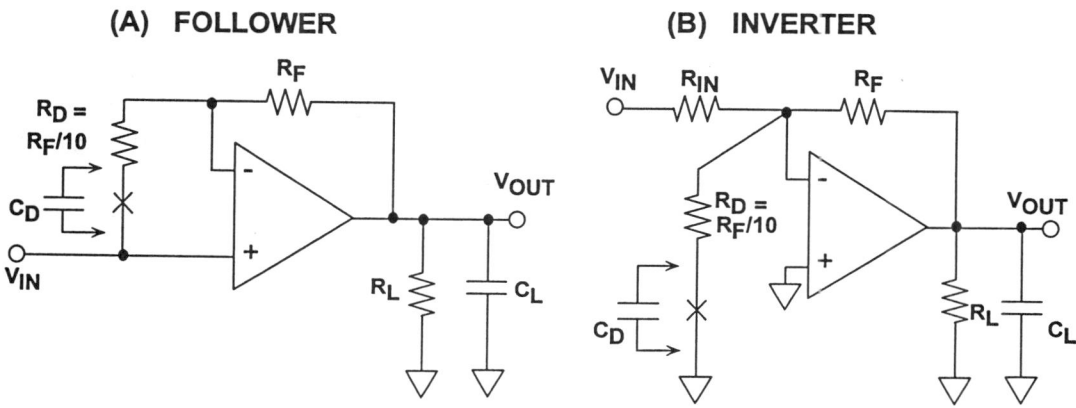

Figure 2.2

Forcing a high noise gain, is shown in Figure 2.1, where the capacitively loaded amplifier with a noise gain of unity at the left is seen to be unstable, due to a $1/\beta$ - open loop rolloff intersection on the Bode diagram in an unstable −12dB/octave region. For such a case, quite often stability can be restored by introducing a higher noise gain to the stage, so that the intersection then occurs in a stable −6dB/octave region, as depicted at the diagram right Bode plot.

To enable a higher noise gain (which does not necessarily need to be the same as the stage's *signal gain*), use is made of resistive or RC pads at the amplifier input, as in Figure 2.2. This trick is more broad in scope than overcompensation, and has the advantage of not requiring access to any internal amplifier nodes. This generally allows use with any amplifier setup, even voltage followers. The technique adds an extra resistor R_D, which works against R_F to force the noise gain of the stage to a level appreciably higher than the signal gain (which is unity in both cases here). Assuming that C_L is a value which produces a parasitic pole near the amplifier's natural crossover, this loading combination would likely lead to oscillation due to the excessive phase lag. However with R_D connected, the higher amplifier noise gain produces a new $1/\beta$ - open loop rolloff intersection, about a decade lower in frequency. This is set low enough that the extra phase lag from C_L is no longer a problem, and amplifier stability is restored.

A drawback to this trick is that the DC offset and input noise of the amplifier are raised by the value of the noise gain, when the optional C_D is *not* present. But, when C_D is used in series with R_D, the offset voltage of the amplifier is not raised, and the gained-up AC noise components are confined to a frequency region above $1/(2\pi \cdot R_D \cdot C_D)$. A further caution is that the technique can be somewhat tricky when separating these operating DC and AC regions, and should be applied carefully with regard to settling time (Reference 2). Note that these simplified examples are generic, and in practice the absolute component values should be matched to a specific amplifier.

"Passive" cap load compensation, shown in Figure 2.3, is the most simple (and most popular) isolation technique available. It uses a simple "out-of-the-loop" series resistor R_X to isolate the cap load, and can be used with any amplifier, current or voltage feedback, FET or bipolar input.

As noted, this technique can be applied to virtually any amplifier, which is a major reason why it is so useful. It is shown here with a current feedback amplifier suitable for high current line driving, the AD811, and it consists of just the simple (passive) series isolation resistor, R_X. This resistor's minimum value for stability will vary from device to device, so the amplifier data sheet should be consulted for other ICs. Generally, information will be provided as to the amount of load capacitance tolerated, and a suggested minimum resistor value for stability purposes.

HIGH SPEED OP AMPS

OPEN-LOOP SERIES RESISTANCE ISOLATES CAPACITIVE LOAD FOR AD811 CURRENT FEEDBACK OP AMP (CIRCUIT BANDWIDTH = 13.5MHz)

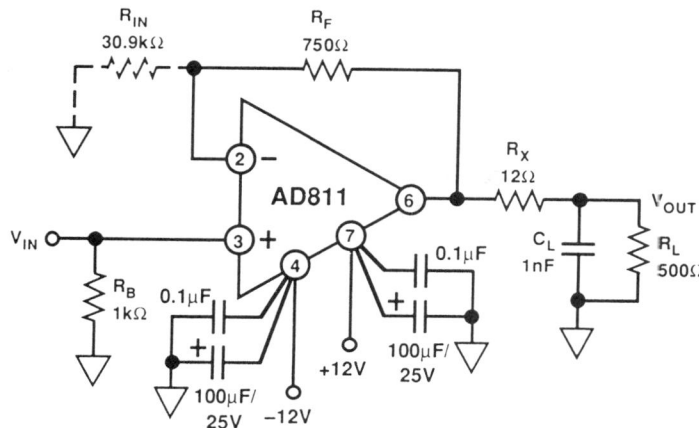

Figure 2.3

Drawbacks of this approach are the loss of bandwidth as R_X works against C_L, the loss of voltage swing, a possible lower slew rate limit due to I_{MAX} and C_L, and a gain error due to the R_X-R_L division. The gain error can be optionally compensated with R_{IN}, which is ratioed to R_F as R_L is to R_X. In this example, a 100mA output from the op amp into C_L can slew V_{OUT} at a rate of 100V/µs, far below the intrinsic AD811 slew rate of 2500V/µs. Although the drawbacks are serious, this form of cap load compensation is nevertheless useful because of its simplicity. If the amplifier is not otherwise protected, then an R_X resistor of 50-100Ω should be used with virtually any amplifier facing capacitive loading. Although a non-inverting amplifier is shown, the technique is equally applicable to inverter stages.

With very speed high amplifiers, or in applications where lowest settling time is critical, even small values of load capacitance can be disruptive to frequency response, but are nevertheless sometimes inescapable. One case in point is an amplifier used for driving ADC inputs. Since high speed ADC inputs quite often look capacitive in nature, this presents an oil/water type problem. In such cases the amplifier *must* be stable driving the capacitance, but it must also preserve its best bandwidth and settling time characteristics. To address this type of cap load case performance, R_s and C_L data for a specified settling time is most appropriate.

Some applications, in particular those that require driving the relatively high impedance of an ADC, do not have a convenient back termination resistor to dampen the effects of capacitive loading. At high frequencies, an amplifier's output impedance is rising with frequency and acts like an inductance, which in combination with C_L causes peaking or even worse, oscillation. When the bandwidth of an amplifier is an appreciable percentage of device f_t, the situation is complicated by the fact that the loading effects are reflected back into its internal stages. In spite of this, the basic behavior of most very wide bandwidth amplifiers such as the AD8001 is very similar.

In general, a small damping resistor (R_S) placed in series with C_L will help restore the desired response (see Figure 2.4). The best choice for this resistor's value will depend upon the criterion used in determining the desired response. Traditionally, simply stability or an acceptable amount of peaking has been used, but a more strict measure such as 0.1% (or even 0.01%) settling will yield different values. For a given amplifier, a family of R_S - C_L curves exists, such as those of Figure 2.4. These data will aid in selecting R_S for a given application.

AD8001 R_S REQUIRED FOR VARIOUS C_L VALUES

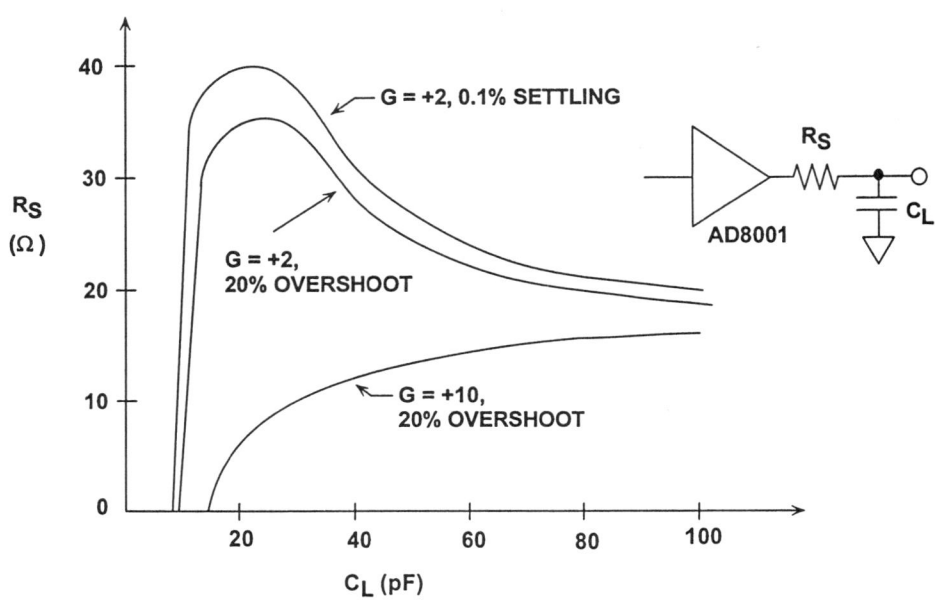

Figure 2.4

The basic shape of this curve can be easily explained. When C_L is very small, no resistor is necessary. When C_L increases to some threshold value an R_S becomes necessary. Since the frequency at which the damping is required is related to the $R_S \cdot C_L$ time constant, the R_S needed will initially increase rapidly from zero, and then will decrease as C_L is increased further. A relatively strict requirement, such as for 0.1%, settling will generally require a larger R_S for a given C_L, giving a curve falling higher (in terms of R_S) than that for a less stringent requirement, such as 20% overshoot. For the common gain condition of +2, these two curves are plotted in the figure for 0.1% settling (upper-most curve) and 20% overshoot (middle curve). It is also worth mentioning that higher closed loop gains lessen the problem dramatically, and will require less R_S for the same performance. The third (lower-most) curve illustrates this, demonstrating a closed loop gain of 10 R_S requirement for 20% overshoot for the AD8001 amplifier. This can be related to the earlier discussion associated with Figure 2.2.

The recommended values for R_S will optimize response, but it is important to note that generally C_L will degrade the maximum bandwidth and settling time performance which is achievable. In the limit, a large $R_S \cdot C_L$ time constant will dominate the response. In any given application, the value for R_S should be taken as a starting point in an optimization process which accounts for board parasitics and other secondary effects.

Active or "in-the-loop" cap load compensation can also be used as shown in Figure 2.5, and this scheme modifies the passive configuration to provide feedback correction for the DC & low frequency gain error associated with R_X. In contrast to the passive form, active compensation can only be used with voltage feedback amplifiers, because current feedback amplifiers don't allow the integrating connection of C_F.

ACTIVE "IN-LOOP" CAPACITIVE LOAD COMPENSATION CORRECTS FOR DC AND LF GAIN ERRORS

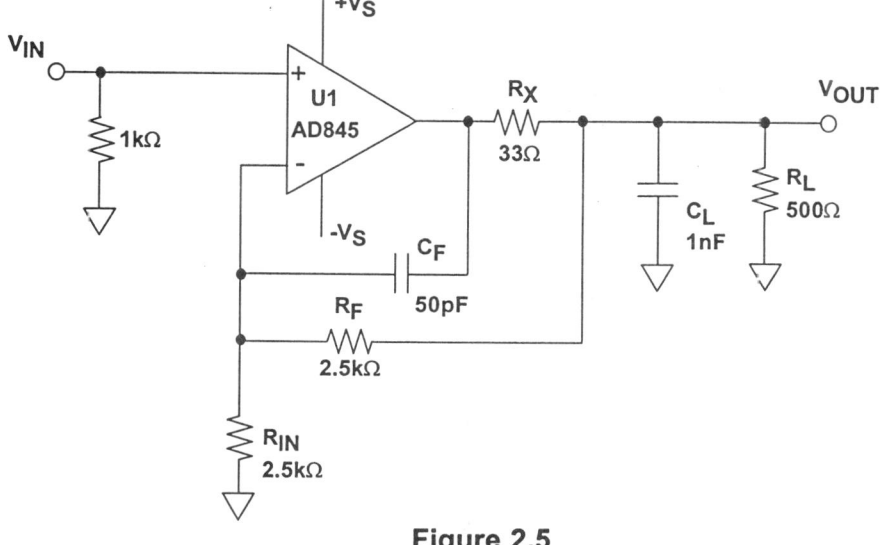

Figure 2.5

HIGH SPEED OP AMPS

This circuit returns the DC feedback from the output side of isolation resistor R_X, thus correcting for errors. AC feedback is returned via C_F, which bypasses R_X/R_F at high frequencies. With an appropriate value of C_F (which varies with C_L, for fixed resistances) this stage can be adjusted for a well damped transient response (Reference 2,3). There is still a bandwidth reduction, a headroom loss, and also (usually) a slew rate reduction, but the DC errors can be very low. A drawback is the need to tune C_F to C_L, as even if this is done well initially, any change to C_L will alter the response away from flat. The circuit as shown is useful for voltage feedback amplifiers only, because capacitor C_F provides integration around U1. It also can be implemented in inverting fashion, by driving the bottom end of R_{IN}.

Internal cap load compensation involves the use of an amplifier which internally has topological provisions for the effects of external cap loading. To the user, this is the most transparent of the various techniques, as it works for any feedback situation, for any value of load capacitance. Drawbacks are that it produces higher distortion than does an otherwise similar amplifier without the network, and the compensation against cap loading is somewhat signal level dependent.

The internal cap load compensated amplifier sounds at first like the best of all possible worlds, since the user need do nothing at all to set it up. Figure 2.6, a simplified diagram of an amplifier with internal cap load compensation, shows how it works. The cap load compensation is the C_F-resistor network shown around the unity gain output stage of the amplifier - note that the dotted connection of this network underscores the fact that it only makes its presence felt for certain load conditions.

AD817 SIMPLIFIED SCHEMATIC ILLUSTRATES INTERNAL COMPENSATION FOR DRIVING CAPACITIVE LOADS

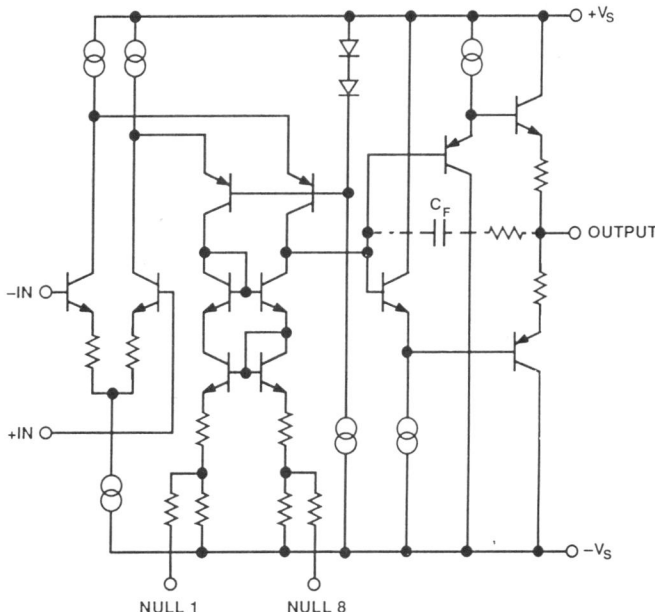

Figure 2.6

HIGH SPEED OP AMPS

Under normal (non-capacitive or light resistive) loading, there is limited input/output voltage error across the output stage, so the C_F network then sees a relatively small voltage drop, and has little or no effect on the amplifier's high impedance compensation node. However when a capacitor (or other heavy) load is present, the high currents in the output stage produce a voltage difference across the C_F network, which effectively adds capacitance to the compensation node. With this relatively heavy loading, a net larger compensation capacitance results, and reduces the amplifier speed in a manner which is adaptive to the external capacitance, C_L. *As a point of reference, note that it requires 6.3mA peak to support a 2Vp-p swing across a 100pF load at 10MHz.*

Since this mechanism is resident in the amplifier output stage and it affects the overall compensation characteristics dynamically, it acts independent of the specific feedback hookup, as well as size of the external cap loading. In other words, it can be transparent to the user in the sense that no specific design conditions need be set to make it work (other than selecting an IC which employs it). Some amplifiers using internal cap load compensation are the AD847 and the AD817, and their dual equivalents, AD827 and AD826.

There are, however, some caveats also associated with this internal compensation scheme. As with the passive compensation techniques, bandwidth decreases as the device slows down to prevent oscillation with higher load currents. Also, this adaptive compensation network has its greatest effect when enough output current flows to produce significant voltage drop across the C_F network. Conversely, at small signal levels, the effect of the network on speed is less, so greater ringing may actually be possible for some circuits for lower-level outputs.

RESPONSE OF INTERNAL CAP LOAD COMPENSATED AMPLIFIER VARIES WITH SIGNAL LEVEL

(A) V_{OUT} = 10V p-p

Vertical Scale: 5V/div

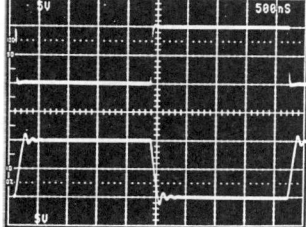

(B) V_{OUT} = 200mV p-p

Vertical Scale, 100mV/div

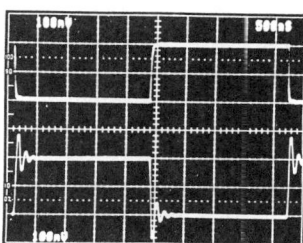

Horizontal Scale: 500ns/div

AD817 INVERTER
R_F = R_{IN} = 1kΩ
R_L = 1kΩ, C_L = 1nF, Vs = ±15V

Figure 2.7

The dynamic nature of this internal cap load compensation is illustrated in Figure 2.7, which shows an AD817 unity gain inverter being exercised at both high and low output levels, with common conditions of V_S = 15V, R_L = 1kΩ, C_L = 1nF, and using 1kΩ input/feedback resistors. In both photos the input signal is on the top trace and the output signal is on the bottom trace, and the time scale is fixed. In the 10Vp-p output (A) photo at the left, the output has slowed down appreciably to accommodate the capacitive load, but settling is still relatively clean, with a small percentage of overshoot. This indicates that for this high level case, the bandwidth reduction due to C_L is most effective. However, in the (B) photo at the right, the 200mVp-p output shows greater overshoot and ringing, for the lower level signal. The point is made that, to some degree at least, the relative cap load immunity of this type of internally cap load compensated amplifier is signal dependent.

Finally, because the circuit is based on a nonlinear principle, the internal network affects distortion and load drive ability, and these factors influence amplifier performance in video applications. Though the network's presence does not by any means make devices like the AD847 or AD817 unusable for video, it does not permit the very lowest levels of distortion and differential gain and phase which are achievable with otherwise comparable amplifiers (for example, the AD818).

While the individual techniques for countering cap loading outlined above have various specific tradeoffs as noted, all of the techniques have a serious common drawback of reducing speed (both bandwidth and slew rate). If these parameters cannot be sacrificed, then a matched transmission line system is the solution, and is discussed in more detail later in the chapter. As for choosing among the cap load compensation schemes, it would seem on the surface that amplifiers using the internal form offer the best possible solution to the problem- just pick the right amplifier and forget about it. And indeed, that would seem the "panacea" solution for all cap load situations - if you use the "right" amplifier you never need to think about cap loading again. Could there be more to it?

Yes! The "gotcha" of internal cap load compensation is subtle, and lies in the fact that the dynamic adaptive nature of the compensation mechanism actually can produce higher levels of distortion, vis-à-vis an otherwise similar amplifier, *without* the C_F-resistor network. Like the old saying about no free lunches, if you care about attaining top-notch levels of high frequency AC performance, you should give the issue of whether to use an internally compensated cap load amplifier more serious thought than simply picking a trendy device.

On the other hand, if you have no requirements for the lowest levels of distortion, then such an amplifier could be a good choice. Such amplifiers are certainly easier to use, and relatively forgiving about loading issues. Some applications of this chapter illustrate the distortion point specifically, quoting performance in a driver circuit with/without the use of an internal cap load compensated amplifiers.

With increased gain bandwidths of ≥100MHz available in today's ICs, layout, grounding and the control of parasitics become much, much more important. In fact, with the fastest available ICs such as the XFCB types, these issues simply cannot be ignored, they are critical and *must* be addressed for stable performance. All high frequency designs can profit from the use of low parasitic construction techniques, such as described in Chapter 9. In the circuit discussions which follow, similar methods should be used for best results, and in the very high frequency circuits (≥100MHz) it is mandatory. Some common pitfalls are covered before getting into specific circuit examples.

As with all wide bandwidth components, good PC board layout is critical to obtain the best dynamic performance with these high speed amplifiers. The ground plane in the area of the op amp and its associated components should cover as much of the component side of the board as possible (or first interior ground layer of a multilayer board).

The ground plane should be removed in the area of the amplifier inputs and the feedback and gain set resistors to minimize stray capacitance at the input. Each power supply trace should be decoupled close to the package with a minimum of 0.1μF ceramic (preferably surface mount), plus a 6.8μF or larger tantalum capacitor within 0.5", as a charge storage reservoir when delivering high peak currents (line drivers, for example). Optionally, larger value conventional electrolytic can be used in place of the tantalum types, if they have a low ESR.

All lead lengths for input, output, and feedback resistor should be kept as short as possible. All gain setting resistors should be chosen for low values of parasitic capacitance and inductance, i.e., microwave resistors (buffed metal film rather than laser-trimmed spiral-wound) and/or carbon resistors.

Microstrip techniques should be used for all input and output lead lengths in excess of one inch (Reference 1). Sockets should be avoided if at all possible because of their parasitic capacitance and inductance. If sockets are necessary, individual *pin sockets* such as AMP p/n 6-330808-3 should be used. These contribute far less stray capacitance and inductance than molded socket assemblies.

The effects of inadequate decoupling on harmonic distortion performance are dramatically illustrated in Figure 2.8. The left photo shows the spectral output of the AD9631 op amp driving a 100Ω load with proper decoupling (output signal is 20MHz, 2V p-p). Notice that the second harmonic distortion at 40MHz is approximately −70dBc. If the decoupling is removed, the distortion is increased, as shown in the right photo of the same figure. Figure 2.8 (right-hand photo) also shows stray RF pickup in the wiring connecting the power supply to the op amp test fixture. Unlike lower frequency amplifiers, the power supply rejection ratio of many high frequency amplifiers is generally fairly poor at high frequencies. For example, at 20MHz, the power supply rejection ratio of the AD9631 is less than 25dB. This is the primary reason for the degradation in performance with inadequate decoupling. The change in output signal produces a corresponding signal-dependent load current change. The corresponding change in power supply voltage due to inadequate decoupling produces a signal-dependent error in the output which manifests itself as an increase in distortion.

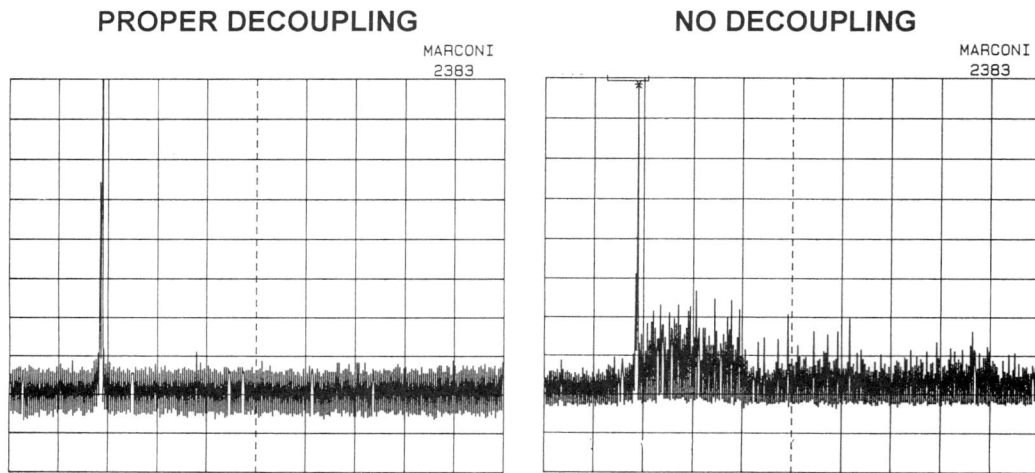

Figure 2.8

Inadequate decoupling can also severely affect the pulse response of high speed amplifiers such as the AD9631. Figure 2.9 shows normal operation and the effects of removing all decoupling capacitors on the AD9631 in its evaluation board. Notice the severe ringing on the pulse response for the poorly decoupled condition, in the right photo. A Tektronix 644A, 500MHz digitizing oscilloscope was used to make the measurement (as well as the pulse responses in Figure 2.10, 2.14, 2.15, 2.16, and 2.17).

The effects of stray parasitic capacitance on the inverting input of such high speed op amps as the AD8001 is shown in Figure 2.10. In this example, 10pF was connected to the inverting input, and the overshoot and ringing increased significantly. (The AD8001 was configured in the inverting mode with a gain of –1, and the feedback and feedforward resistors were equal to 649Ω). In some cases, low-amplitude oscillation may occur at frequencies of several hundred megahertz when there is significant stray capacitance on the inverting input. Unfortunately, you may never actually observe it unless you have a scope or spectrum analyzer which has sufficient bandwidth. Unwanted oscillations at RF frequencies will probably be rectified and averaged by devices to which the oscillating signal is applied. This is referred to as *RF rectification* and will create small unexplained dc offsets which may even be a function of moving your hand over the PC board. It is absolutely essential when building circuits using high frequency components to have high bandwidth test equipment and use it to check for oscillation at frequencies well beyond the signals of interest.

EFFECT OF INADEQUATE DECOUPLING ON PULSE RESPONSE OF AD9631 OP AMP

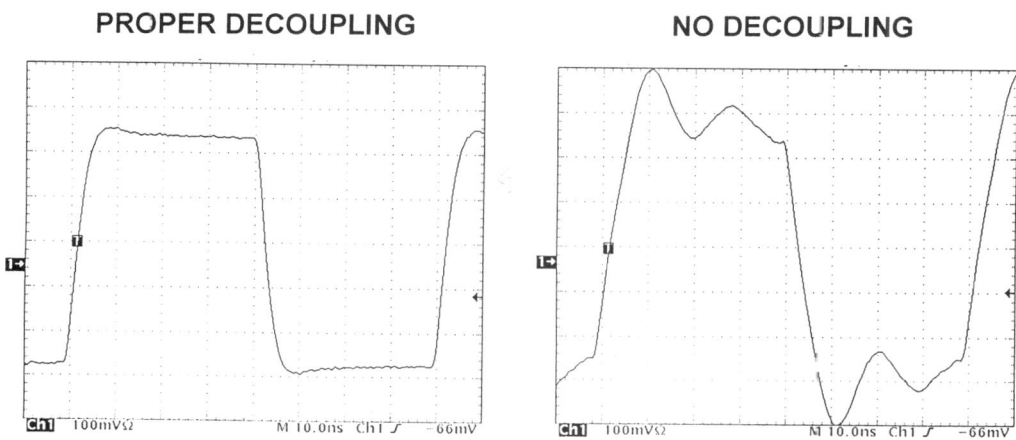

VERTICAL SCALE: 100mV/div
HORIZONTAL SCALE: 10ns/div

Figure 2.9

EFFECT OF 10pF STRAY INVERTING INPUT CAPACITANCE ON PULSE RESPONSE OF AD8001 OP AMP

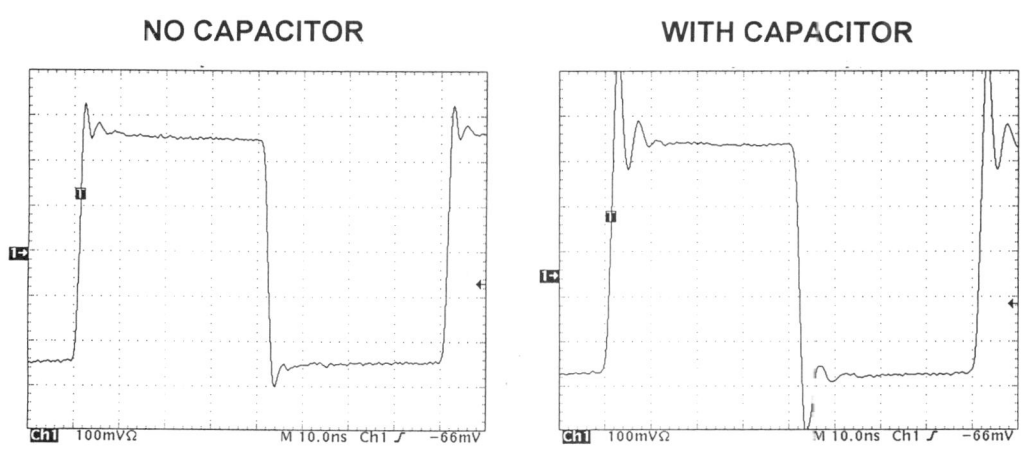

VERTICAL SCALE: 100mV/div
HORIZONTAL SCALE: 10ns/div

Figure 2.10

HIGH SPEED OP AMPS

Many of these problems occur in the prototype phase due to a disregard for high frequency layout and decoupling techniques. The solutions to them lie in rigorous attention to such details as above, and those described in Chapter 9.

CABLE DRIVING

For a number of good reasons, wide bandwidth amplifier systems traditionally use transmission line interconnections, such as that shown in the basic diagram of Figure 2.11. This system uses a drive amplifier A, matched in terms of output impedance by the 75Ω source termination R_T to the transmission line connecting stages A and B. In this particular case the line is a 75Ω coax, but in general it is a wideband line matched at both ends, and can alternately be of twisted pair or stripline construction. It is followed immediately by the differential receiver circuit, B, which terminates the line with a load R_{TERM}, equal to its 75Ω impedance. The receiver stage recovers a noise-free 1V signal which is referenced to system ground B.

SINGLE-ENDED DRIVER AND DIFFERENTIAL RECEIVER

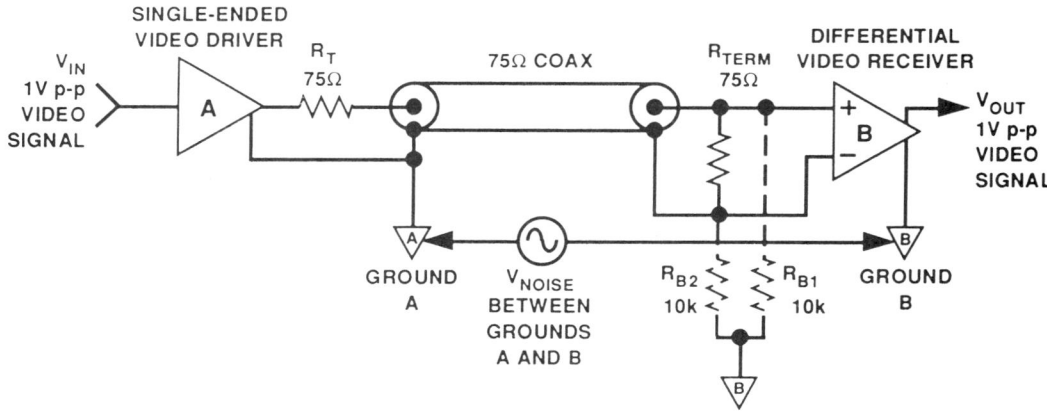

Figure 2.11

When properly implemented (i.e., the line is source and load terminated in its characteristic impedance), this system presents resistive-only loading to drive amplifier A. This factor makes it near ideal from the mutual viewpoints of amplifier stability, distortion and frequency response, as well as minimizing line reflections and associated time domain aberrations. There is an intrinsic 2/1 (6dB) signal loss associated with the line's source and load terminations, but this is easily made up by a 2× driver stage gain.

It is very important to understand that the capacitive-load compensation techniques described above are hardly the perfect solution to the line-driving problem. The most foolproof way to drive a long line (which could otherwise present a substantial capacitive load) is to use a transmission line, a standard for signal distribution in video and RF systems for years. Figure 2.12 summarizes several important cable characteristics.

CABLE CAPACITANCE

- All Interconnections are Really Transmission Lines Which Have a Characteristic Impedance (Even if Not Controlled)

- The Characteristic Impedance is Equal to $\sqrt{L/C}$, where L and C are the Distributed Inductance and Capacitance

- Correctly Terminated Transmission Lines Have Impedances Equal to Their Characteristic Impedance

- Unterminated Transmission Lines Behave Approximately as Lumped Capacitance at Frequencies $<< 1/t_p$, where t_p = Propagation Delay of Cable

Figure 2.12

A transmission line correctly terminated with pure resistance (no reactive component) does *not* look capacitive. It has a controlled distributed capacitance per foot (C) and a controlled distributed inductance per foot (L). The characteristic impedance of the line is given by the equation $Z_0 = \sqrt{(L/C)}$. Coaxial cable is the most popular form of single-ended transmission line and comes in characteristic impedances of 50Ω, 75Ω, and 93Ω.

Because of skin effect, it exhibits a loss which is a function of frequency as shown in Figure 2.13 for several popular coaxial cables (Reference 5). Skin effect also affects the pulse response of long coaxial cables. The response to a fast pulse will rise sharply for the first 50% of the output swing, then taper off during the remaining portion of the edge. Calculations show that the 10 to 90% waveform risetime is 30 times greater than the 0 to 50% risetime when the cable is skin effect limited (Reference 5).

COAXIAL CABLE ATTENUATION VERSUS FREQUENCY

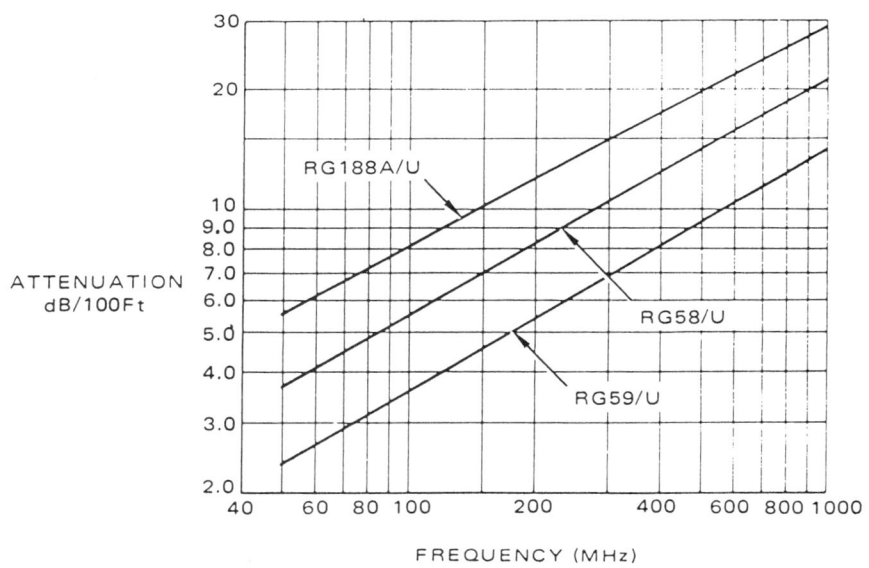

Copyright of Motorola, Inc., Used by Permission

Figure 2.13

It is useful to examine what happens for conditions of proper and improper cable source/load terminations. To illustrate the behavior of a high speed op amp driving a coaxial cable, consider the circuit of Figure 2.14. The AD8001 drives 5 feet of 50Ω coaxial cable which is load-end terminated in the characteristic impedance of 50Ω. No termination is used at the amplifier (driving) end. The pulse response is also shown in the figure.

The output of the cable was measured by connecting it directly to the 50Ω input of a 500MHz Tektronix 644A digitizing oscilloscope. The 50 resistor termination is actually the input of the scope. The 50Ω load is not a perfect termination (the scope input capacitance is about 10pF), so some of the pulse is reflected out of phase back to the source. When the reflection reaches the op amp output, it sees the closed-loop output impedance of the op amp which, at 100MHz, is approximately 100Ω. Thus, it is reflected back to the load with no phase reversal, accounting for the negative-going "blip" which occurs approximately 16ns after the leading edge. This is equal to the round-trip delay of the cable (2·5ft·1.6 ns/ft=16ns). In the frequency domain (not shown), the cable mismatch will cause a loss of bandwidth flatness at the load.

HIGH SPEED OP AMPS

PULSE RESPONSE OF AD8001 DRIVING 5 FEET OF LOAD-TERMINATED 50Ω COAXIAL CABLE

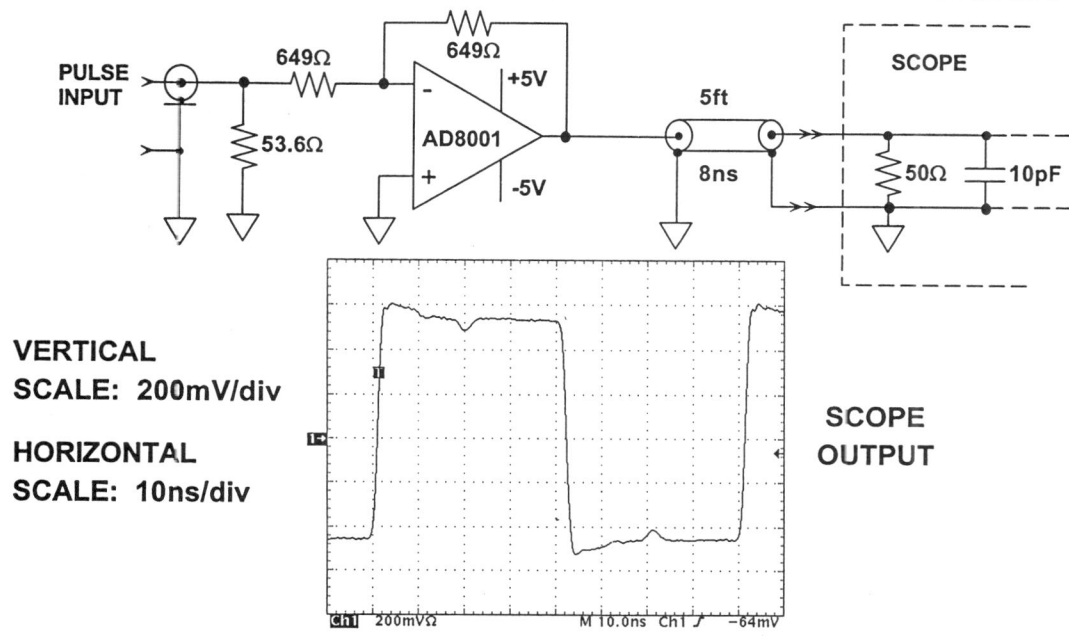

Figure 2.14

Figure 2.15 shows a second case, the results of driving the same coaxial cable, but now used with both a 50Ω source-end as well as a 50Ω load-end termination. This case is the preferred way to drive a transmission line, because a portion of the reflection from the load impedance mismatch is absorbed by the amplifier's source termination resistor. The disadvantage is that there is a 2× gain reduction, because of the voltage division between the equal value source/load terminations. However, a major positive attribute of this configuration, with matched source and load terminations in conjunction with a low-loss cable, is that the best bandwidth flatness is ensured, especially at lower operating frequencies. In addition, the amplifier is operated under near optimum loading conditions, i.e., a resistive load.

HIGH SPEED OP AMPS

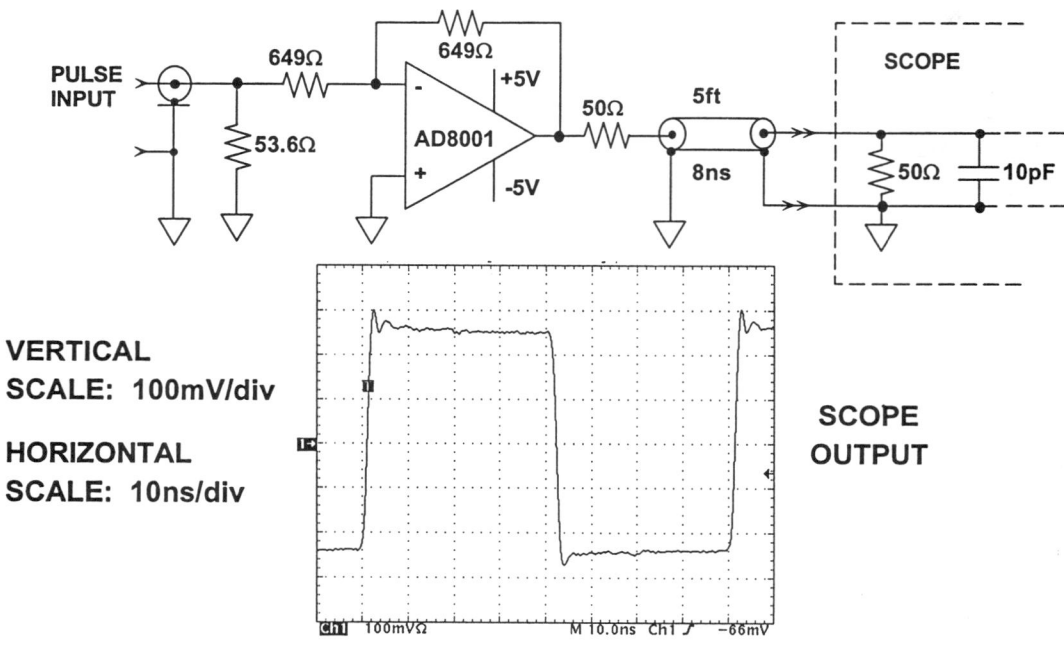

Figure 2.15

Source-end (only) terminations can also be used as shown in Figure 2.16, where the op amp is source terminated by the 50Ω resistor which drives the cable. The scope is set for 1MΩ input impedance, representing an approximate open circuit. The initial leading edge of the pulse at the op amp output sees a 100Ω load (the 50Ω source resistor in series with the 50Ω coax impedance. When the pulse reaches the load, a large portion is reflected in phase because of the high load impedance, resulting in a full-amplitude pulse at the load. When the reflection reaches the source-end of the cable, it sees the 50Ω source resistance in series with the op amp closed loop output impedance (approximately 100Ω at the frequency represented by the 2ns risetime pulse edge). The reflected portion remains in phase, and appears at the scope input as the positive-going "blip" approximately 16ns after the leading edge.

HIGH SPEED OP AMPS

PULSE RESPONSE OF AD8001 DRIVING 5 FEET OF SOURCE-TERMINATED 50Ω COAXIAL CABLE

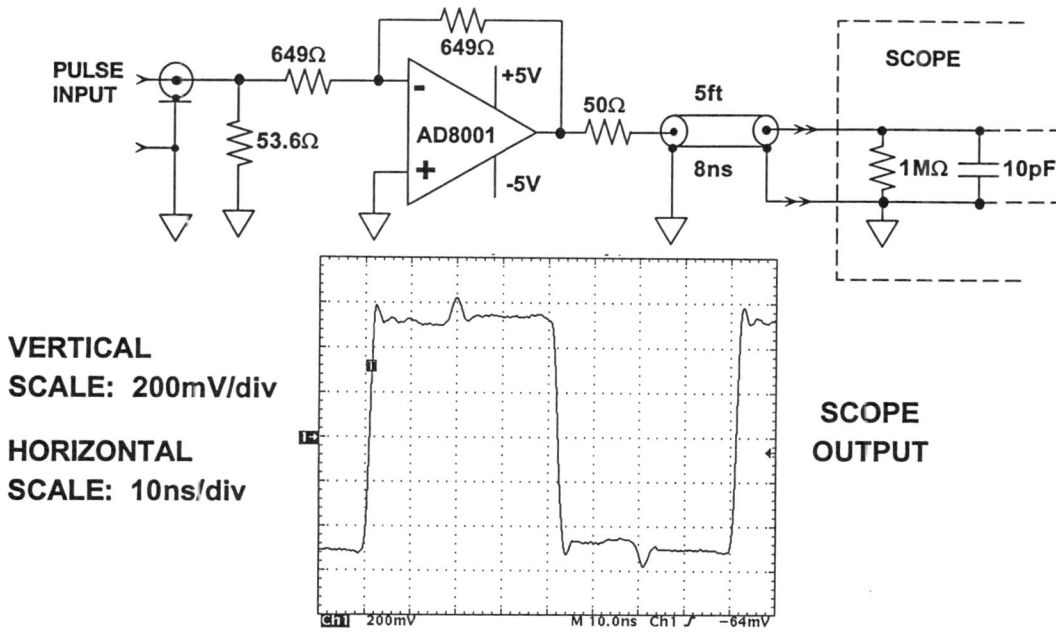

VERTICAL SCALE: 200mV/div

HORIZONTAL SCALE: 10ns/div

SCOPE OUTPUT

Figure 2.16

From these experiments, one can easily see that the preferred method for minimum reflections (and therefore maximum bandwidth flatness) is to use both source and load terminations and try to minimize any reactance associated with the load. The experiments represent a worst-case condition, where the frequencies contained in the fast edges are greater than 100MHz. (Using the rule-of-thumb that bandwidth = 0.35/risetime). At video frequencies, either load-only, or source-only terminations may give acceptable results, but the data sheet should always be consulted to determine the op amp's closed-loop output impedance at the maximum frequency of interest. A major disadvantage of the source-only termination is that it requires a truly high impedance load (high resistance and minimal parasitic capacitance) for minimum absorption of energy.

Now, for a truly worst case, let us replace the 5 feet of coaxial cable with an uncontrolled-impedance cable (one that is largely capacitive with little inductance). Let us use a capacitance of 150pF to simulate the cable (corresponding to the total capacitance of 5 feet of coaxial cable whose distributed capacitance is about 30pF/foot). Figure 2.17 shows the output of the AD8001 driving a lumped 160pF capacitance (including the scope input capacitance of 10pF). Notice the overshoot and ringing on the pulse waveform due to the capacitive loading. This example illustrates the need to use good quality controlled-impedance coaxial cable in the transmission of high frequency signals.

HIGH SPEED OP AMPS

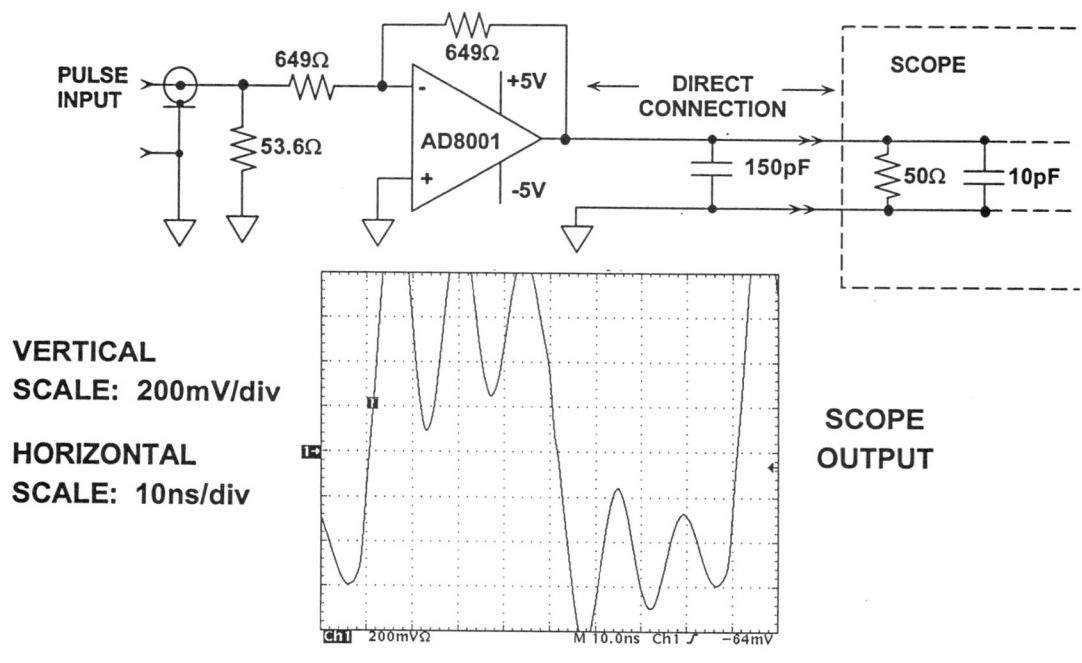

Figure 2.17

Line Drivers

The single-ended line driver function complements the receiver at the transmission end (below). This driver type driver is usually non-inverting, and accepts a signal from a high impedance board level source, scales it, and buffers it to drive a source matched coaxial transmission line of 50-100Ω. Typically, the driver has a gain of 2 times, which complements the 2/1 attenuation of system source/load terminations. A gain-of-two stage seems simple on the surface, but actually many factors become involved in optimizing device and power supply selection to meet overall system performance criteria. Among these are bandwidth/distortion levels versus load impedance, supply voltage/power consumption, and device type.

Fortunately, modern ICs have become more complete in their specifications, as well as more flexible in terms of supply voltage, so there is much from which to choose. For high performance in demanding such applications as wideband video, designers need a fully specified circuit environment, so that the best choice can be easy. For NTSC video systems, distortion is usually rated in terms of differential gain and differential phase, expressed as Δ% for gain and Δ° for phase, driving a rated load at the 3.58MHz subcarrier frequency. While traditional 3dB bandwidth is important, a more stringent specification of 0.1dB bandwidth is also often used.

VIDEO LINE DRIVER USING THE AD818 VOLTAGE FEEDBACK OP AMP HAS 50MHz BANDWIDTH (–0.1dB)

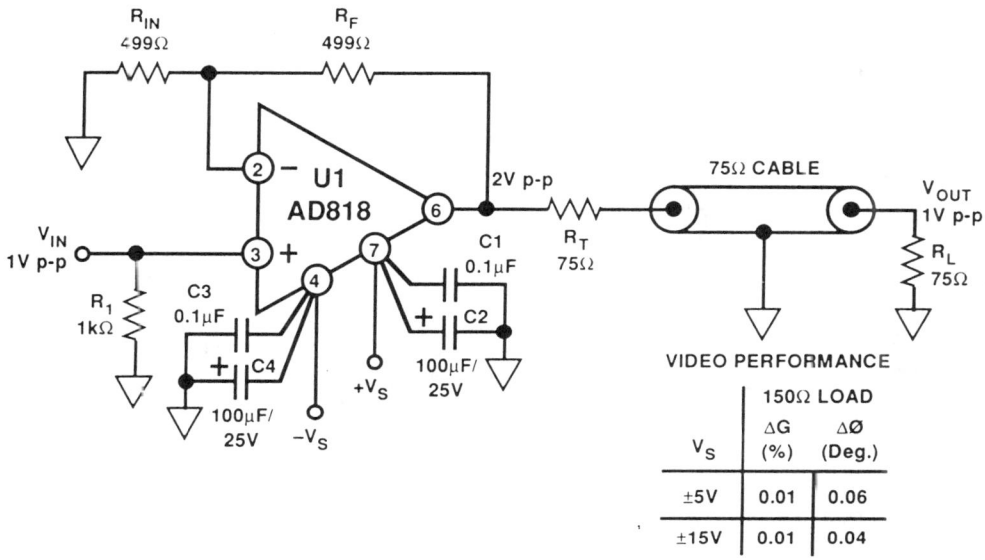

Figure 2.18

An excellent single-ended, high performance line driver meeting these guidelines is shown in Figure 2.18. Although this circuit uses inexpensive amplifiers, an AD818 (or 1/2 an AD828), it is still has excellent performance. Stage gain is set at 2 times by the equal R_1- R_2 values, which are also relatively low (500-1kΩ), to minimize the feedback time constant. This voltage feedback op amp has maximum effective bandwidth when operated at G=2, and has been optimized for this specific application, thus it is able to achieve a 50MHz 0.1dB bandwidth. For highest linearity, it does not use internal capacitive load compensation. This factor, plus a high current (50mA) output stage provides the gain linearity required for high performance video, and line driving applications.

The NTSC video differential gain/phase performance of this circuit is quite good with the 150Ω loading presented by the 75Ω source termination R_T plus the 75 load, R_L, and is typically on the order of 0.01%/0.05° for a 2Vp-p V_{OUT} video swing while operating at ±15V. These figures do degrade for operation at V_S= ±5V, but can be maintained for supplies of ±10V or more. Thus to minimize distortion, supplies of ±10V to ±15V should be used. Other video grade op amps can also be used for U1, but illustrate the potential for distortion tradeoffs. For example, the AD817 (a similar op amp with internal cap load compensation, see above) achieves NTSC video distortion figures of 0.04%/0.08° at V_S=±15V for the same 150Ω loading.

Supply bypassing for line drivers such as this should include both local low inductance caps C1- C3, as well as larger value electrolytics, for a charge reservoir to buffer heavy load currents.

HIGH SPEED OP AMPS

Tantalum types can be used for C2 - C4, and tend to have both lower ESR and small physical size (both desirable), but the 100μF aluminum types shown can also be used. Quiescent current of this circuit is 7mA, which equates to power dissipations of 210 and 70mW for $V_S=\pm15V$ and $\pm5V$, respectively.

Figure 2.19 shows another high-performance video line driver using the AD810 current feedback amplifier. This circuit is also inexpensive and has higher slewrate and higher output current. The AD810 has a 3dB bandwidth of 65MHz, and a 0.1dB bandwidth of 20MHz, while the quiescent current is only 8mA.

A unique feature of the AD810 is its power-down mode. The DISABLE pin is active-low to shut the device down to a standby current drain of 2mA, with 60dB input isolation at 10MHz. This permits on/off control of a single amplifier, or "wire or-ing" the outputs of a number of devices to achieve a multiplexing function. *Note: If the AD810's disabling function is not required, then the DISABLE pin can float, and it operates conventionally.*

Note that when the AD810 is used on different power supplies, the optimum R_F will change (see table... this also will be true for other current feedback amplifiers). Other current feedback amplifiers suitable for this driver function are the single AD811, the dual AD812, and the triple AD813 (with the AD813 also featuring a DISABLE function).

VIDEO LINE DRIVER USING THE AD810 CURRENT FEEDBACK OP AMP HAS DISABLE MODE, 65MHz BANDWIDTH (–3dB) AND 20MHz BANDWIDTH (–0.1dB)

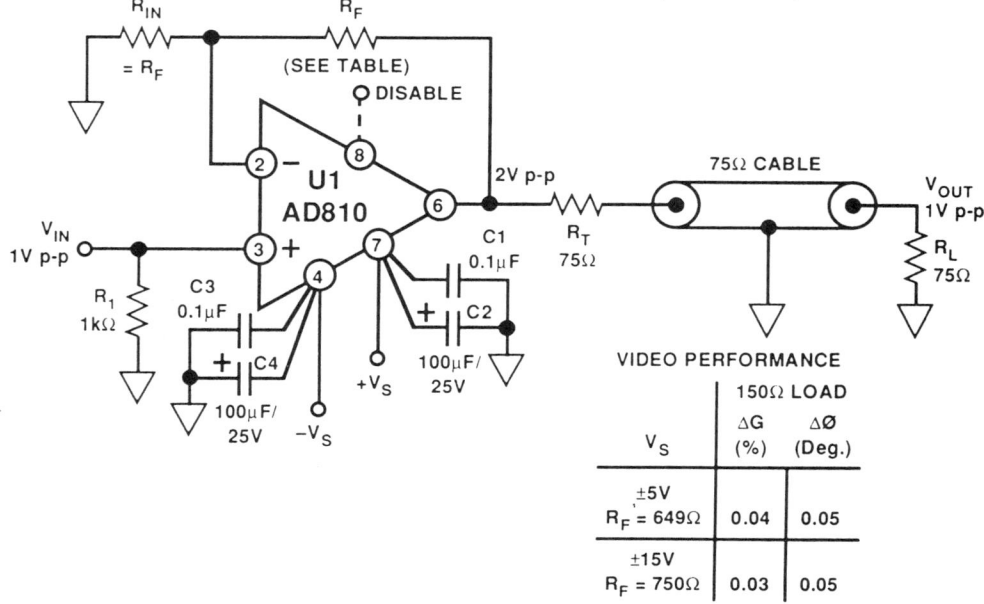

Figure 2.19

HIGH SPEED OP AMPS

VERY HIGH PERFORMANCE ±5V VIDEO LINE DRIVER/ DISTRIBUTION AMPLIFIER HAS 440MHz BANDWIDTH (-3dB) AND 110MHz BANDWIDTH (-0.1dB)

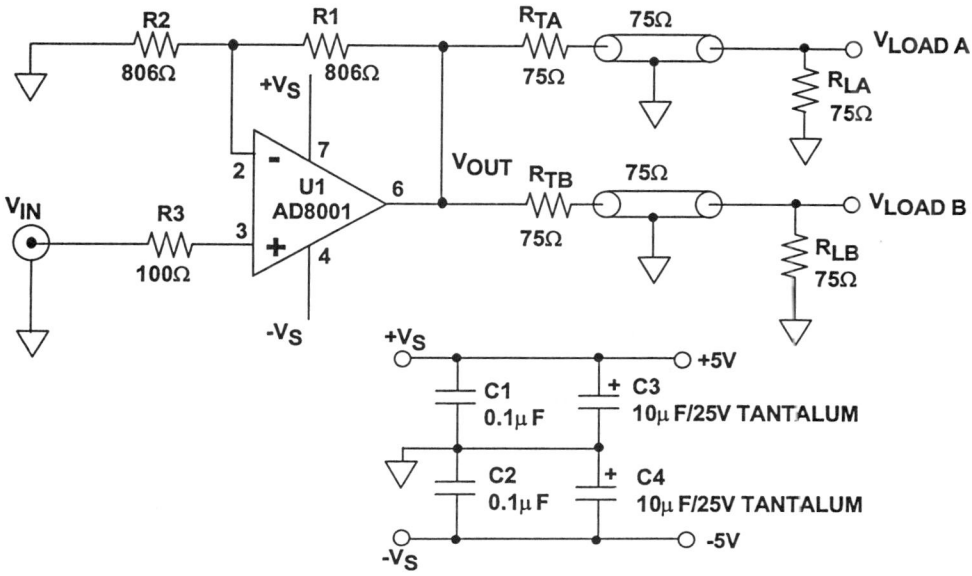

Figure 2.20

Figure 2.20 illustrates a very high performance video line driver, which has optional distribution amplifier features. This circuit uses the AD8001 current feedback amplifier as a gain-of-2 dual 75Ω line driver, generally similar to the above. Some key performance differences which set this circuit apart lie in the fact that the op amp employs a complementary UHF process. It is capable of extremely wideband response, due to the NPN/PNP f_ts of 3-5GHz. This allows higher frequency response at a lower power dissipation than the 500-600MHz f_t complementary process parts, such as the AD818 or AD810 above. The extended frequency response and lower power helps achieve low distortion at higher frequencies, but operating at a much lower quiescent power on supply voltages up to 12V.

Because the higher frequency response per mW of power, a higher output drive is possible. Here, this leads to a performance difference in that this circuit also doubles as a distribution amplifier, that is it can drive two 75Ω output lines if desired. Operated from ±5V supplies as shown, the circuit has 3dB bandwidth of 440MHz. Video distortion for differential gain/phase is 0.01%/0.025° with one line driven (R_L = 150Ω). Driving two lines (R_L = 75Ω), the differential gain errors are essentially the same, while the differential phase errors rise to about 0.07°. The 0.1dB bandwidth of this circuit is 110MHz, and the quiescent power is 50mW with the ±5V supplies. Supply bypassing should follow the same guidelines as the previous drivers, and the general physical layout should follow the RF construction techniques described above and in Chapter 9.

SINGLE-SUPPLY CONSIDERATIONS

The term *single-supply* has various implications, some of which are often further confused by marketing hype, etc. As mentioned above, there is a distinct trend toward systems which run on lower supply voltages. For high speed designs where 2Vp-p swings are often the norm, ±5V power supplies have become standard. There are many obvious reasons for lower power dissipation, such as the ability to run without fans, reliability issues, etc. There are, therefore, many applications for single-supply ADCs other than in systems which have only one supply voltage. In many instances, the lower power drain of a single-supply ADC can be the reason for its selection, rather than the fact that it requires just one supply.

Then, there are also systems which truly operate on a single power supply. In such cases, it can often be difficult to maintain DC coupling from a source all the way to the ADC. In fact, AC coupling is quite often used in single-supply systems, with a DC restoration circuit preceding the ADC. This is necessary to prevent the loss of dynamic range which could otherwise occur, because of a need to provide maximum headroom to an AC coupled signal of arbitrary duty cycle. In the AC-coupled portions of such systems, a "false ground" is often created, usually centered between the rails.

This introduces the question of an optimum input voltage range for a single-supply ADC. At first it would seem that a zero-volt referenced input might be desirable. But in fact, this places severe constraints on the ADC driving amplifier in DC coupled systems, as it must maintain full linearity at or near 0V out. In actuality, there is no such thing as a true rail-rail output amplifier, since all output stage types will have finite saturation voltage(s) to the $+V_s$ rail or ground. Bipolar stages come the closest, and can go as low as an NPN V_{CESAT} of ground (or, a complementary PNP can go within a V_{CESAT} of the + rail). The exact saturation voltage is current dependent, and while it may be only a few mV at light currents, it can be several hundred mV for higher load currents. The traditional CMOS rail-rail output stage looks like a resistor to ground for zero-volt outputs (or to the + rail, for + outputs), so substantial load currents create proportionally higher voltage drops across these resistors, thereby limiting the output swing.

A more optimum ADC input range is thus one which includes neither ground nor the positive supply, and a range centered around $V_s/2$ is usually optimum. For example, an input range of 2Vp-p around +2.5V is bounded by +1.5V and +3.5V. A complementary common-emitter type single-supply output stage is quite capable of handling this range.

For design and process reasons however, the ADC input common-mode range may be offset from the ideal $V_s/2$ voltage midpoint. Single-supply op amps dynamic specifications such as distortion, settling time, slew rate, bandwidth, etc. are typically stated for a $V_s/2$ output bias condition. Distortion and other dynamics can degrade if the signal is offset substantially in either direction from this nominal range.

The output voltage range of an op amp is most often given for DC or low frequency output signals. Distortion may increase as the signal approaches either high or low saturation voltage limits. For instance, a +5V op amp can have a distortion specification of −60dBc for a 3Vp-p output signal. If not stated otherwise, the implication is that the signal is centered around +2.5V (or $V_s/2$), i.e., the sinewave is bounded by +1V and +4V. If the signal is offset from +2.5, distortion may very well increase.

Low distortion op amp designs typically use a complementary emitter follower output stage. This limits their output swing to slightly greater than 1 diode drop of the rails, in general more like 1V of the rails. In order to maintain low distortion at high frequencies, even more headroom may be required, reducing the available peak-to-peak swing.

The complementary common-emitter output stage allows the output to approach within V_{CESAT} of the rails. However, it does have a higher open-loop output impedance than that of the follower type of stage, and is more likely to distort when driving such non-linear loads as flash converters. When driving constant impedance loads, distortion performance can be equal to or better that the follower type of stage, but such a generalization obviously has its limitations.

The input voltage range of a single-supply op amp may also be restricted. The ability to handle zero-volt input signals can be realized by either a PNP bipolar transistor or N-channel JFET differential input stage. Including both supply rails is rarely required for high frequency signal processing. Of course, the positive rail can be included, if the op amp design uses NPN bipolars, or P-channel JFETs. For true rail-rail input capability, two amplifier input stages must operate in parallel, and the necessary bias crossover point between these stages can result in distortion and reduced CMRR. This type of input stage serves DC and lower frequency amplifiers best.

In many cases, an amplifier may be AC-specified for low voltage single-supply operation, but neither its input nor its output can actually swing very close to the rails. Such devices must use applications designed so that both the input and output common-mode restrictions are not violated. This generally involves offsetting the inputs using some sort of a false ground reference scheme.

To summarize, there are many tradeoffs involved in single-supply high speed designs. In many cases using devices specified for operation on +5V, but without true rail inclusive input/output operation, can give best available performance. As more high speed devices which are truly single-supply become available, they can be added to the designer's bag of tricks.

Direct Coupling Requires Careful Design and Controlled Levels

A design which operates on a single +5V supply with DC coupling is illustrated in Figure 2.21. This type of circuit is useful when the input signal maximum amplitude is known, and a specified output bias level is required to interface to the next stage (such as the ADC input range mentioned).

CAREFUL BIASING AND CONTROLLED LEVELS ALLOW DC COUPLING WITHIN SINGLE-SUPPLY SYSTEMS

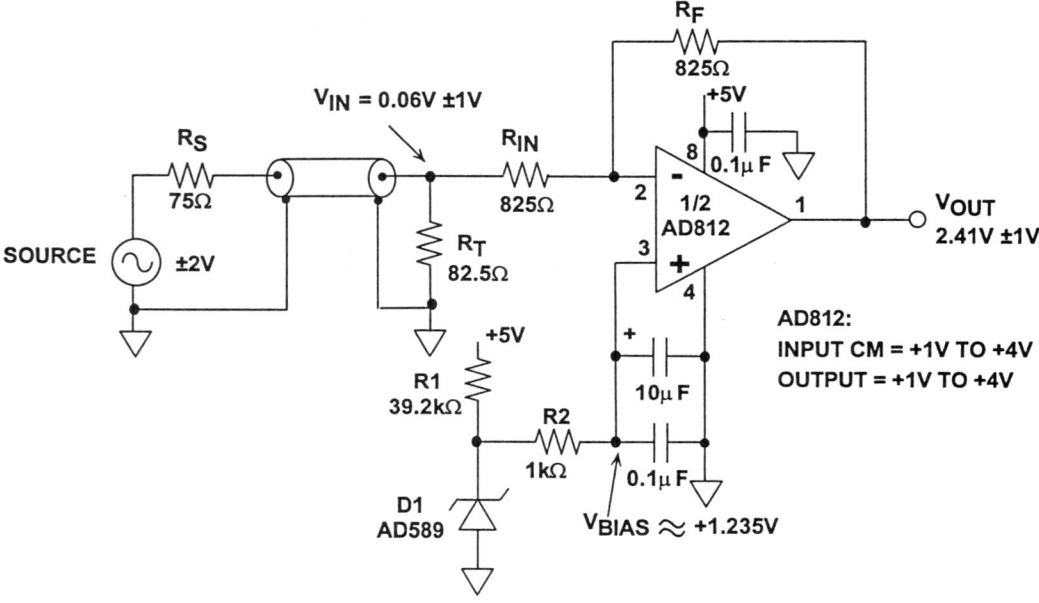

Figure 2.21

Here the source voltage is a ±2V 75Ω source, which is DC coupled and terminated by R_T. An AD812 amplifier is used, which has an input CM range of +1V to +4V (when operating on +5V), and a minimum output swing of 3Vp-p into 1kΩ on 5V. It can easily swing the required 2Vp-p at the output, so there is some latitude for biasing it around an output level of $V_s/2$. For unity signal gain, equal value feedback and input resistors are used. The R_F and R_{IN} values chosen are a slight compromise to maximum bandwidth (the optimum value is 715Ω), but the input line is terminated properly at 75Ω (the parallel equivalent value of R_T and R_{IN}).

The resulting resistor values provide a gain of about 1.95 to the DC voltage applied to pin 3, V_{BIAS}. A voltage of 1.235V from a stable reference source such the AD589 will fix the static output DC level at 1.95•1.235V, or 2.41V. Because of the inverting mode signal operation, pin 2 of the op amp does not change appreciably with signal, therefore this node effectively operates at a fixed DC level, equal to V_{BIAS}. As long as this bias voltage is well above the amplifier's minimum CM range of +1V, there should be no problem. The RC network at pin 3 provides a noise filter for the diode. If a more precise output DC level is required from this stage, then V_{BIAS} can be adjusted to provide it without change to the stage's signal gain.

Single-Supply Line Drivers

By choosing a high frequency op amp which is specified for operation on low voltage supplies, single-ended line drivers can also be adapted for single 5V supply operation. An example is the 5V supply line driver circuit of Figure 2.22, which also illustrates AC coupling in single-supply design.

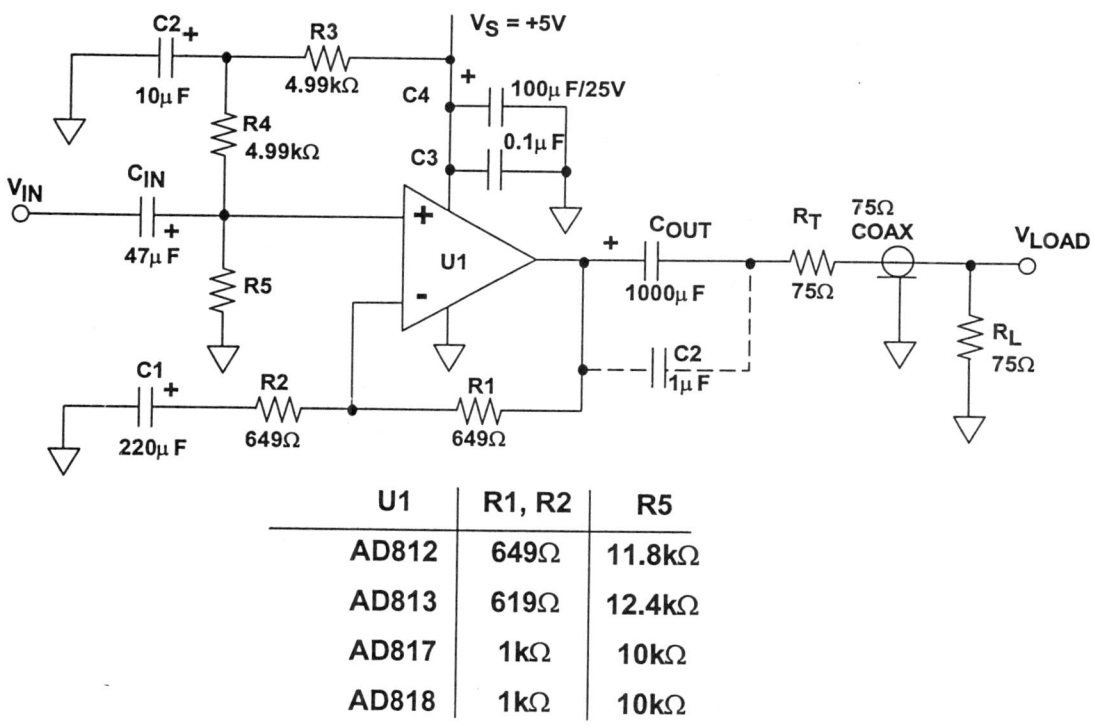

Figure 2.22

Speaking generally, this circuit can use a number of op amps, both voltage and current feedback types. The output of the AD812 and AD813 devices can swing to within about 1V of either rail, allowing 3Vp-p outputs to be delivered into 150Ω loads on 5V. While bandwidth of these current feedback amplifiers does reduces with low voltage operation, they are still capable of a small signal 0.1dB bandwidth on the order of 10MHz, and good differential gain/ phase performance, about 0.07%/0.06°, quite good for a simple circuit. The AD817 and AD818 are not as clean in performance, and not necessarily recommended for 5V line drivers, but are included as examples of more general purpose voltage feedback types operable on 5V. For optimizing the bias and bandwidth of any of these amplifiers, use the resistor values from the table.

As would be expected, headroom is critical on such low supplies, so if nothing else, biasing should be optimized for the voltage in use, via the R5 value as noted. R3 and R4 are equal values, AC bypassed for minimum noise coupling from the supply line. All input and output coupling capacitors are large in value, and are so chosen for a minimum of low frequency phase shift, for composite video uses. They can be reduced for applications with higher low frequency cutoffs. C_{OUT} is potentially a problem in the large value shown, as large electrolytics can be inductive. C5, a non-critical optional low inductance shunt, can minimize this problem.

Obviously, the stage cannot be driven beyond 3Vp-p at V_{OUT} without distortion, so operating levels need to maintained conservatively below this. The next section discusses AC coupling issues and headroom in more detail.

The AC coupling of arbitrary waveforms can actually introduce problems which don't exist at all in DC coupled or DC restored systems. These problems have to do with the waveform duty cycle, and are particularly acute with signals which approach the rails, as they can in low supply voltage systems which are AC coupled.

In an amplifier circuit such as that of Figure 2.22, the output bias point will be equal to the DC bias as applied to the op amp's (+) input. For a symmetric (50% duty cycle) waveform of a 2Vp-p output level, the output signal will swing symmetrically about the bias point, or nominally 2.5V± 1V. If however the pulsed waveform is of a very high (or low) duty cycle, the AC averaging effect of C_{IN} and R4||R5 will shift the effective peak level either high or low, dependent upon the duty cycle. This phenomenon has the net effect of reducing the working headroom of the amplifier, and is illustrated in Figure 2.23.

In Figure 2.23 (A), an example of a 50% duty cycle square wave of about 2Vp-p level is shown, with the signal swing biased symmetrically between the upper and lower clip points of a 5V supply amplifier. This amplifier, for example, (an AD817 biased similarly to Figure 2.22) can only swing to the limited DC levels as marked, about 1V from either rail. In cases (B) and (C), the duty cycle of the input waveform is adjusted to both low and high duty cycle extremes *while maintaining the same peak-to-peak input level*. At the amplifier output, the waveform is seen to clip either negative or positive, in (B) and (C), respectively.

HIGH SPEED OP AMPS

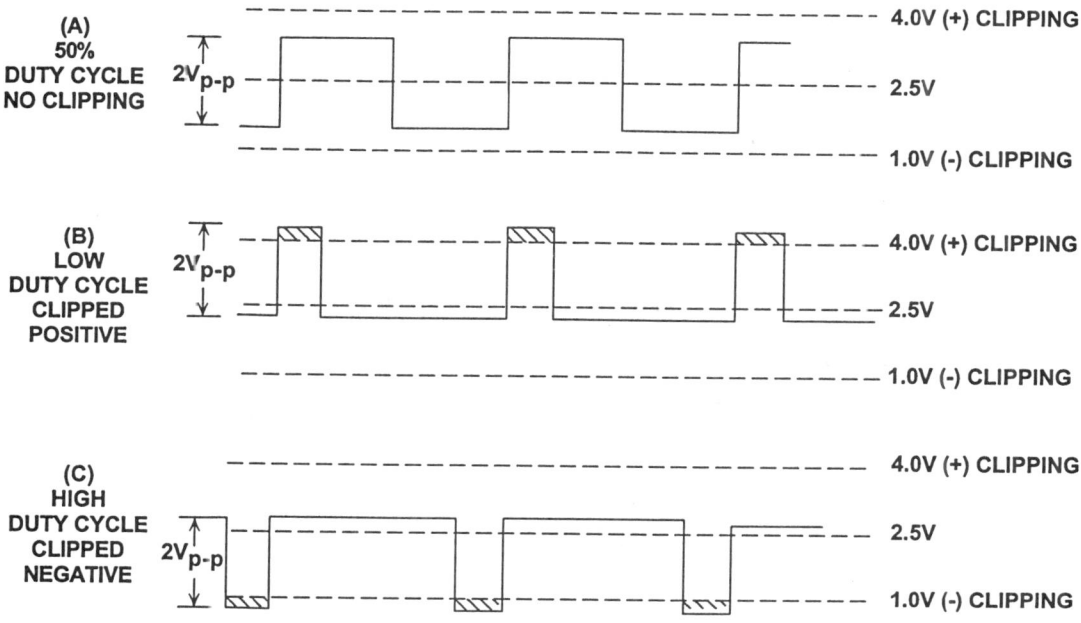

Figure 2.23

Since standard video waveforms *do* vary in duty cycle as the scene changes, the point is made that low distortion operation on AC coupled single supply stages must take the duty cycle headroom degradation effect into account. If a stage has a 3Vp-p output swing available before clipping, and it must cleanly reproduce an arbitrary waveform, then the maximum allowable amplitude is less than 1/2 of this 3Vp-p swing, that is <1.5Vp-p. An example of violating this criteria is contained the 2Vp-p waveform of Figure 2.23, which is clipping for both the high and low duty cycles. Note that the criteria set down above is based on avoiding hard clipping, while subtle distortion increases may in fact take place at lower levels. This suggests an even more conservative criteria for lowest distortion operation such as composite NTSC video amplifiers.

Of course, amplifiers designed with rail-rail outputs and low distortion in mind address these problems most directly. One such device is the AD8041, an XFCB single-supply op amp designed for video applications, and summarized briefly by Figure 2.24. As these data show, this part is designed to provide low NTSC video distortion while driving a single 75Ω source terminated load (in a circuit such as Figure 2.22).

2 - 29

RAIL-RAIL OUTPUT VIDEO OP AMPS ALLOW LOW DISTORTION OUTPUT AND GREATEST FLEXIBILITY

Typical specifications for AD8041 op amp @ V_S = +5V, T_A = 25°C

- Common Mode Range: −0.2V to +4V
- Offset Voltage: 2mV
- Bias Current: 1.2 µA
- Bandwidth: 80MHz
- Slew Rate: 160V/µs
- Differential Gain/Phase: 0.03% / 0.03°
  (V_{out} = 2Vp-p, R_L = 150Ω)
- Output Current: 50mA (0.5V from rails)
- Quiescent Current: 5mA
- Disable Feature Allows Multiplexing

Figure 2.24

Application Circuits

A common video circuit requirement is the multiplexer, a stage which selects one of "N" video inputs, and transmits a buffered version of the selected signal to an output transmission line. Video amplifiers which can operate internally in a switched mode, such as the AD810 and AD813, allow this operation to be performed directly in the video signal path with no additional hardware. This feature is activated with the use of the device's *disable* pin, which when pulled low, disables the amplifier and drops power to a low state. The AD810 is a single channel current feedback amplifier with this disable feature, while the AD813 offers similar functionality, in a 14 pin, three channel format. The high performance of the AD813 on low voltages allows it to achieve high performance on 5V supplies, and to be directly interfaced with standard 5V logic drivers.

2:1 Video Multiplexer

The outputs of two AD810s can be wired together to form a 2:1 multiplexer without degrading the flatness of the gain response. Figure 2.25 shows a recommended configuration, which results in a 0.1dB bandwidth of 20MHz and OFF channel isolation of 77dB at 10MHz on ±5V supplies. The time to switch between channels is about 750ns when the disable pins are driven by open drain output logic. With the use of the recommended 74HC04 as shown, the switching time is about 180ns. The switching time is only slightly affected by the signal level.

A 2:1 VIDEO MULTIPLEXER USING AD810s HAS −0.1dB BANDWIDTH OF 20MHz AND SWITCHES IN 180ns

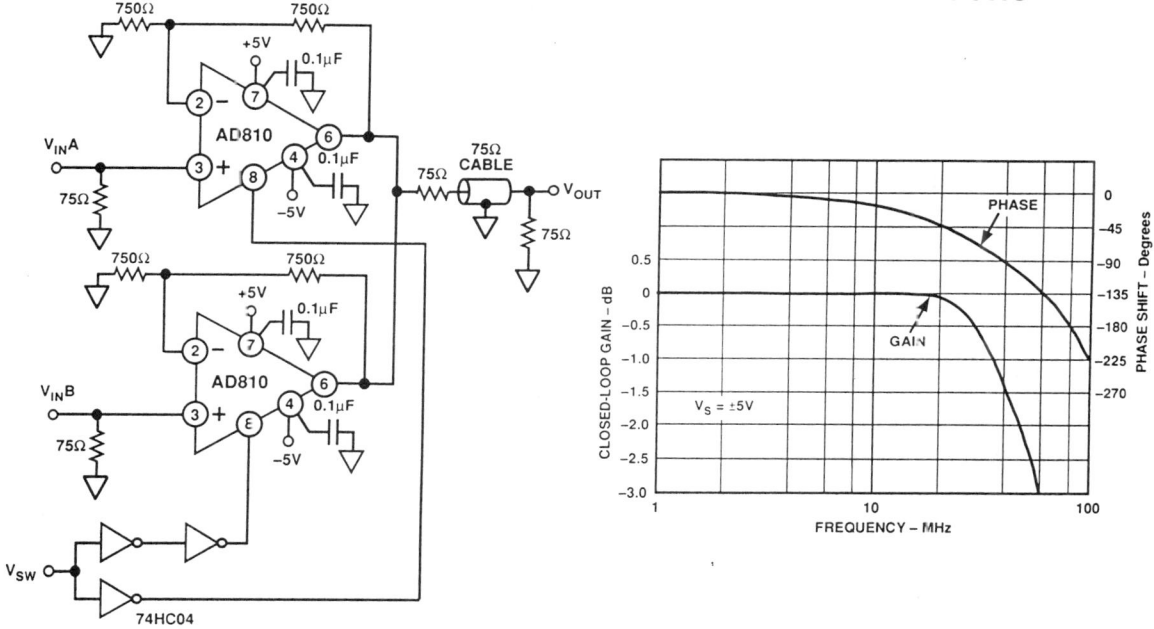

Figure 2.25

3:1 Video Multiplexer

A 3:1 video multiplexer circuit using the triple AD813 is shown in Figure 2.26, and features relative simplicity and high performance while operating from 5V power supplies. The 3 standard 1Vp-p video input signals V_{IN1} - V_{IN3} drive the 3 channels of the AD813, one of which is ON at a given moment. If say channel 1 is selected, amplifier section 1 is enabled, by virtue of a logic HIGH signal on the SELECT1 line driving the ENABLE input of the first amplifier. The remaining two amplifier channels appear as open circuits looking back into them, but their feedback networks do appear as a load to the active channel. Control logic decoding is provided by U2, a 74HC238 1 of 8 logic decoder. The control lines A0 and A1 are decoded as per the truth table, which provides selection between the 3 input signals and OFF, as noted.

3:1 VIDEO MULTIPLEXER USING AD813 TRIPLE OP AMP

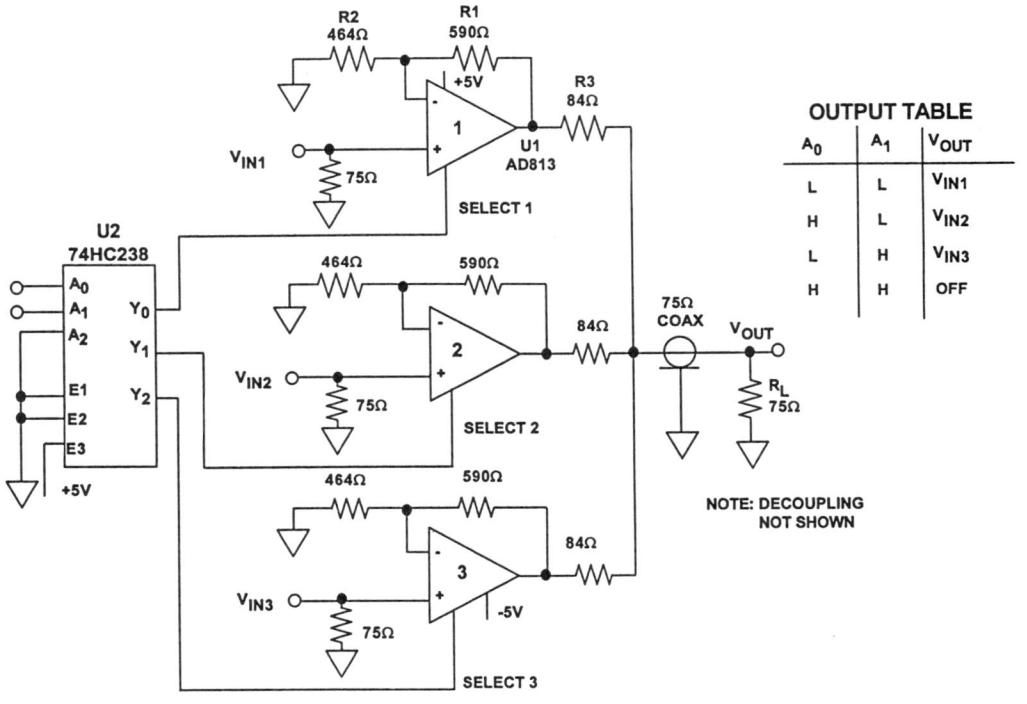

Figure 2.26

Some design subtleties of the circuit come about because of necessity to account for several design criteria. One is the 590Ω value of feedback resistor R1, to provide optimum response to the AD813 current feedback amplifier; another is the parasitic loading of the two unused gain resistor networks; a third is the source termination of the line, 75Ω in this case. While any given channel is ON, it drives not only load resistor R_L, but also the net dummy resistance $R_X/2$, where R_X is an equivalent series resistance equal to R1 + R2 + R3. To provide a net overall gain of unity plus and effective 75Ω source impedance, this sets the resistance values of R1 + R2 + R3 as shown.

HIGH SPEED OP AMPS

SWITCHING CHARACTERISTICS OF 3:1 VIDEO MULTIPLEXER

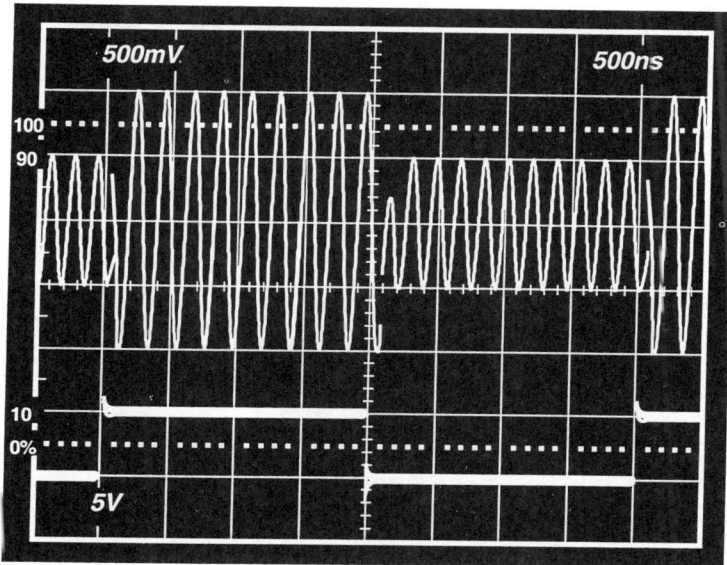

Vertical Scale: 500mV/div (Top Waveform), 5V/div (Bottom Waveform)
Horizontal Scale: 500ns/div

Figure 2.27

Performance of the circuit is excellent, with 0.1dB bandwidth of 20MHz, and an OFF state isolation of 60dB at 10MHz. Switching time is about 180ns, and is shown in Figure 2.27 switching between two different inputs (top trace) with the control input also shown (bottom trace).

Video Programmable Gain Amplifier

Closely related to the 3:1 multiplexer of Figure 2.26 is a programmable gain video amplifier, or PGA, as shown in Figure 2.28. With a similarly configured 2 line digital control input, this circuit can be set up to provide 3 different gain settings. This makes it a useful tool in various systems which can employ signal normalization or gain ranging prior to A/D conversion, such as CCD systems, ultrasound, etc. The gains can be binary related as here, or they can be arbitrary. An extremely useful feature of the AD813 current feedback amplifier to this application is the fact that the bandwidth does not reduce in inverse proportion, as gain is increased. Instead, it stays relatively constant as gain is raised. Thus more useful bandwidth is available at the higher programmed gains than would be true for a fixed gain-bandwidth product amplifier type.

2 - 33

HIGH SPEED OP AMPS

GAIN OF 1, 2, 4 PROGRAMMABLE GAIN VIDEO AMPLIFIER

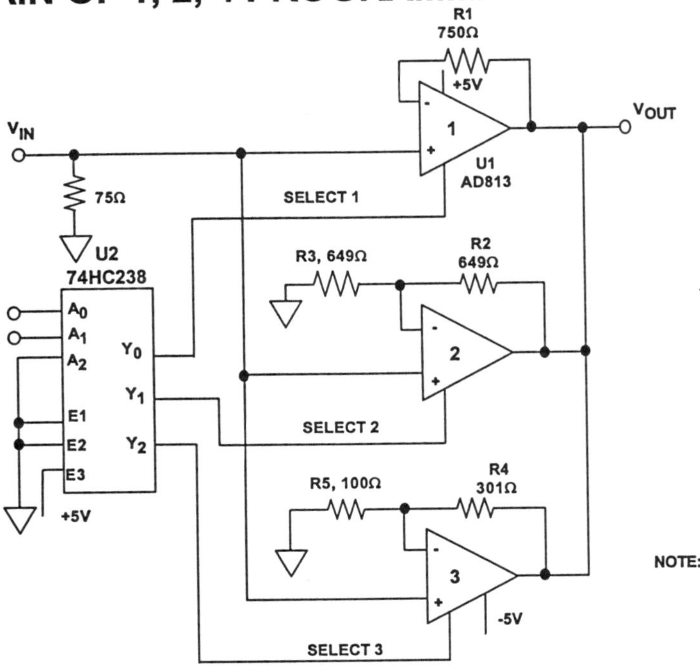

OUTPUT TABLE

A_0	A_1	V_{OUT}/V_{IN}
L	L	1
H	L	2, (1 + R2/R3)
L	H	4, (1 + R4/R5)
H	H	0, (OFF)

NOTE: DECOUPLING NOT SHOWN

Figure 2.28

In the circuit, channel 1 of the AD813 is a unity gain channel, channel 2 has a gain of 2, and channel 3 a gain of 4, while the fourth control state is OFF. As is indicated by the table, these gains can varied by adjustment of the R2/R3 or R4/R5 ratios. For the gain range and values shown, the PGA will be able to maintain a 3dB bandwidth of about 50MHz or more for loading as shown (a high impedance load of 1kΩ or more is assumed). Fine tuning of the bandwidth for a given gain setting can be accomplished by tweaking the absolute values of the feedback resistors (most applicable at higher gains).

Differential Drivers

Many applications require gain/phase matched complementary or differential signals. Among these are analog-digital-converter (ADC) input buffers, where differential operation can provide lower levels of 2nd harmonic distortion for certain converters. Other uses include high frequency bridge excitation, and drivers for balanced transmission twisted pair lines such as UTP-5. While various topologies can be employed to derive differential drive signals, many circuit details as well as the topologies themselves are important as to how accurate two outputs can be maintained.

HIGH SPEED OP AMPS

Inverter-Follower Differential Driver

The circuit of Figure 2.29 is useful as a high speed differential driver for driving high speed 10-12 bit ADCs, differential video lines, and other balanced loads at levels of 1-4Vrms. As shown it operates from ±5V supplies, but it can also be adapted to supplies in the range of ±5 to ±15V. When operated directly from ±5V as here, it minimizes potential for destructive ADC overdrive when higher supply voltage buffers drive a ±5V powered ADC, in addition to minimizing driver power.

DIFFERENTIAL DRIVER USING INVERTER/FOLLOWER

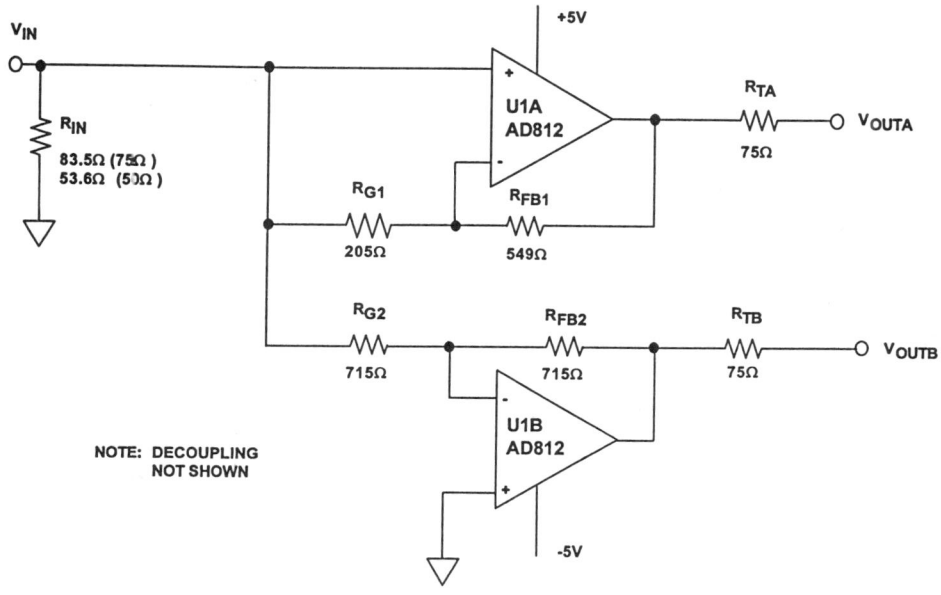

Figure 2.29

In many of these differential drivers the performance criteria is high. In addition to low output distortion, the two signals should maintain gain/phase flatness. In this driver, two sections of an AD812 dual current feedback amplifier are used for the channel A & B buffers, U1A & U1B. This measure can provide inherently better open-loop bandwidth matching than will the use of two individual same part number singles (where bandwidth varies between devices from different manufacturing lots).

The two buffers here operate with precise gains of ±1, as defined by their respective feedback and input resistances. Channel B buffer U1B is conventional, and uses a matched pair of 715Ω resistors- the value for using the AD812 on ±5V supplies.

In channel A, non-inverting buffer U1A has an inherent signal gain of 1, by virtue of the bootstrapped feedback network R_{FB1} and R_{G1} (Reference 5). It also has a higher noise gain, for phase matching. Normally a current feedback amplifier operating as a simple unity gain follower would use one (optimum) resistor R_{FB1}, and no gain resistor at all. Here, with input resistor R_{G1} added, a U1A noise gain like that of U1B results. Due to the bootstrap connection of R_{FB1}-R_{G1}, the signal gain is maintained at unity. Given the matched open loop bandwidths of U1A and U1B, similar noise gains in the A-B channels provide closely matched output bandwidths between the driver sides, a distinction which greatly impacts overall matching performance.

In setting up a design for the driver, the effects of resistor gain errors should be considered for R_{G2}-R_{FB2}. Here a worst case 2% mis-match will result in less than 0.2dB gain error between channels A and B. This error can be improved simply by specifying tighter resistor ratio matching, avoiding trimming.

If desired, phase matching is trimmed via R_{G1}, so that the phase of channel A closely matches that of B. This can be done for new circuit conditions, by using a pair of closely matched (0.1% or better) resistors to sum the A and B channels, as R_{G1} is adjusted for the best null conditions at the sum node. The A-B gain/phase matching is quite effective in this driver, with test results of the circuit as shown 0.04dB and 0.1° between the A and B output signals at 10MHz, when operated into dual 150Ω loads. The 3dB bandwidth of the driver is about 60MHz.

Net input impedance of the circuit is set to a standard line termination value such as 75Ω (or 50Ω), by choosing R_{IN} so that the desired value results with R_{IN} in parallel with R_{G2}. In this example, an R_{IN} value of 83.5Ω provides a standard input impedance of 75Ω when paralleled with 715Ω. For the circuit just as shown, dual voltage feedback amplifier types with sufficiently high speed and low distortion can also be used. This allows greater freedom with regard to resistor values using such devices as the AD826 and AD828.

Gain of the circuit can be changed if desired, but this is not totally straightforward. An easy step to satisfy diverse gain requirements is to simply use a triple amplifier such as the AD813, with the third channel as a variable gain input buffer.

HIGH SPEED OP AMPS

CROSS-COUPLED DIFFERENTIAL DRIVER PROVIDES BALANCED OUTPUTS AND 250MHz BANDWIDTH

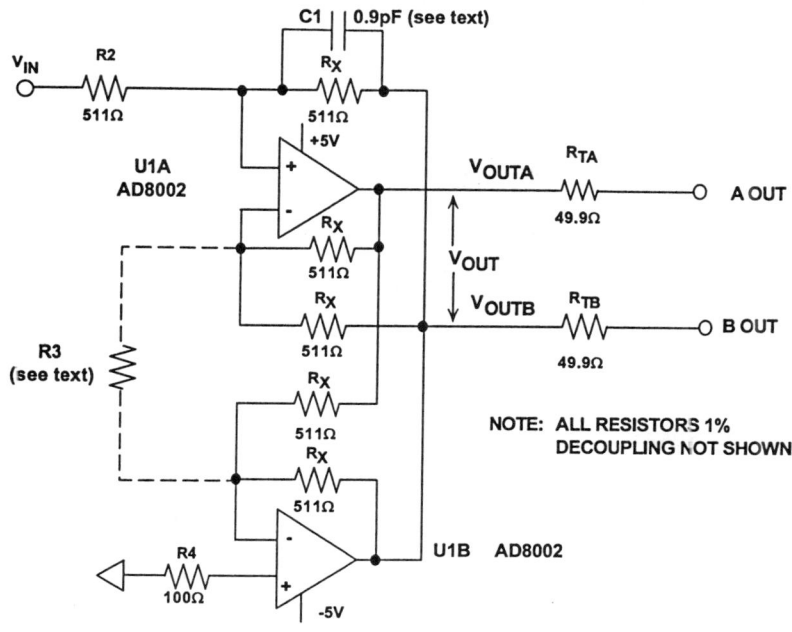

Figure 2.30

Cross-Coupled Differential Driver

Another differential driver approach uses cross-coupled feedback to get very high CMR and complementary outputs at the same time. In Figure 2.30, by connecting AD8002 dual current feedback amplifier sections as cross-coupled inverters, their outputs are forced equal and opposite, assuring zero output common mode voltage. The gain cell which results, U1A and U1B plus cross-coupling resistances R_X, is fundamentally a differential input and output topology, but it behaves as a voltage feedback amplifier with regard to the feedback port at the U1A (+) node. The gain of the stage from V_{IN} to V_{OUT} is:

$$G = \frac{V_{OUT}}{V_{IN}} = \frac{2R1}{R2}$$

where V_{OUT} is the differential output, equal to $V_{OUTA} - V_{OUTB}$.

This circuit has some unique benefits. First, differential gain is set by a single resistor ratio, so there is no necessity for side-side resistor matching with gain changes, as is the case for conventional differential amplifiers (see line receivers, below). Second, because the (overall) circuit emulates a voltage feedback amplifier, these gain resistances are not as restrictive as is true in the case of a conventional current feedback amplifier. Thus they are not highly critical as to absolute value. This is unlike standard applications using current feedback amplifiers, and will be true as long as the equivalent resistance seen by U1A is reasonably low (≤1kΩ in this case). Third, the cell bandwidth can be optimized to the desired gain by a single optional resistor, R3, as follows. If for instance, a net gain of 20 is desired (R1/R2=10), the bandwidth would otherwise be reduced by roughly this amount, since without R3 the cell operates with a constant gain-bandwidth product. With R3 present however, advantage can be taken of the AD8002 current feedback amplifier characteristics. Additional internal gain is added by the connection of R3, which, given an appropriate value, effectively raises gain-bandwidth to a level so as to restore the bandwidth which would otherwise be lost by the higher closed loop gain.

In the circuit as shown, no R3 is necessary at the low working gain of 2 times differential, since the 511Ω R_X resistors are already optimized for maximum bandwidth. Note that these four matched resistances are somewhat critical, and will change in absolute value with the use of another current feedback amplifier. At higher gain closed loop gains as set by R1/R2, R3 can be chosen to optimize the working transconductance in the input stages of U1A and U1B, as follows:

As in any high speed inverting feedback amplifier, a small high-Q chip type feedback capacitance, C1, may be needed to optimize flatness of frequency response. In this example, a 0.9pF value was found optimum for minimizing peaking. In general, provision should be made on the PC layout for an NPO chip capacitor in the range of 0.5-2pF. This capacitor is then value selected at board characterization for optimum frequency response.

For the dual trace, 1-500MHz swept frequency response plot of Figure 2.31, output levels were 0dBm into matched 50Ω loads, through back termination resistances R_{TA} and R_{TB}, at V_{OUTA} and V_{OUTB}. In this plot the vertical scale is 2dB/div, and it shows the 3dB bandwidth of the driver measuring about 250MHz, with peaking about 0.1dB. The four R_X resistors along with R_{TA} and R_{TB} control low frequency amplitude matching, which was within 0.1dB in the lab tests, using 511Ω 1% resistor types. For tightest amplitude matching, these resistor ratios can be more closely controlled.

Due to the high gain-bandwidths involved with the AD8002, the construction of this circuit should follow RF rules, with the use of a ground plane, chip bypass capacitors of zero lead length at the ±5V supply pins, and surface mount resistors for lowest inductance.

$$R3 \cong \frac{Rx}{(R1/R2) - 1}$$

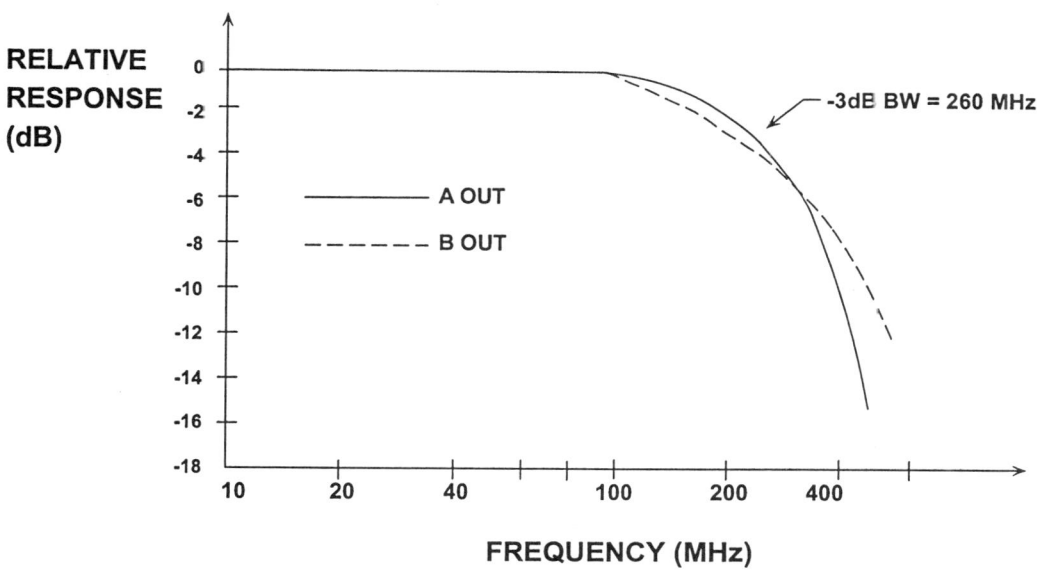

Figure 2.31

Differential Receivers

Another standard system application is the line receiver function, a stage which accepts signals from a single-ended (or differential) transmission line, and converts them into a buffered version for local processing (referring back to the system diagram of Figure 2.11). In a typical system, there is a single ended driving signal V_{IN} at the origination point, a coaxial transmission line with the signal terminated in the characteristic line impedance at the local end, and finally, a differential input receiver. The system operates the same in principle for standard impedances of 50-100Ω, as long as the source and load impedances match that of the line, under which conditions bandwidth is maximized. In the receiver, input impedance is assumed high in relation to line impedance, and high common-mode rejection (CMR) allows rejection of spurious noise appearing between driver/receiver grounds. Noise is rejected in proportion to the CMR of the receiver amplifier, and typical CMR performance goal for such a stage is 60-70dB or better for frequencies up to 10MHz.

Functionally, this system delivers at the output of the receiver stage a signal V_{OUT}, a replica of the original driving signal V_{IN}. The receiver may also scale the received signal, and may also be called on to drive another transmission line. Critical performance parameters for the receiver are signal bandwidth and distortion specifications. For line receivers, video distortion is rated in terms of differential gain and differential phase at the 3.58 MHz subcarrier frequency.

4 Resistor Differential Line Receiver

Figure 2.32 shows a low cost, medium performance line receiver using a high speed op amp rated for video use. It is actually a standard 4 resistor instrumentation amplifier optimized for high speed, with a differential to single-ended gain of R2/R1. Using low value, DC accurate/AC trimmed resistances for R1-R4 and a high speed, high CMR op amp provides the good performance. Practically speaking however, at low frequencies resistor matching can be more critical to overall CMR than the rated CMR of the op amp. For example, the worst case CMR (in dB) of this circuit due to resistor effects is:

$$\text{CMR} = 20\log_{10}\left(\frac{1+\frac{R2}{R1}}{4Kr}\right)$$

In this expression the term "Kr" is a single resistor tolerance in fractional form (1%=0.01, etc.), and it is assumed the amplifier has significantly higher CMR (≥100dB). Using discrete 1% metal films for R1/R2 and R3/R4 yields a worst case CMR of 34dB, 0.1% types 54dB, etc. Of course 4 random 1% resistors will on the average yield a CMR better than 34dB, but not dramatically so. A single substrate dual matched pair thin film network is preferred, for reasons of best noise rejection and simplicity. One type suitable is the Ohmtek 1005, (Reference 6) which has a ratio match of 0.1%, which will provide a worst case low frequency CMR of 66 dB.

SIMPLE VIDEO LINE RECEIVER USING THE AD818 OP AMP

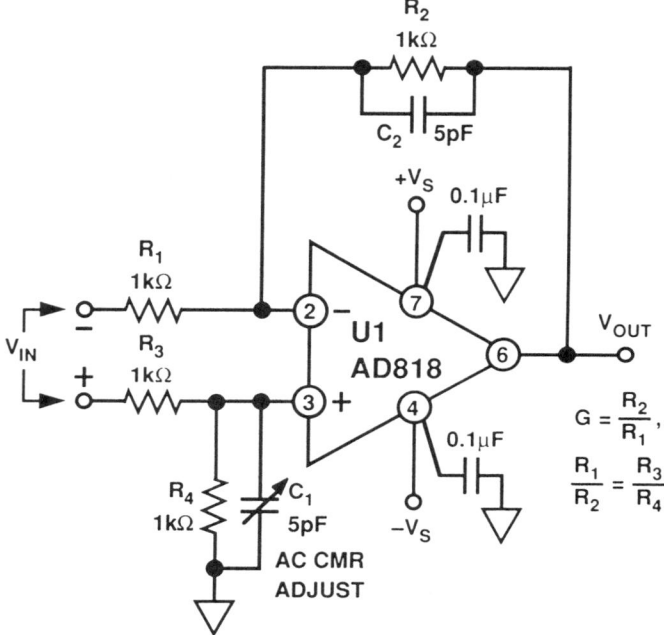

Figure 2.32

This circuit has an interesting and desirable side property. Because of the resistors it divides down the input voltage, and the amplifier is protected against overvoltage. This allows CM voltages to exceed ±5V supply rails in some cases without hazard. Operation at ±15V should constrain the inputs within the rails.

At frequencies above 1MHz, the bridge balance is dominated by AC effects, and a C1-C2 capacitive balance trim should be used for best performance. The C1 adjustment is intended to allow this, providing for the cancellation of stray layout capacitance(s) by electrically matching the net C1-C2 values. In a given PC layout with low and stable parasitic capacitance, C1 is best adjusted once in 0.5pF increments, for best high frequency CMR. Using designated PC pads, production values then would use the trimmed value. Good AC matching is essential to achieving good CMR at high frequencies. C1-C2 should be types similar physically, such as NPO (or other stable) ceramic chip style capacitors.

While the circuit as shown has unity gain, it can be gain-scaled in discrete steps, as long as the noted resistor ratios are maintained. In practice, this means using taps on a multi-ratio network for gain change, so as to raise both R2 and R4, in identical proportions. There is no other simple way to change gain in this receiver circuit. Alternately, a scheme for continuous gain control without interaction with CMR is to follow this receiver with a scaling amplifier/driver with adjustable gain. The similar AD828 dual amplifier allows this with the addition of only two resistors.

Video gain/phase performance of this stage is dependent upon the device is used for U1 and the operating supply voltages. Suitable voltage feedback amplifiers work best at supplies of ±10 - ±15V, which maximizes op amp bandwidth. And, while many high speed amplifiers function in this circuit, those expressly designed with low distortion video operation perform best. The circuit as shown can be used with supplies of ±5 to ±15V, but lowest NTSC video distortion occurs for supplies of ±10V or more, where differential gain/differential phase errors are less than 0.01%/0.05°. Operating at ±5V the distortion rises somewhat, but the lowest power drain of 70mW occurs.

One drawback to this circuit is that it does load a 75Ω video line to some extent, and so should be used with this loading taken into account. On the plus side, it has wide dynamic range for both signal and CM voltages, plus the inherent overvoltage protection.

Active Feedback Differential Line Receiver

Fully integrating the line receiver function eliminates the resistor-related drawbacks of the 4 resistor line receiver, improving CMR performance, ease of use, and overall circuit flexibility. An IC designed for this function is the AD830 active feedback amplifier (Reference 7,8). Its use as a differential line receiver with gain is illustrated in Figure 2.33.

VIDEO LOOP-THROUGH CONNECTION USING THE AD830

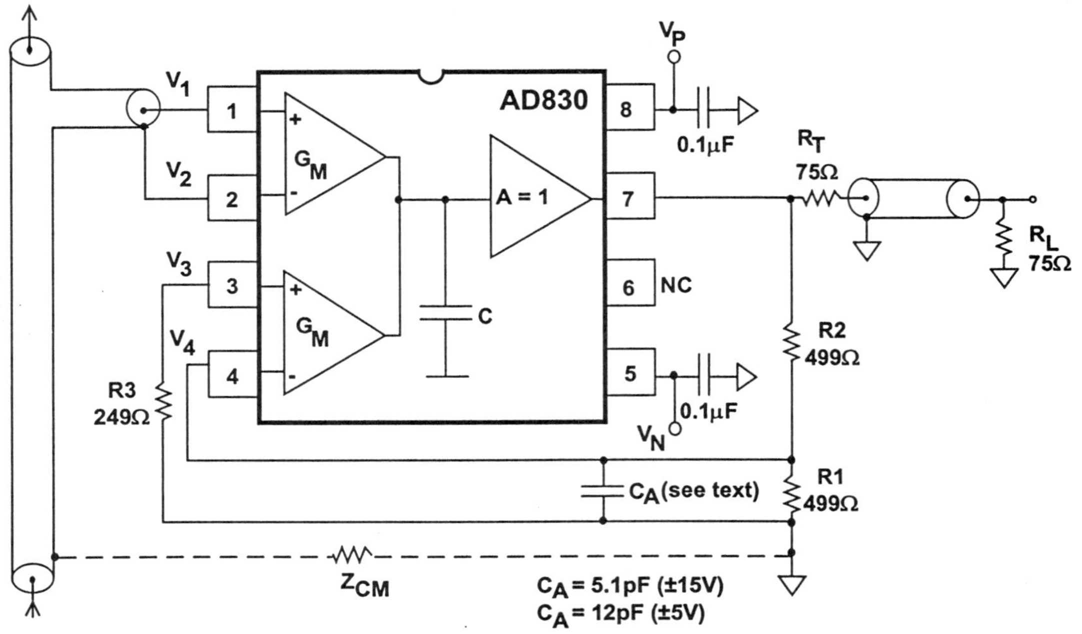

Figure 2.33

$C_A = 5.1pF\ (\pm15V)$
$C_A = 12pF\ (\pm5V)$

The AD830 operates as a feedback amplifier with two sets of fully differential inputs, available at pins 1-2 and 3-4, respectively. Internally, the outputs of the two stages are summed and drive a buffer output stage. Both input stages have high CMR, and can handle differential signals up to ±2V, and CM voltages can range up to $-V_S+3V$ or $+V_S-2.1V$, with a 1V differential input applied. While the AD830 does not normally need protection against CM voltages, if sustained transient voltage beyond the rails is encountered, an optional pair of equal value ($\cong 200\Omega$) resistances can be used in series with pins 1-2.

In this device the overall feedback loop operates so that the differential voltages V_{1-2} and V_{3-4} are forced to be equal. Feedback is taken from the output back to one input differential pair, while the other pair is driven by a differential input signal. An important point of this architecture is that high CM rejection is provided by the two differential input pairs, so CMR isn't dependent on resistor bridges and their associated matching problems. The inherently wideband balanced circuit and the quasi-floating operation of the driven input provide the high CMR, which is typically 100dB at DC.

The general expression for the U1 stage's gain "G" is like a non-inverting op amp, or:

$$G = \frac{V_{OUT}}{V_{IN}} = 1 + \frac{R2}{R1}$$

For lowest DC offset, balancing resistor R3 is used (equal to R1 || R2).

In this example of a video "loop-through" connection, the input signal tapped from a coax line and applied to one input stage at pins 1-2, with the scaled output signal tied to the second input stage between pins 3-4. With the R1-R4 feedback attenuation of 2/1, the net result is that the output of U1, is then equal to $2 \cdot V_{IN}$, i.e., a gain of 2.

Functionally, the input and local grounds are isolated by the CMR of the AD830, which is typically 75dB at frequencies below 1MHz, 60dB at 4.43MHz, and relatively supply independent.

With the addition of an output source termination resistor R_T, this circuit has an overall loaded gain of unity at the load termination, R_L. It is a ground isolating video repeater, driving the terminated 75Ω output line, delivering a final output equal to the original input, V_{IN}.

NTSC video performance will be dependent upon supplies. Driving a terminated line as shown, the circuit has optimum video distortion levels for V_s ±15V, where differential gain is typically 0.06%, and differential phase 0.08°. Bandwidth can be optimized by the optional 5.1pF (or 12pF) capacitor, C_A, which allows a 0.1dB bandwidth of 10MHz with ±15V operation. The differential gain and phase errors deteriorate about 2 or more times at ±5V.

REFERENCES:

1. Walt Kester, *Maintaining Transmission Line Impedances on the PC Board*, within Chapter 11 of **System Applications Guide**, Analog Devices, 1993.

2. Joe Buxton, *Careful Design Tames High-Speed Op Amps*, **Electronic Design**, April 11, 1991.

3. Walt Jung, *Op Amps in Line-Driver and Receiver Circuits, Part 1*, **Analog Dialogue**, Vol. 26-2, 1992.

4. William R. Blood, Jr., **MECL System Design Handbook** (HB205, Rev.1), Motorola Semiconductor Products, Inc., 1988.

5. Dave Whitney, Walt Jung, *Applying a High-Performance Video Operational Amplifier*, **Analog Dialogue**, 26-1, 1992.

6. Ohmtek, Niagara Falls, NY, (716) 283-4025.

7. Walt Kester, *Video Line Receiver Applications Using the AD830 Active Feedback Amplifier Topology*, within Chapter 11 of **System Applications Guide**, Analog Devices, 1993.

8. Walt Jung, *Analog-Signal-Processing Concepts Get More Efficient*, **Electronic Design Analog Applications Issue**, June 24, 1993.

9. Walt Jung, Scott Wurcer, *Design Video Circuits Using High-Speed Op-Amp Systems*, **Electronic Design Analog Applications Issue**, November 7, 1994.

10. Thomas M. Frederiksen, **Intuitive Operational Amplifiers**, McGraw Hill, 1988.

11. D. Stout, M. Kaufman, **Handbook of Operational Amplifier Circuit Design**, New York, McGraw-Hill, 1976.

12. J. Dostal, **Operational Amplifiers**, Elsevier Scientific Publishing, New York, 1981.

13. **1992 Amplifier Applications Guide**, 1992, Analog Devices.

14. Walter G. Jung, **IC Op Amp Cookbook, Third Edition**, SAMS (Division of Macmillan, Inc.), 1986.

15. W. A. Kester, *PCM Signal Codecs for Video Applications*, **SMPTE Journal**, No. 88, November 1979, pp. 770-778.

16. *IEEE Standard for Performance Measurements of A/D and D/A Converters for PCM Television Circuits*, **IEEE Standard 746-1984**.

17. Tim Henry, *Analysis and Design of the Op Amp Current Source*, **Application Note AN-587**, Motorola, Inc., 1973.

HIGH SPEED OP AMPS

SECTION 3

HIGH RESOLUTION SIGNAL CONDITIONING ADCs

- Sigma-Delta ADCs

- High Resolution, Low Frequency Measurement ADCs

High Resolution Signal Conditioning ADCs

SECTION 3

HIGH RESOLUTION SIGNAL CONDITIONING ADCs
Walt Kester, James Bryant, Joe Buxton

The trend in ADCs and DACs is toward higher speeds and higher resolutions at reduced power levels. Modern data converters generally operate on ±5V (dual supply) or +5V (single supply). There are now a few converters which operate on a single +3V supply. This trend has created a number of design and applications problems which were much less important in earlier data converters, where ±15V supplies were the standard.

Lower supply voltages imply smaller input voltage ranges, and hence more susceptibility to noise from all potential sources: power supplies, references, digital signals, EMI/RFI, and probably most important, improper layout, grounding, and decoupling techniques. Single-supply ADCs often have an input range which is not referenced to ground. Finding compatible single-supply drive amplifiers and dealing with level shifting of the input signal in direct-coupled applications also becomes a challenge.

In spite of these issues, components are now available which allow extremely high resolutions at low supply voltages and low power. This section discusses the applications problems associated with such components and shows techniques for successfully designing them into systems.

LOW POWER, LOW VOLTAGE ADC DESIGN ISSUES

- Low Power ADCs typically run on ±5V, +5V, +5/+3V, or +3V
- Lower Signal Swings Increase Sensitivity to All Types of Noise (Device, Power Supply, Logic, etc.)
- Device Noise Increases at Low Quiescent Currents
- Bandwidth Suffers as Supply Current Drops
- Input Common-Mode Range May be Limited
- Selection of Zero-Volt Input/Output Amplifiers is Limited
- Auto-Calibration Modes Highly Desirable at High Resolutions

Figure 3.1

Sigma-Delta (Σ-Δ) ADCs
(Courtesy of James M. Bryant)

Because Sigma-Delta (Σ –Δ) is such an important and popular architecture for high resolution (16 to 24 bits) ADCs, the section begins with a basic description of this type of converter.

Sigma-Delta Analog-Digital Converters (Σ –Δ ADCs) have been known for nearly thirty years, but only recently has the technology (high-density digital VLSI) existed to manufacture them as inexpensive monolithic integrated circuits. They are now used in many applications where a low-cost, low-bandwidth, low-power, high-resolution ADC is required.

There have been innumerable descriptions of the architecture and theory of Σ –Δ ADCs, but most commence with a maze of integrals and deteriorate from there. In the Applications Department at Analog Devices, we frequently encounter engineers who do not understand the theory of operation of Σ –Δ ADCs and are convinced, from study of a typical published article, that it is too complex to comprehend easily.

There is nothing particularly difficult to understand about Σ –Δ ADCs, as long as you avoid the detailed mathematics, and this section has been written in an attempt to clarify the subject. A Σ –Δ ADC contains very simple analog electronics (a comparator, a switch, and one or more integrators and analog summing circuits), and quite complex digital computational circuitry. This circuitry consists of a digital signal processor (DSP) which acts as a filter (generally, but not invariably, a low pass filter). It is not necessary to know precisely how the filter works to appreciate what it does. To understand how a Σ –Δ ADC works one should be familiar with the concepts of *over-sampling, noise shaping, digital filtering,* and *decimation.*

An ADC is a circuit whose digital output is proportional to the ratio of its analog input to its analog reference. Often, but by no means always, the scaling factor between the analog reference and the analog signal is unity, so the digital signal represents the normalized ratio of the two.

Figure 3.4 shows the transfer characteristic of an ideal 3-bit unipolar ADC. The input to an ADC is analog and is not quantized, but its output is quantized. The transfer characteristic therefore consists of eight horizontal steps (when considering the offset, gain and linearity of an ADC we consider the line joining the midpoints of these steps).

SIGMA-DELTA (Σ-Δ) ADCs

- Sigma-Delta ADCs are low-cost and have high resolution, excellent DNL, low-power, although limited input bandwidth

- A Σ-Δ ADC is Simple

- The Mathematics, however, is Complex

- This section concentrates on What Actually Happens!

Figure 3.2

SIGMA-DELTA ADC KEY CONCEPTS

- Oversampling
- Noise Shaping
- Digital Filtering
- Decimation

Figure 3.3

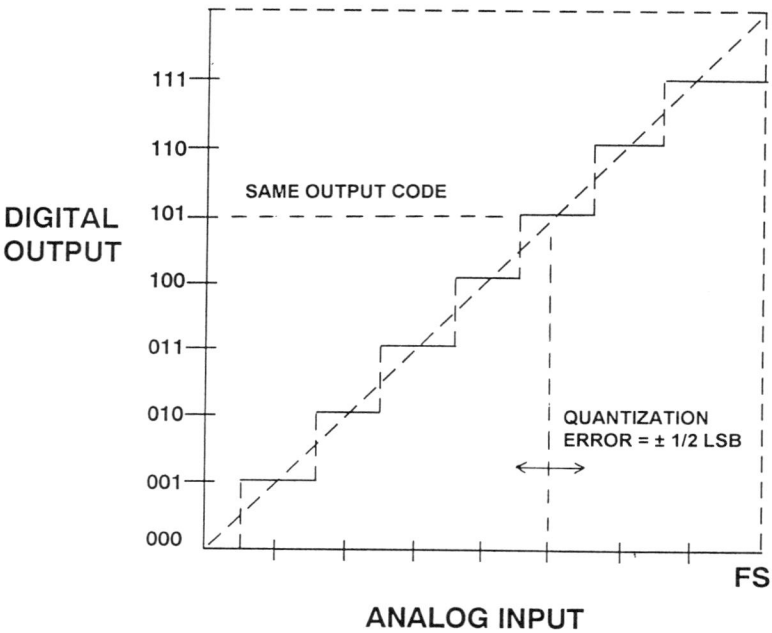

Figure 3.4

Digital full scale (all "1"s) corresponds to 1 LSB below the analog full scale (the reference or some multiple thereof). This is because, as mentioned above, the digital code represents the *normalized* ratio of the analog signal to the reference, and if this were unity, the digital code would be all "0"s and "1" in the bit *above* the MSB.

The (ideal) ADC transitions take place at ½ LSB above zero and thereafter every LSB, until 1½ LSB below analog full scale. Since the analog input to an ADC can take any value, but the digital output is quantized, there may be a difference of up to ½ LSB between the actual analog input and the exact value of the digital output. This is known as the *quantization error* or *quantization uncertainty*. In AC (sampling) applications, this quantization error gives rise to quantization noise.

If we apply a fixed input to an ideal ADC, we will always obtain the same output, and the resolution will be limited by the quantization error.

Suppose, however, that we add some AC (dither) to the fixed signal, take a large number of samples, and prepare a histogram of the results. We will obtain something like the result in Figure 3.5. If we calculate the mean value of a large number of samples, we will find that we can measure the fixed signal with greater resolution than that of the ADC we are using. This procedure is known as *over-sampling*.

OVERSAMPLING WITH DITHER ADDED TO INPUT

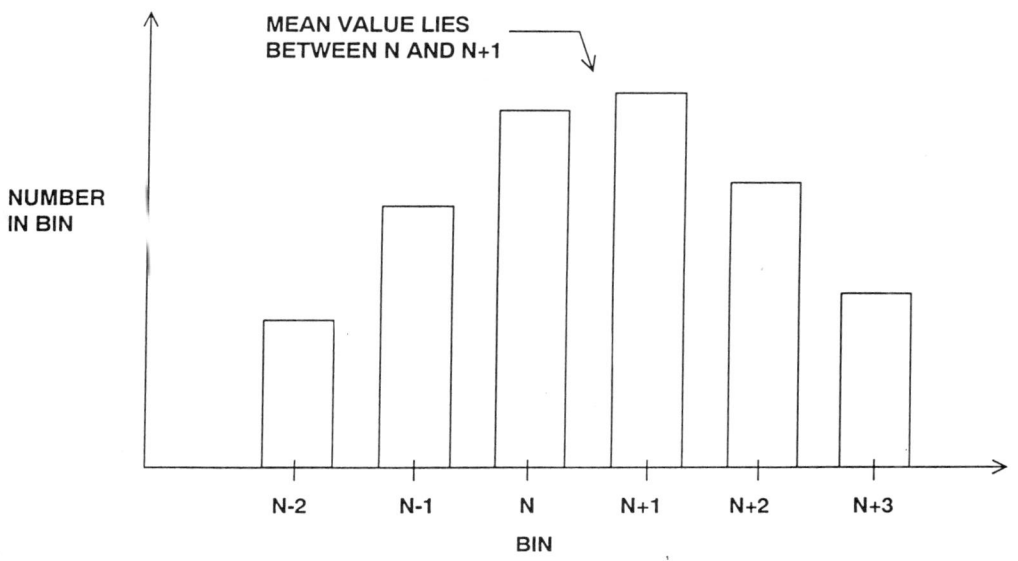

Figure 3.5

The AC (dither) that we add may be a sine-wave, a tri-wave, or gaussian noise (but *not* a square wave) and, with some types of sampling ADCs (including Σ–Δ ADCs), an external dither signal is unnecessary, since the ADC generates its own. Analysis of the effects of differing dither waveforms and amplitudes is complex and, for the purposes of this section, unnecessary. What we do need to know is that with the simple oversampling described here, the number of samples must be doubled for each ½ bit of increase in effective resolution.

If, instead of a fixed DC signal, the signal that we are over-sampling is an AC signal, then it is not necessary to add a dither signal to it in order to over-sample, since the signal is moving anyway. (If the AC signal is a single tone harmonically related to the sampling frequency, dither may be necessary, but this is a special case.)

Let us consider the technique of oversampling with an analysis in the frequency domain. Where a DC conversion has a *quantization error* of up to ½ LSB, a sampled data system has *quantization noise*. As we have already seen, a perfect classical N-bit sampling ADC has an rms quantization noise of $q/\sqrt{12}$ uniformly distributed within the Nyquist band of DC to $f_s/2$ (where q is the value of an LSB and f_s is the sampling rate). Therefore, its SNR with a full-scale sinewave input will be $(6.02N + 1.76)$ dB. If the ADC is less than perfect, and its noise is greater than its theoretical minimum quantization noise, then its *effective* resolution will be less than N-bits. Its actual resolution (often known as its Effective Number of Bits or ENOB) will be defined by

$$\text{ENOB} = \frac{\text{SNR} - 1.76\text{dB}}{6.02\text{dB}}.$$

If we choose a much higher sampling rate, the quantization noise is distributed over a wider bandwidth as shown in Figure 3.7. If we then apply a digital low pass filter (LPF) to the output, we remove much of the quantization noise, but do not affect the wanted signal - so the ENOB is improved. We have accomplished a high resolution A/D conversion with a low resolution ADC.

Since the bandwidth is reduced by the digital output filter, the output data rate may be lower than the original sampling rate and still satisfy the Nyquist criterion. This may be achieved by passing every Mth result to the output and discarding the remainder. The process is known as "decimation" by a factor of M. Despite the origins of the term (*decem* is Latin for ten), M can have any integer value, provided that the output data rate is more than twice the signal bandwidth. Decimation does not cause any loss of information (see Figure 3.8).

If we simply use over-sampling to improve resolution, we must over-sample by a factor of 2^{2N} to obtain an N-bit increase in resolution. The $\Sigma-\Delta$ converter does not need such a high over-sampling ratio because it not only limits the signal passband, but also shapes the quantization noise so that most of it falls outside this passband.

SAMPLING ADC QUANTIZATION NOISE

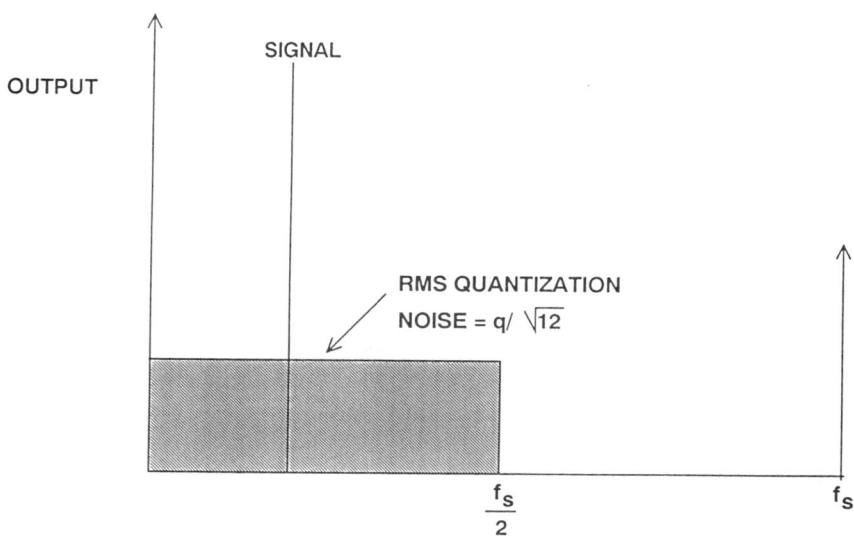

Figure 3.6

HIGH RESOLUTION SIGNAL CONDITIONING ADCs

OVERSAMPLING FOLLOWED BY DIGITAL FILTERING AND DECIMATION IMPROVES SNR AND ENOB

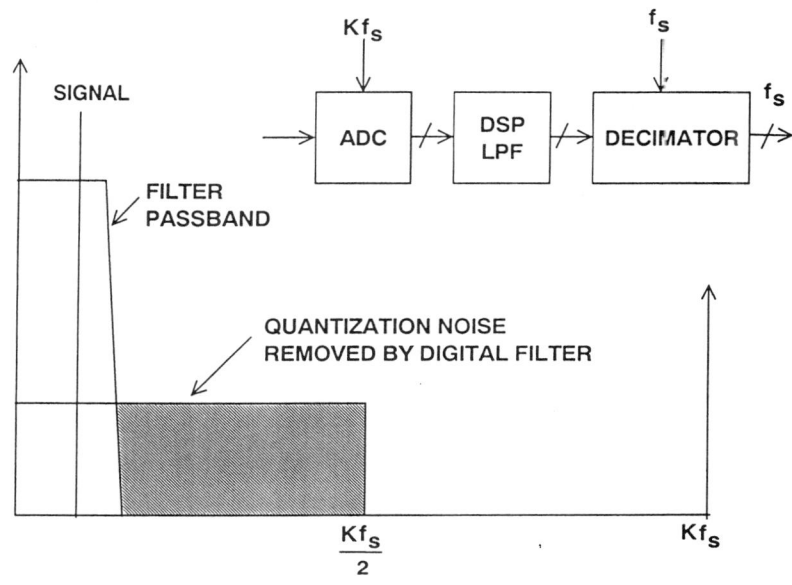

Figure 3.7

DECIMATION

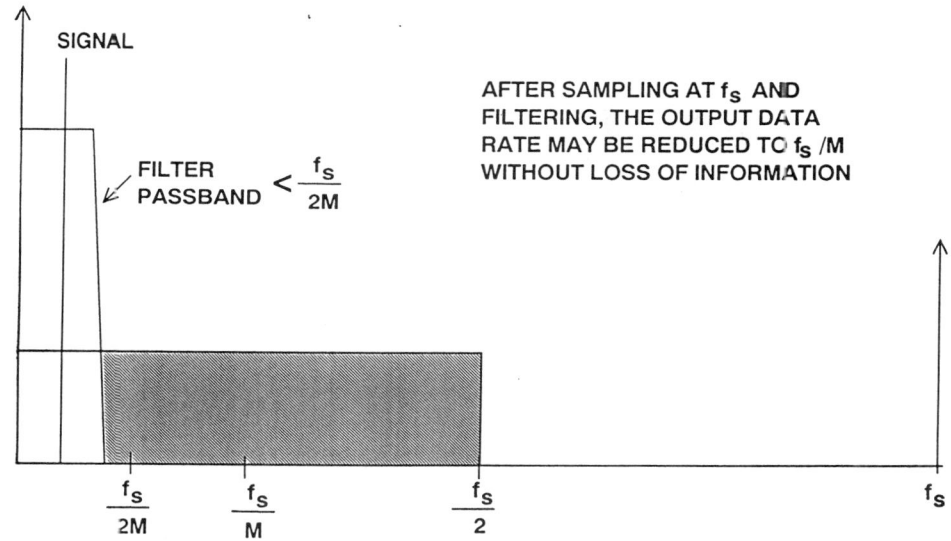

AFTER SAMPLING AT f_S AND FILTERING, THE OUTPUT DATA RATE MAY BE REDUCED TO f_S/M WITHOUT LOSS OF INFORMATION

Figure 3.8

If we take a 1-bit ADC (generally known as a comparator), drive it with the output of an integrator, and feed the integrator with an input signal summed with the output of a 1-bit DAC fed from the ADC output, we have a first-order Σ–Δ modulator as shown in Figure 3.9. Add a digital low pass filter (LPF) and decimator at the digital output, and we have a Σ–Δ ADC: the Σ–Δ modulator shapes the quantization noise so that it lies above the passband of the digital output filter, and the ENOB is therefore much larger than would otherwise be expected from the over-sampling ratio.

By using more than one integration and summing stage in the Σ–Δ modulator, we can achieve higher orders of quantization noise shaping and even better ENOB for a given over-sampling ratio as is shown in Figure 3.10 for both a first and second-order Σ–Δ modulator. The block diagram for the second-order Σ–Δ modulator is shown in Figure 3.11. Third, and higher, order Σ–Δ ADCs were once thought to be potentially unstable at some values of input — recent analyses using *finite* rather than infinite gains in the comparator have shown that this is not necessarily so, but even if instability does start to occur, it is not important, since the DSP in the digital filter and decimator can be made to recognize incipient instability and react to prevent it.

FIRST-ORDER SIGMA-DELTA ADC

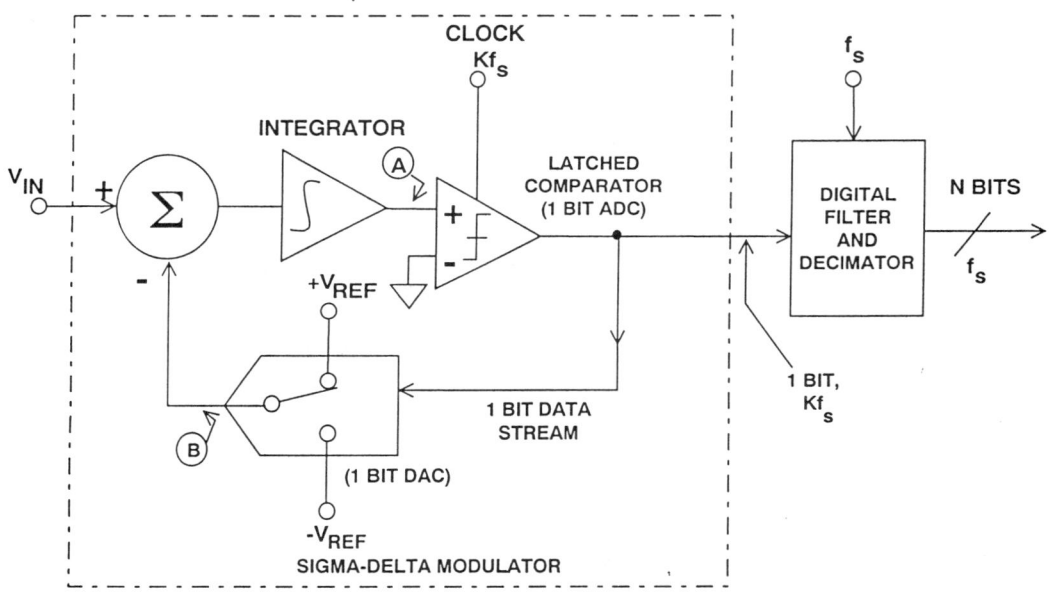

Figure 3.9

SIGMA-DELTA MODULATORS SHAPE QUANTIZATION NOISE

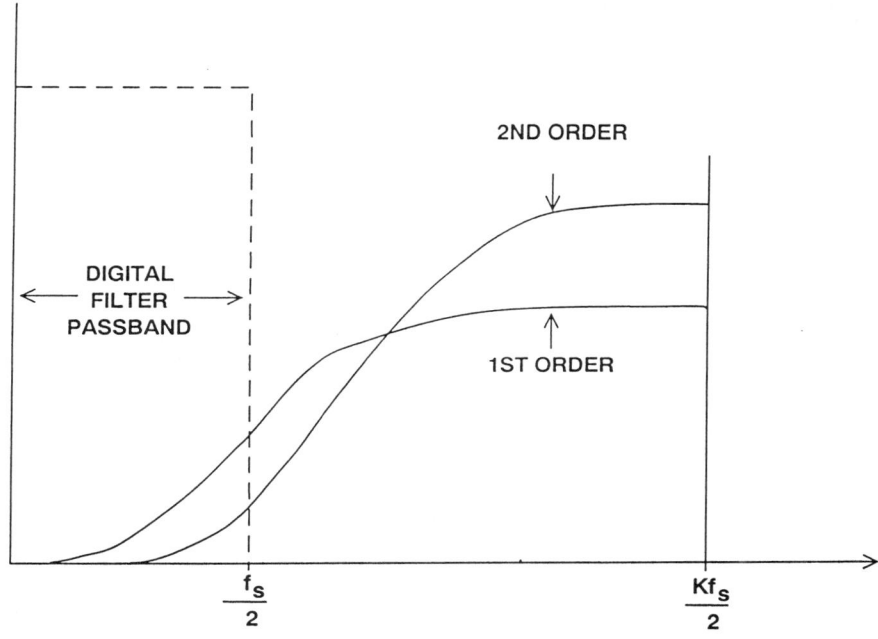

Figure 3.10

SECOND-ORDER SIGMA-DELTA ADC

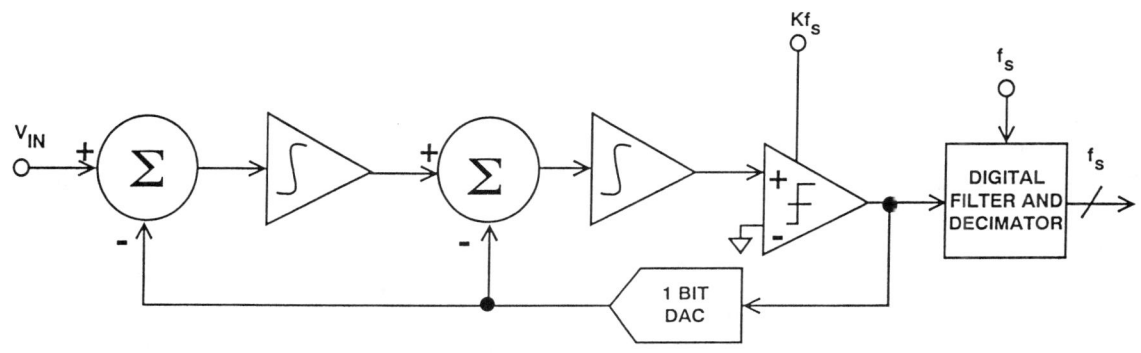

Figure 3.11

HIGH RESOLUTION SIGNAL CONDITIONING ADCs

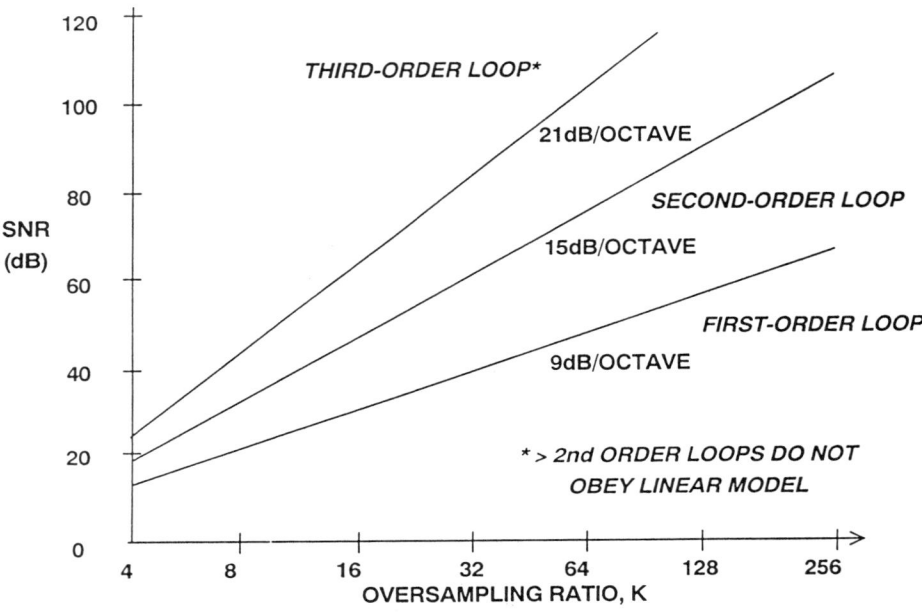

Figure 3.12

Figure 3.12 shows the relationship between the order of the $\Sigma-\Delta$ modulator and the amount of over-sampling necessary to achieve a particular SNR.

The $\Sigma-\Delta$ ADCs that we have described so far contain integrators, which are low pass filters, whose passband extends from DC. Thus, their quantization noise is pushed up in frequency. At present, all commercially available $\Sigma-\Delta$ ADCs are of this type (although some which are intended for use in audio or telecommunications applications contain bandpass rather than lowpass digital filters to eliminate any system DC offsets). $\Sigma-\Delta$ ADCs are available with resolutions up to 24-bits for DC measurement applications (AD7710, AD7711, AD7712, AD7713, AD7714), and with resolutions of 18-bits for high quality digital audio applications (AD1879).

But there is no particular reason why the filters of the $\Sigma-\Delta$ modulator should be LPFs, except that traditionally ADCs have been thought of as being baseband devices, and that integrators are somewhat easier to construct than bandpass filters. If we replace the integrators in a $\Sigma-\Delta$ ADC with bandpass filters (BPFs), the quantization noise is moved up and down in frequency to leave a virtually noise-free region in the pass-band (see Reference 1). If the digital filter is then programmed to have its pass-band in this region, we have a $\Sigma-\Delta$ ADC with a bandpass, rather than a low pass characteristic (see Figure 3.13). Although studies of this architecture are in their infancy, such ADCs would seem to be ideally suited for use in digital radio receivers, medical ultrasound, and a number of other applications.

REPLACING INTEGRATORS WITH RESONATORS GIVES A BANDPASS SIGMA-DELTA ADC

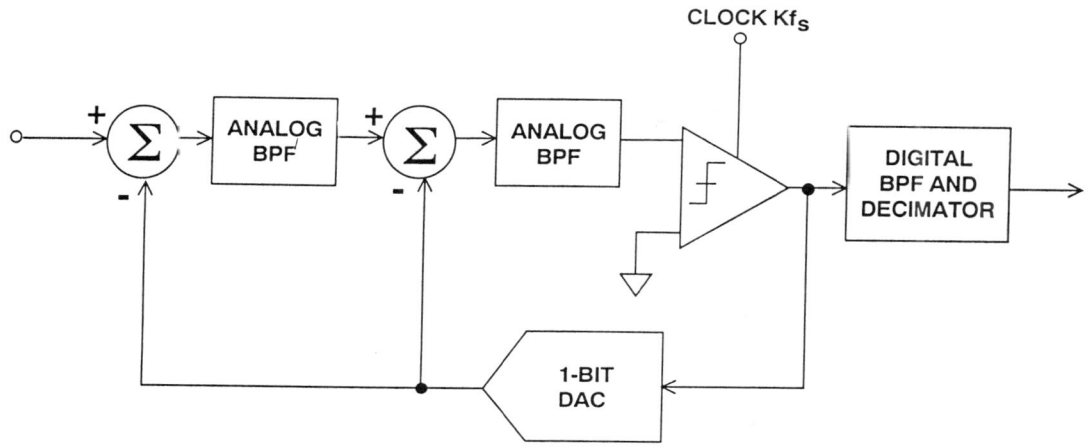

Figure 3.13

A Σ–Δ ADC works by over-sampling, where simple analog filters in the Σ–Δ modulator shape the quantization noise so that the SNR *in the bandwidth of interest* is much lower than would otherwise be the case, and by using high performance digital filters and decimation to eliminate noise outside the required passband. Because the analog circuitry is so simple and undemanding, it may be built with the same digital VLSI process that is used to fabricate the DSP circuitry of the digital filter. Because the basic ADC is 1-bit (a comparator), the technique is inherently linear.

Although the detailed analysis of Σ–Δ ADCs involves quite complex mathematics, their basic design can be understood without the necessity of any mathematics at all. For further discussion on Σ–Δ ADCs, refer to References 2 and 3.

SIGMA-DELTA SUMMARY

- Linearity is Inherently Excellent

- High Resolutions (16 - 24 Bits)

- Ideal for Mixed-Signal IC Processes, no Trimming

- No SHA Required

- Charge Injection at Input Presents Drive Problems

- Upper Sampling Rate Currently Limits Applications to Measurement, Voiceband, and Audio, but Bandpass Sigma-Delta Techniques Will Change This

- Analog Multiplexing Applications are Limited by Internal Filter Settling Time. Consider One Sigma-Delta ADC per Channel.

Figure 3.14

High Resolution, Low-Frequency Measurement ADCs

The AD7710, AD7711, AD7712, AD7713, and AD7714 are members of a family of sigma-delta converters designed for high accuracy, low frequency measurements. They have no missing codes at 24-bits and useful resolution of up to 21.5-bits (AD7710, AD7711, AD7712, and AD7713), and 22.5 bits (AD7714). They all use similar sigma-delta cores, and their main differences are in their analog inputs, which are optimized for different transducers. The AD7714 is the newest member of the family and is fully specified for either +5V (AD7714-5) or +3V (AD7714-3) operation.

The digital filter in the sigma-delta core may be programmed by the user for output update rates between 10Hz and 1kHz (AD7710, AD7711, AD7712), 2Hz and 200Hz (AD7713), and 2Hz and 1kHz (AD7714). The effective resolution of these ADCs is inversely proportional to the bandwidth. For example, for 22.5-bits of effective resolution, the output update rate of the AD7714 cannot exceed 10Hz. The AD771X family is ideal for such sensor applications as those shown in Figure 3.15.

The AD771X family has a high level of integration which simplifies the design of data acquisition systems. For example, the AD7710 (Figures 3.16 and 3.17) has two high impedance differential inputs that can be interfaced directly to many different sensors, includ-

SIGNAL CONDITIONING, TRANSDUCER INPUT ADCs: THE AD7710, AD7711, AD7712, AD7713, AD7714

- ■ Ultra-High Resolution Measurement Systems
- ■ Implemented Using $\Sigma\Delta$ Conversion
- ■ Ideal for Applications Such As:
  - ◆ Weigh Scales
  - ◆ RTDs
  - ◆ Thermocouples
  - ◆ Strain Gauges
  - ◆ Process Control
  - ◆ Smart Transmitters
  - ◆ Medical

Figure 3.15

ing resistive bridges. The two inputs are selected by the internal multiplexer, which passes the signal to a programmable gain amplifier (PGA). The PGA has a digitally programmable gain range of 1 to 128 to accommodate a wide range of signal inputs. After the PGA, the signal is digitized by the sigma-delta modulator. The digital filter notch frequency may be adjusted from 10Hz to 1kHz, which allows various input bandwidths.

To achieve this high accuracy, the AD771X family has four different internal calibration modes, including system and background calibration. All of these functions are controlled via a serial interface. A benefit of this serial interface is that the AD771X-family fits into a 24-pin package, giving a small footprint for the high level of integration. All of the parts except the AD7713 and AD7714 can operate on either a single +5V or dual ±5V supplies. The AD7713 is designed exclusively for single supply (+5V) operation. The AD7714 is the newest member of the family and is designed for either single +3V (AD7714-3) or single +5V (AD7714-5) low power applications. The AD771X family has <0.0015% non-linearity.

HIGH RESOLUTION SIGNAL CONDITIONING ADCs

THE AD771X-SERIES PROVIDES A HIGH LEVEL OF INTEGRATION IN A 24-PIN PACKAGE

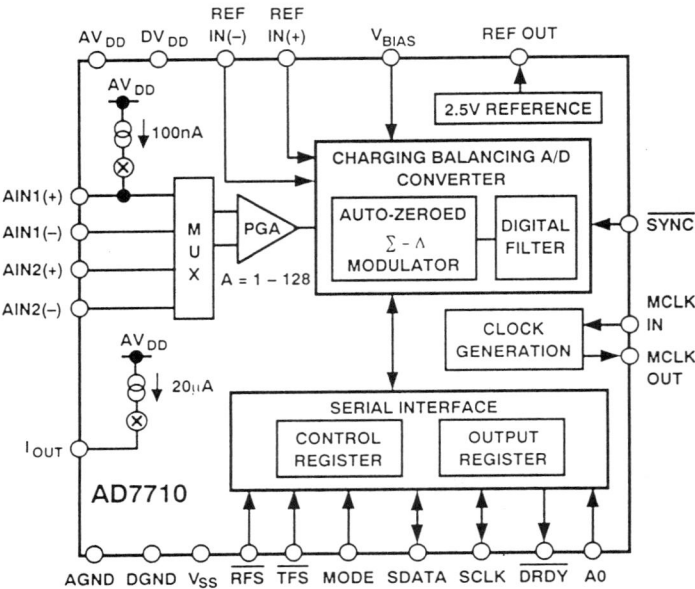

Figure 3.16

KEY FEATURES OF THE AD7710

- ±0.0015% Nonlinearity
- Two Channels with Differential Inputs
- Programmable Gain Amplifier (G = 1 to 128)
- Programmable Low Pass Filter
- System or Self-Calibration Options
- Single or Dual 5V Supply Operation
- Microcontroller Serial Interface

Figure 3.17

The AD7710, AD7711, AD7712, and AD7713 have identical structures of PGA, sigma-delta modulator, and serial interface. Their main differences are in their input configurations. The AD7710 has two low level differential inputs, the AD7711 two low level differential inputs with excitation current sources which make it ideal for RTD applications, the AD7712 has one low level differential input and a single ended high level input that can accommodate signals of up to four times the reference voltage, and the AD7713 is designed for loop-powered applications where power dissipation is important. The AD7713 consumes only 3.5mW of power from a single +5V supply.

The AD7714 is designed for either +3V (AD7714-3) or +5V (AD7714-5) single-supply, low power applications. It has a buffer between the multiplexer and the PGA which can be enabled or bypassed using a control line. When the buffer is active, it isolates the analog inputs from the transient currents and variable impedance of the switched-capacitor PGA.

Because of the differences in analog interfaces, each device is best suited to a particular sensor or system application. In other words, the sensor and the system requirements (i.e. type of sensor, single versus dual supply, power consumption, etc.) determine which converter should be used. Figure 3.19 lists the converters, and the sensors and applications to which they are best suited.

SUMMARY TABLE OF AD771X DIFFERENCES

- **AD7710:**
  - 2-Channel Low-Level Differential Inputs

- **AD7711:**
  - 1-Channel Low-Level Differential Input
  - 1-Channel Low-Level Single-Ended Input
  - Excitation Current Sources for 3 or 4-Wire RTDs

- **AD7712:**
  - 1-Channel Low-Level Differential Input
  - 1-Channel High-Level Single-Ended Input

- **AD7713:**
  - 2-Channel Low-Level Differential Inputs
  - 1-Channel High-Level Single-Ended Input
  - Excitation Current Sources for 3 or 4-Wire RTDs
  - Single 5V Operation Only
  - Low Power (3.5mW)
  - No Internal Reference

- **AD7714**
  - 3-Channel Low-Level Differential Inputs or 5-Channel Pseudo-Differential Inputs
  - Single +3V (AD7714-3) or Single +5V (AD7714-5)
  - Low Power (1.5mW: AD7714-3)
  - No Internal Reference

Figure 3.18

HIGH RESOLUTION SIGNAL CONDITIONING ADCs

AD771X APPLICATIONS

- **AD7710:**
  - Weigh Scales
  - Thermocouples
  - Chromatography
  - Strain Gauge

- **AD7711:**
  - RTD Temperature Measurement

- **AD7712:**
  - Smart Transmitters
  - Process Control

- **AD7713:**
  - Loop-Powered Smart Transmitters
  - RTD Temperature Measurement
  - Process Control
  - Portable Industrial Instruments

- **AD7714:**
  - Single +3V Supply Applications
  - Portable Industrial Instruments
  - Portable Weigh Scales

Figure 3.19

AD7714 SINGLE-SUPPLY MEASUREMENT ADC

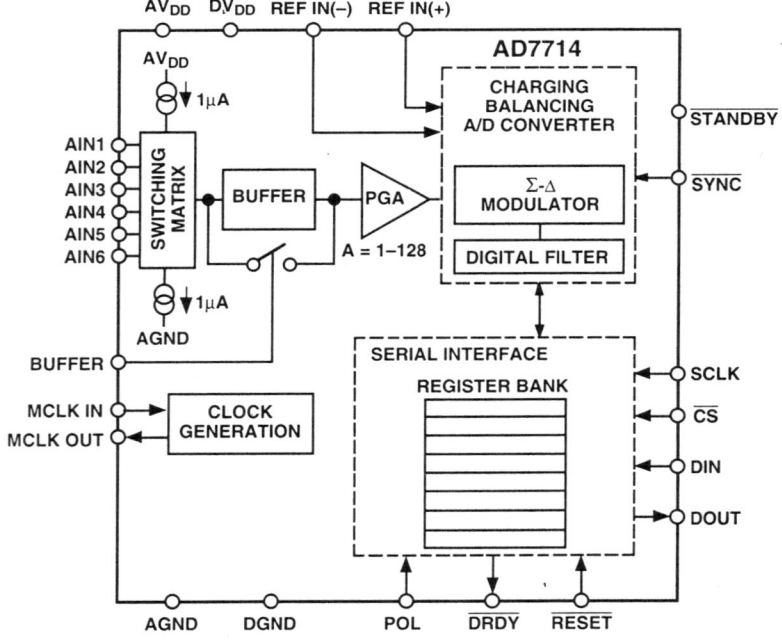

Figure 3.20

Although the AD7714 is often used as an example, the following discussion applies to all the converters in the family, with some minor exceptions. The basic AD7714 ADC (see Figure 3.20) is a switched capacitor sigma-delta converter which operates as has previously been discussed in this section. The signals on the input channels pass through a switching matrix (multiplexer) and into a bypassable buffer. The buffer (available only in the AD7714) allows the input signals to be isolated from the PGA switching transients and variable impedance (PGA operation will be described shortly). The PGA gain is programmable from 1 to 128, thereby allowing low level signals to be converted without the need for external amplification.

The functional block diagram of the AD7714 shows the PGA as separate from the sigma-delta modulator. In fact, it is part of the sigma-delta integrator (see Figure 3.21). The differential signal input charges C2, which is then discharged into the integrator summing node. This is done by closing S1 and S2, and then, after opening them, closing S3 and S4. When the PGA has a gain of 1 this happens once per cycle of the basic 19.2kHz clock, but for gains of 2, 4, and 8 respectively it happens 2, 4, or 8 times per cycle. The integrator charge is balanced by switching charge in the same way from the reference into C1, and thence to the integrator summing node. The polarity of reference switched depends on the state of the comparator output.

THE AD7714 Σ-Δ MODULATOR INCLUDES A PGA FUNCTION

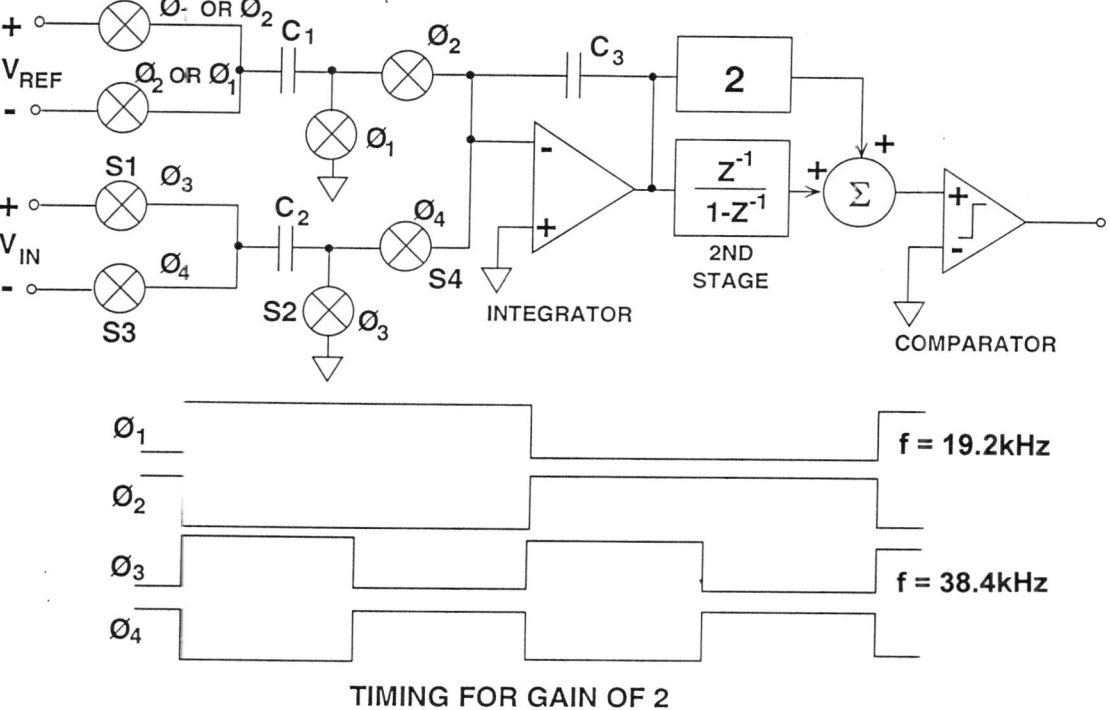

Figure 3.21

At a gain of 8, the sampling rate is 153.6kHz. Higher switching rates than this would not allow C2 sufficient time to charge, so for PGA gains greater than 8, the value of the reference capacitor, C1, is reduced, rather than the sampling rate being increased. Each time C1 is halved the gain of the system is doubled. The original value of C1 for gains of 1-8 is about 7pF in the AD7714 and 20pF for the other members of the AD771X family.

The internal digital filter has the sinc^3 response illustrated in Figure 3.22. The first notch in the filter response is programmable according to the formula:

$$f_{notch} = \left(\frac{f_{clkin}}{128}\right)\left(\frac{1}{\text{Decimal Value of Digital Code}}\right)$$

where f_{clkin} is normally either 2.4576MHz or 1MHz for the AD7714. The decimal value of the digital code in the above equation is loaded into the appropriate register in the AD7714. The master clock f_{clkin} frequency of 2.4576MHz is chosen because 50Hz and 60Hz may be obtained by direct division as well as the popular communications frequencies of 19.2kHz and 9.6kHz.

AD7714 DIGITAL FILTER FREQUENCY RESPONSE

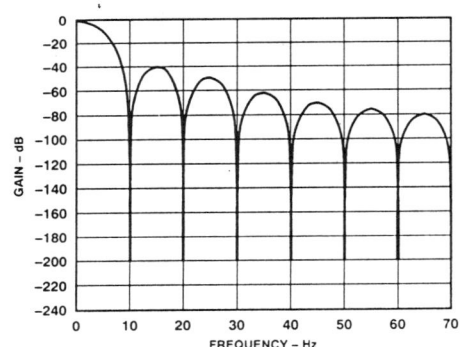

- Response Follows a sinc^3 = $\left(\dfrac{\sin x}{x}\right)^3$
- First Notch Frequency is Programmable and given by:
$$f_{notch} = \left(\frac{f_{clkin}}{128}\right)\left(\frac{1}{\text{Decimal Value of Digital Code}}\right)$$
- For f_{clkin} = 2.4576MHz, 5Hz $\leq f_{notch} \leq$ 1kHz

Figure 3.22

The first notch frequency is 3.82 times the -3 dB frequency, so the notch frequency must be chosen so that the maximum signal frequency falls within the filter passband.

The lower the notch frequency, the lower the noise bandwidth, and therefore the higher the effective resolution of the converter. Moreover, the PGA gain will also set limits on the achievable resolution. With a 2.5V span, 1 LSB in a 24-bit system is only 150nV - with a gain of 128 it is 1.2nV!

As is evident from their pipeline architecture, sigma-delta ADCs have a conversion time which is related to the bandwidth of the digital filter:- the narrower the bandwidth, the longer the conversion. For a 10Hz notch frequency, the AD7714 has a 10Hz output data rate.

When the input to a sigma-delta ADC changes by a large step, the entire digital filter must fill with the new data before the output becomes valid, which is a slow process. This is why sigma-delta ADCs are sometimes said to be unsuitable for multi-channel multiplexed systems - they are not, but the time taken to change channels can be inconvenient. In the case of the AD771X-series, four conversions must take place after a channel change before the output data is again valid (Figure 3.23). The $\overline{\text{SYNC}}$ input pin resets the digital filter, and if it used, data is valid on the third output afterwards, saving one conversion cycle (when the internal multiplexer is switched, the $\overline{\text{SYNC}}$ is automatically operated). The $\overline{\text{SYNC}}$ input also allows several AD771X ADCs to be synchronized.

THE RATE OF CONVERSION AND SETTLING TIME DEPENDS ON THE FILTER SETTING

	FILTER NOTCH FREQUENCY (Hz)								
	10	25	30	50	60	100	250	500	1k
CONVERSION TIME (ms)	100	40	33.3	20	16.7	10	4	2	1
MUX SWITCHING OR FULLSCALE WITH $\overline{\text{SYNC}}$, SETTLING TIME (ms)	300	120	100	60	50	30	12	6	3
ASYNCHRONOUS FULLSCALE SETTLING TIME (ms)	400	160	133.3	80	66.7	40	16	8	4

- Conversion Time = $\dfrac{1}{\text{Filter Notch Frequency}}$
- Digital Filter Requires Settling Time for Input Step Changes
- Use $\overline{\text{SYNC}}$ Input to Decrease Settling Time

Figure 3.23

High Resolution Signal Conditioning ADCs

Although the AD771X sigma-delta ADCs are 24-bit devices, it is not possible to obtain 24 bits of useful resolution from a single sample because internal ADC noise limits the accuracy of the conversion. We thus introduce the concept of "Effective Resolution," or "Effective Number of Bits," ENOB. This is a measure of the useful signal-noise ratio of an ADC.

Noise in the ADC is generated by unwanted signal coupling and by components such as resistors and active devices. There is also intrinsic *quantization noise* which is inescapably linked to the analog-digital conversion process. As discussed in the first part of this section, sigma-delta ADCs use special techniques to shape their quantization noise and thus reduce their oversampling ratio for a given ENOB, but they cannot eliminate quantization noise entirely.

In an ideal noise-free ADC, it is possible to position a dc input signal so that the ADC digital output is always the same code from sample to sample. There is no quantization noise present, because only one code is being exercised. In a real-world high resolution ADC, however, there are internal noise sources which can cause the output code to change from sample to sample for a constant-value dc input signal. Figure 3.24 shows the comparison between an ideal ADC and an one which has internal noise. The results are plotted as a histogram, where the vertical axis represents the number of occurrences of each code out of the total number of 5000 samples used in this example. In the ideal ADC, all 5000 samples result in the same output code. In the practical ADC, however, internal noise generally results in a distribution of codes, centered around the primary code. In most cases, the noise is gaussian, and a normal distribution can be fitted to the points on the histogram. The standard deviation of this distribution, σ, represents the rms value of the sum of all internal noise sources reflected to the ADC output, measured in LSBs. This of course assumes a noise-free input. The rms value in LSBs can be converted easily to an effective rms voltage noise.

The signal-to-noise ratio can then be computed by dividing the full scale ADC input range by the rms noise computed from the histogram. The full scale input range for the AD771X-series is equal to twice the reference voltage divided by the gain of the PGA, and the equation for calculating the *effective resolution* in bits is given by:

$$\text{Effective Resolution} = \log_2\left(\frac{2 \times V_{REF}}{GAIN} \bullet \frac{1}{RMS\ NOISE}\right)$$

HIGH RESOLUTION SIGNAL CONDITIONING ADCs

HISTOGRAM SHOWS THE EFFECT OF INTERNAL ADC NOISE FOR A DC INPUT SIGNAL

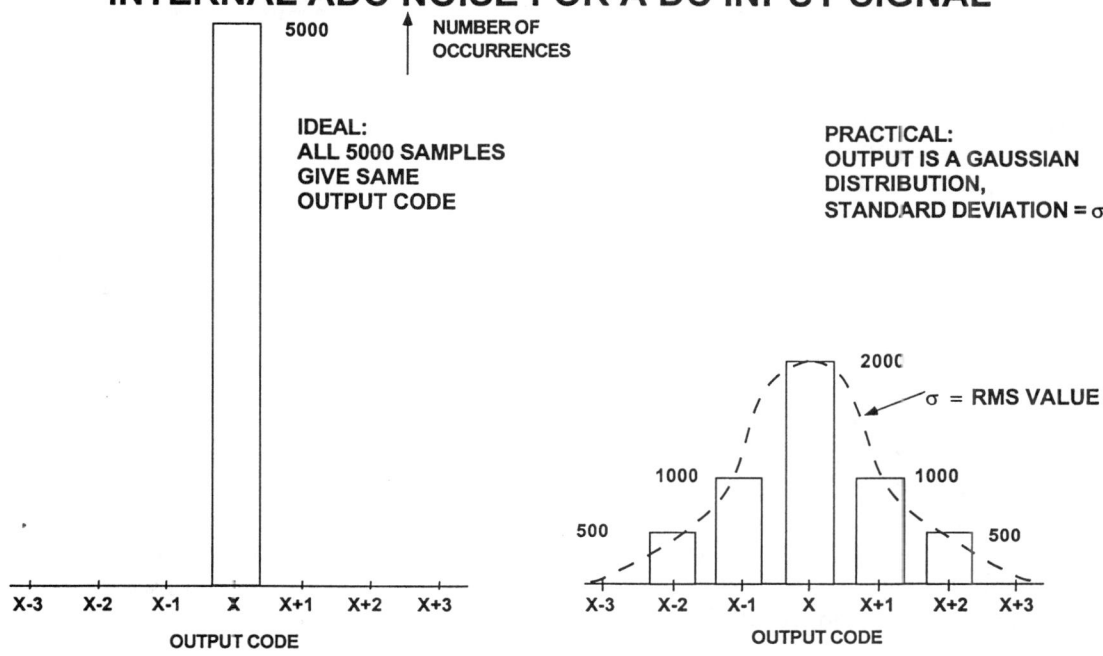

Figure 3.24

DETERMINING EFFECTIVE RESOLUTION

- Effective Resolution (ENOB) = $\log_2\left(\dfrac{\text{Full Scale Signal}}{\text{RMS Noise}}\right)$

  $= \log_2\left(\dfrac{2 \times V_{REF}}{\text{Gain} \times \text{RMS Noise}}\right)$

- Output RMS Noise = Effective Noise in the Digital Output Code

- ENOBs is Greatest at Low Filter Frequency and Low Gain

Figure 3.25

Figure 3.26 shows how RMS noise in an AD7714 varies with gain and notch frequency. Figure 3.27 gives the same results in terms of effective resolution, or ENOB using the previous equation. The effective output noise comes from two sources. the first is the electrical noise in the semiconductor devices used in the implementation of the ADC front end and the modulator (device noise). Secondly, when the analog input signal is converted into the digital domain, quantization noise is added.

The device noise is at a low level, and is largely independent of frequency. The quantization noise starts at an even lower level, but rises rapidly with increasing frequency to become the dominant noise source. Consequently, lower filter notch settings (below 100Hz approximately for f_{clkin} = 2.4576MHz tend to be device-noise dominated, while higher notch settings are dominated by quantization noise. Reducing the filter notch and cutoff frequency in the quantization-noise dominated region results in a more dramatic improvement in noise performance than it does in the device-noise dominated region. Furthermore, quantization noise is added after the PGA, so effective resolution is largely independent of gain for the higher filter notch frequencies. Meanwhile, device noise is added in the PGA, and therefore effective resolution suffers a little at high gains for lower notch frequencies.

Additionally, in the device-noise dominated region, the output noise (in µV) is largely independent of reference voltage, while in the quantization-noise dominated region, the noise is proportional to the value of the reference.

NOISE VARIES AS A FUNCTION OF GAIN AND FILTER CUTOFF FREQUENCY (AD7714-5, UNBUFFERED MODE, CLOCK = 2.4576MHz)

First Notch of Filter and O/P Data Rate	-3dB Frequency	Typical Output RMS Noise (µV)							
		Gain of 1	Gain of 2	Gain of 4	Gain of 8	Gain of 16	Gain of 32	Gain of 64	Gain of 128
5 Hz	1.31 Hz	0.87	0.48	0.24	0.2	0.18	0.17	0.17	0.17
10 Hz	2.62 Hz	1.0	0.78	0.48	0.33	0.25	0.25	0.25	0.25
25 Hz	6.55 Hz	1.8	1.1	0.63	0.5	0.44	0.41	0.38	0.38
30 Hz	7.86 Hz	2.5	1.31	0.84	0.57	0.46	0.43	0.4	0.4
50 Hz	13.1 Hz	4.33	2.06	1.2	0.64	0.54	0.46	0.46	0.46
60 Hz	15.72 Hz	5.28	2.36	1.33	0.87	0.63	0.62	0.6	0.56
100 Hz	26.2 Hz	12.1	5.9	2.86	1.91	1.06	0.83	0.82	0.76
250 Hz	65.5 Hz	127	58	29	15.9	6.7	3.72	1.96	1.5
500 Hz	131 Hz	533	267	137	66	38	20	8.6	4.4
1kHz	262 Hz	2,850	1,258	680	297	131	99	53	28

Figure 3.26

EFFECTIVE RESOLUTION VERSUS GAIN AND FIRST NOTCH FREQUENCY
(AD7714-5, UNBUFFERED MODE, CLOCK = 2.4576MHz)

First Notch of Filter and O/P Data Rate	-3dB Frequency	Effective Resolution (ENOBs)							
		Gain of 1	Gain of 2	Gain of 4	Gain of 8	Gain of 16	Gain of 32	Gain of 64	Gain of 128
5 Hz	1.31 Hz	22.5	22.5	22.5	21.5	20.5	20	19	18
10 Hz	2.62 Hz	22.5	21.5	21.5	21	20.5	19.5	18.5	17.5
25 Hz	6.55 Hz	21.5	21	21	20	19.5	18.5	17.5	16.5
30 Hz	7.86 Hz	21	21	20.5	20	19.5	18.5	17.5	16.5
50 Hz	13.1 Hz	20	20	20	20	19	18.5	17.5	16.5
60 Hz	15.72 Hz	20	20	20	19.5	19	18	17	16
100 Hz	26.2 Hz	18.5	18.5	19	18.5	18	17.5	16.5	15.5
250 Hz	65.5 Hz	15.5	15.5	15.5	15.5	15.5	15.5	15.5	14.5
500 Hz	131 Hz	13	13	13	13	13	13	13	13
1kHz	262 Hz	11	11	11	11	11	10.5	10.5	10.5

- Effective Resolution = $\log_2 \left[\dfrac{2 \times V_{REF}}{GAIN} \cdot \dfrac{1}{RMS\ NOISE} \right]$
- Highest resolution occurs at low gains and low frequency

Figure 3.27

It is important to distinguish between RMS and peak-to-peak noise. Noise in a sigma-delta ADC has a gaussian (or near gaussian) distribution. This means that if one waits long enough, any value of peak noise will eventually occur, and it is not possible to write a specification *absolutely* prohibiting a specified value of noise peak. For practical purposes, the peak-to-peak noise is defined as 6.6 times the RMS noise, since such peaks occur less than 0.1% of the time. The noise specified in the ENOB table in Figure 3.27 is expressed in RMS terms. If a figure for "noise-free" code resolution is required, it will be approximately 3-bits worse: 20-bits ENOB becomes 17-bits noise-free code, etc. Since most applications are concerned with noise *power*, however, the RMS ENOB figure is the more commonly used.

This does not mean that the original 24-bit resolution has no value, however. Additional external filtering, to narrower bandwidths than the internal filter, can further improve the resolution and ENOB at the expense of longer conversion times. The histogram approach using a large number of samples can also be used to more accurately define the input signal.

The results in Figures 3.27 and 3.28 assume the use of a low noise, heavily decoupled external reference and a noise-free analog input. Noisy inputs (and the reference is an input) reduce the effective resolution. For this reason, careful attention must be paid to external noise sources. Figure 3.29 lists aspects of board layout which may affect system noise, and hence the ENOB of the AD771X-series.

ESTIMATING NOISE-FREE CODE RESOLUTION

- Determined Using Peak-to-Peak Noise
- Output RMS Noise × 6.6 = Peak-to-Peak Noise
- Factor of 6.6 is Approximately Equal to 3 bits:

  $\log_2 (6.6) = 2.72$

- Therefore, subtract 3 bits from Effective Resolution Given in Figure 3.27 to Determine Noise Free Code Resolution

Figure 3.28

OPTIMIZING NOISE PERFORMANCE

- Pay Attention to Layout!
- Use Ground Planes
- Keep Analog PCB Tracks Short
- Interface Directly with Transducer
- Use Low Noise Amplifiers (AD797, OP-213, OP-177, AD707) (Only if Required)
- Connect Analog and Digital Grounds of Converters Together at the Device, and Connect them to Analog Ground Plane
- Route Digital PCB Tracks Clear of Analog Tracks
- Filter Signal and Reference Inputs
- Minimize Reference Noise
- The Evaluation Board is an Example of Good Layout

Figure 3.29

The AD771X-series is designed to interface directly with most transducers without the need for external buffering or amplification. If external amplifiers are used, however, low noise devices such as the OP-213 and AD797 should be chosen. To determine if external amplifiers will lower the AD771X system resolution, the total additional noise (in the bandwidth 0.1 Hz to the cutoff frequency set in the AD771X) should be calculated and compared with the RMS noise figures given in Figure 3.26. (Uncorrelated noise adds by root sum of squares, so if the additional noise is <50% of the AD7710 noise, it may be ignored; but if it exceeds this level, its effect on system performance must be studied carefully.)

An external filter on the input of the AD771X-series can improve its noise performance, because the modulator does not reject noise at integer multiples of the sampling frequency. This means that there are frequency bands $\pm f_{3dB}$ wide (f_{3dB} is the cutoff frequency of the internal digital filter) where noise passes unattenuated to the output. However, due to the AD771X high oversampling ratio, these bands occupy only a small fraction of the spectrum, and most broadband noise is filtered. The internal analog front end provides some filtering at these frequencies (the attenuation at 19.2kHz is approximately 70dB), but high level wideband noise can degrade system ENOB. A simple external RC low-pass filter is generally sufficient to minimize the effects of this noise, but the resistor and capacitor must be carefully chosen so that the gain accuracy of the AD771X is not affected. If the AD7714 is used in the buffered mode (i.e. the internal buffer is active), this restriction does not apply.

INPUT FILTER HELPS REDUCE WIDEBAND NOISE

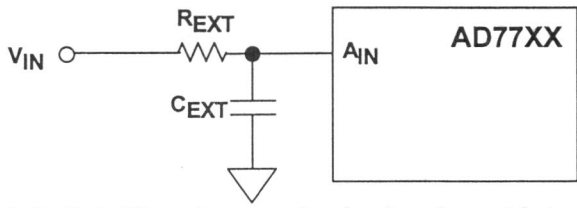

- Internal digital filter does not reject noise at integer multiples of sampling frequency (n x 19.2kHz, where n = 1, 2, 3, ...)

- RC low pass filter on the inputs limits broadband noise.

- RC combination must be such that gain accuracy is maintained

- Tables on data sheet show allowable combinations for 20-bit and 16-bit gain errors

- The cutoff frequency of such a filter is much too high to act as a true antialiasing filter, but can reduce high level wideband noise

- These restrictions do not apply to AD7714 in the buffered mode

Figure 3.30

High Resolution Signal Conditioning ADCs

A simplified model of the analog input of the AD7714 in the unbuffered mode is shown in Figure 3.31 (The AD7710, AD7711, AD7712, and AD7713 have similar structures). It consists of a resistor of approximately 7kΩ (input multiplexer on-resistance) connected to the input terminal and to an analog switch which switches a 7pF sampling capacitor between the resistor and ground, with a mark-space ratio of 50%. The switching frequency depends on f_{clkin} and the gain which is being used: with a gain of unity and the standard clock frequency of 2.4576MHz, the switching frequency is 19.2kHz, and at gains of 2, 4, and 8 or more it is 38.4, 76.8, and 153.6kHz respectively.

If the converter is working to an accuracy of 20-bits, the capacitor must charge with an accuracy of 20-bits. The input RC time constant due to the switch on-resistance (7kΩ) and the sampling capacitor (7pF) is 49ns. If the charge is to achieve 20-bit accuracy, it must charge for at least 14× the time constant, or 686ns. Any external resistor in series with the input will increase the time constant, and the chart in Figure 3.31 shows acceptable values of series resistance necessary to maintain 20-bit performance.

To determine the minimum charge time for 20-bit performance with an external resistance R_{ext} we use the equation:

Minimum Charge Time = $14(R_{ext} + 7k\Omega) \times 7pF$

The minimum charge time must be less than half the period of the switching signal used (it has a 50% duty cycle). The fastest switching frequency with the standard 2.4576MHz clock is 153.6kHz (for a gain of 8 or greater), and half of that clock period is 3.3µs, which allows a maximum R_{ext} of 26.8kΩ. At lower gains R_{ext} may be larger.

AD7714 ANALOG INPUT STRUCTURE (UNBUFFERED MODE)

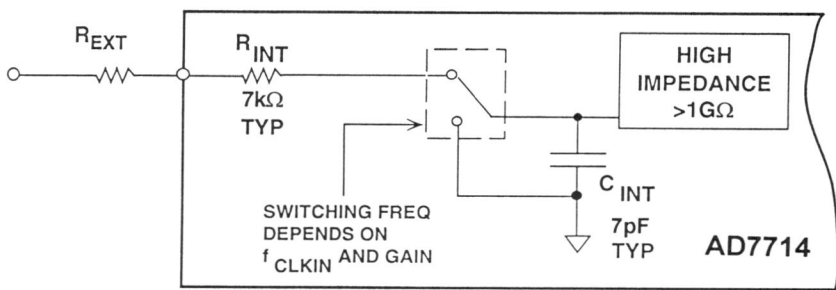

- R_{EXT} increases C_{INT} charge time and may result in gain error
- Charge time dependent on the input sampling rate and therefore gain
- Use following R_{EXT} values to maintain 20 bit accuracy:

GAIN:	1	2	4	8-128
R_{EXT}:	<290kΩ	<141kΩ	<63.6kΩ	<26.8kΩ

Figure 3.31

HIGH RESOLUTION SIGNAL CONDITIONING ADCs

It is not practical to use R_{ext} in conjunction with a capacitor to ground from the input pin of the AD7710, AD7711, AD7712, or AD7713 (or the AD7714 operating in the unbuffered mode) to make an anti-aliasing filter (with a cutoff frequency less than one-half the input sampling frequency), unless the capacitor is dramatically larger than the 7pF C_{int}. This is because C_{int} is discharged on every sampling clock cycle and will recharge from the filter capacitor. Therefore, either the filter capacitor must be so large that charging C_{int} from it changes its voltage by less than an LSB at 20-bits (i.e. it is larger than 7μF), or the time constant $R_{ext}C_{ext}$ must be short enough for C_{ext} to recharge before the next clock cycle - in which case the cutoff frequency due to R_{ext} and C_{ext} is not low enough to make an anti-aliasing filter with respect to the sampling frequency. There may, however, be some benefit in such a filter if there is input noise at high frequencies. The data sheets for the AD771X-series contain tables which give the allowable external capacitor and resistor values as a function of PGA gain for 16-bit and 20-bit gain accuracy (see Figure 3.32).

Note that if an external R_{ext} and C_{ext} are used, the capacitor type must have low non-linearities and dielectric absorption. Film types such as polystyrene or polypropylene are recommended.

AD7714 EXTERNAL FILTER RESTRICTIONS ON R AND C FOR NO 20-BIT GAIN ERROR (UNBUFFERED MODE ONLY)

GAIN	EXTERNAL CAPACITOR (pF)					
	0	50	100	500	1000	5000
1	290 kΩ	69 kΩ	40.8 kΩ	10.4 kΩ	5.6 kΩ	1.4 kΩ
2	141 kΩ	33.8 kΩ	20 kΩ	5 kΩ	2.8 kΩ	700 Ω
4	63.6 kΩ	16 kΩ	9.6 kΩ	2.4 kΩ	1.34 kΩ	340 Ω
8 - 128	26.8 kΩ	7.2 kΩ	4.4 kΩ	1.1 kΩ	600 Ω	160 Ω

Figure 3.32

High Resolution Signal Conditioning ADCs

The advantage of the AD7714 operating in the buffered mode is that a true input antialiasing filter can be used without affecting the gain accuracy. The penalty is only a slight increase in noise (approximately 10%) and a small reduction in input common-mode voltage range. If the value of the series resistor is large, the effects of the input bias current must be considered, but this error can be removed using the calibration modes.

Some successive approximation and subranging ADCs draw large transient currents at their analog and reference inputs which load their respective drive circuitry and cause errors. Often, special drive amplifiers with low output impedance at frequencies well above the conversion clock frequency are necessary to avoid this problem, but these problems do not occur with the very small transient loads of the AD771X devices. The oscilloscope photograph in Figure 3.33 shows the transient current in an AD7710. It was taken with a 1kΩ resistor in series with the input to measure the change in current. This circuit produces a 15mV spike of less than 1μs duration. The corresponding peak pulse current is only 15μA, which permits the use of quite high impedance signal sources with no risk of degrading the ENOB. As discussed earlier, the AD7714 operating in the buffered mode has no significant transients on its analog input.

INPUT TRANSIENT LOADING IS MINIMAL IN THE AD77XX FAMILY

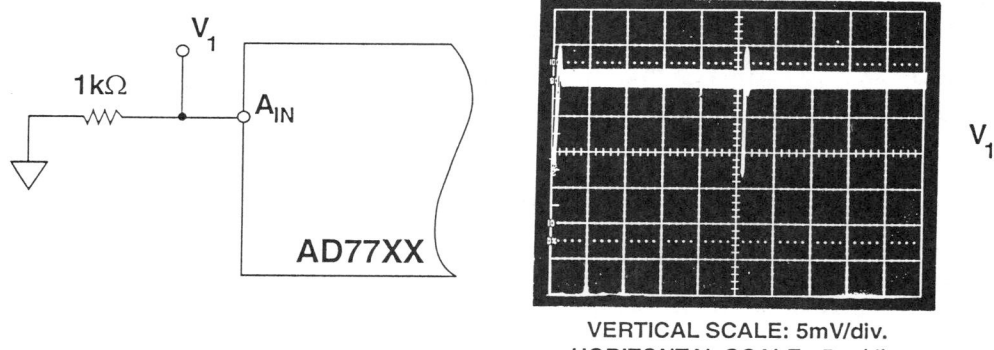

VERTICAL SCALE: 5mV/div.
HORIZONTAL SCALE: 5μs/div.

- Transient settles quickly and does not affect conversion
  Peak amplitude ≈ 15mV across 1kΩ (15μA),
  Duration less than 1μs

- Inputs can accommodate high bridge resistances

- No measurable transient for AD7714 in buffered mode

Figure 3.33

The AD771X family was designed to simplify transducer interfacing. Many types of transducers can be connected directly to the input of one of the AD771X family without additional circuitry, but some care is necessary to achieve the best possible accuracy:- noise needs to be minimized (a simple capacitor across a resistive sensor may be all the filtering that is needed, but this must be checked — noise is particularly important, because noise cannot be removed by the system calibration which eliminates gain and offset errors); transducer source impedance may affect charge times (as mentioned above); and bias currents flowing in high impedance transducers may cause errors, although these can be removed by system calibration. In general, system calibration can remove most dc errors in systems using the AD771X family.

TRANSDUCER CONNECTION CONSIDERATIONS

- Filter Noisy Signals

- Use Shielded, Twisted Pair Cable (Shield Grounded at AGND/DGND Connection Point at ADC, floating at transducer)

- DC Leakage (bias) Current = 10pA can cause offset with R_{source}

- This causes drift over temperature

- To Maintain Accuracy:
    - Minimize R_{source}
    - Use differential inputs and balance R_{source}
    - Use system calibration techniques

Figure 3.34

HIGH RESOLUTION SIGNAL CONDITIONING ADCs

Circuitry connected to transducers must generally be protected against over-voltage from ESD, noise pickup, or accidental shorts. If signals are likely to go outside the positive or negative supplies, some form of clamp is necessary to keep them within them. Figure 3.35 shows a suitable circuit for protecting AD771X devices. The AD7710 has internal ESD protection diodes between the input and both supplies which conduct when the input exceeds either supply by more than about 0.6V. Excessive current in these diodes will vaporize metal tracks on the chip and damage the circuit, so an external resistor, R_p, is necessary to limit current to a safe 5mA during over-voltage events.

R_p may be determined by a simple calculation:

$$R_p = \frac{V_{max} - V_{supply}}{5\,mA}$$

R_p will contribute noise to the system (the basic Johnson noise equation applies:

$$e_n = \sqrt{4kTBR_p}$$

where k is Boltzmann's Constant, T is the absolute temperature and B is the bandwidth). If the noise due to R_p is too high, R_p can be reduced if external Schottky diodes are used in addition to the diodes on the chip.

INPUT OVERVOLTAGE PROTECTION

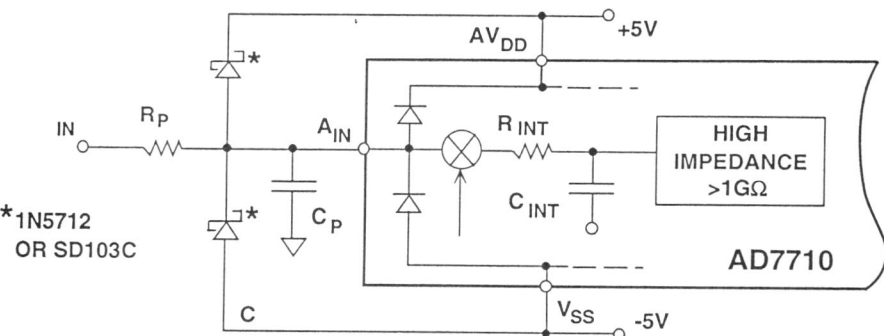

- Internal ESD protection diodes clamp input to within 0.6V of either supply.
- Limit current due to overvoltage to less than 5mA.

$$R_P = \frac{A_{inmax} - AV_{DD}}{5mA}, \text{ or } \frac{|A_{inmin}| - |V_{SS}|}{5mA}, \text{ whichever is greater.}$$

- External Schottkys can be added to reduce the size of R_P.
- Include C_P to filter noise due to R_P.

Figure 3.35

HIGH RESOLUTION SIGNAL CONDITIONING ADCs

There are no special requirements for these Schottky diodes, as long as they have low leakage current and can handle the necessary fault current levels while maintaining a low turn-on voltage.

As important as the analog signal input is the reference input. Figure 3.36 shows a simplified model of the reference input, which is very similar to that of the analog input (with the exception of the AD7714 operating in the buffered mode). The series resistor is 5kΩ, and the value of the capacitor depends on the gain setting and the particular device. For gains of 1-8, the capacitor is approximately 7pF for the AD7714. Above 8, the capacitor's value is halved for each doubling of gain. The value of the capacitor (G = 1-8) for other members of the AD771X series is 20pF.

An important consideration in choosing a reference for the AD771X-series is noise. Many references have output noise which exceeds that of the AD771X and cause reduced accuracy. Filtering may help in such cases, but a low noise reference should be selected wherever possible.

Although the AD7710, AD7711, and AD7712 have internal +2.5V references which may be connected to their positive reference input, their use will degrade the effective resolution of the ADCs by approximately 1 bit for filter cut-off frequencies of 60Hz or less. The noise calculations using the AD7710 internal reference are shown in Figure 3.37. For optimum noise performance, a low noise external reference such as the AD780 should be used as shown in Figure 3.38.

REFERENCE VOLTAGE CONSIDERATIONS

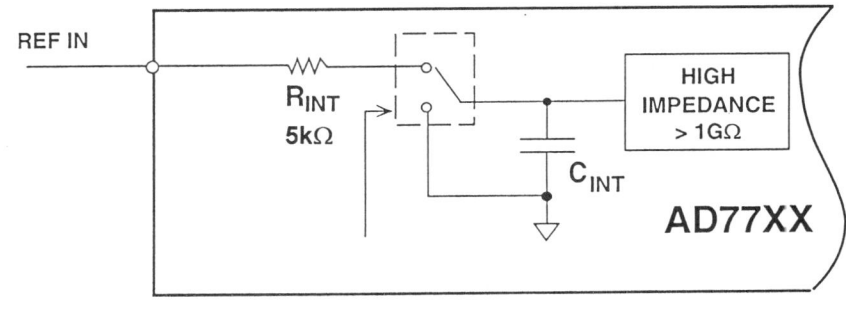

PGA GAIN	1	2	4	8	16	32	64	128
C_{INT}	C	C	C	C	C/2	C/4	C/8	C/16

- C = 20pF for AD7710, AD7711, AD7712, AD7713
- C = 7pF for AD7714
- Heavy decoupling on REF IN required
- Minimal transient loading

Figure 3.36

INTERNAL REFERENCE VOLTAGE NOISE CONSIDERATIONS FOR AD7710, AD7711, AD7712 ADCs

- Specified Output Noise   = 8.3µV rms typical (0.1 to 10Hz)
  of AD7710, AD7711, AD7712
  Internal +2.5V Reference

  = 3.4µV rms (0.1 to 2.62Hz, for 10Hz Output Rate)

- Reference Noise adds to intrinsic ADC noise:

  AD7710 Noise = 1.7µV rms (G = +1)

  Total Noise = $\sqrt{(1.7\mu V)^2 + (3.4\mu V)^2}$ = 3.8µV rms, or 25µV p–p

- This reduces effective resolution from 21.5 to 20.5 bits

- Use low-noise external reference for highest resolution at low input bandwidths

Figure 3.37

USE A LOW NOISE EXTERNAL REFERENCE FOR OPTIMUM NOISE PERFORMANCE AND RESOLUTION

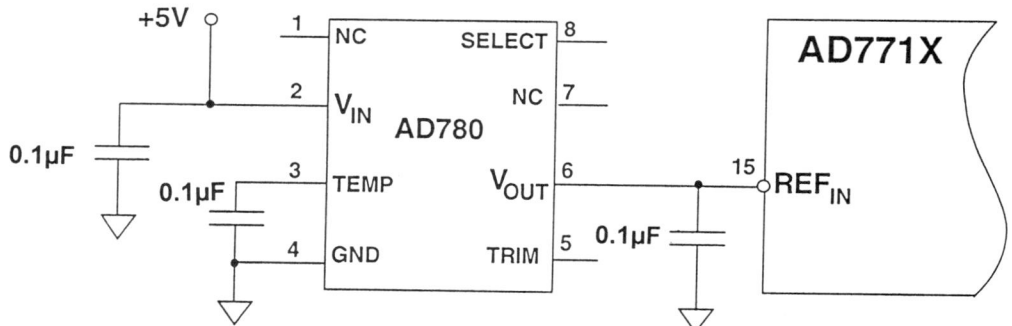

- AD780 has 4µV peak-to-peak noise is 0.1 to 10Hz bandwidth.

- In a 2.62Hz bandwidth (notch frequency = 10Hz), the rms value of the noise is 0.31µV rms

- Well below the noise of the AD771X

Figure 3.38

The AD780 2.5V reference has noise of 4µV p-p in the range 0.1 to 10Hz. This is equivalent to 0.606µV rms (obtained by dividing the peak-to-peak value by 6.6). This gives 0.31µV rms noise in the 2.62Hz bandwidth associated with a 10Hz update rate. This is negligible compared to the AD7710 inherent noise of 1.7µV rms (G = 1). With output rates of 1kHz or more, and cutoff frequency of 262Hz, the noise of the AD780 may be reduced by 50% if a 0.1µF capacitor is connected to its output (unlike many IC voltage references, the AD780 is stable with all values of capacitive load).

CALCULATING REFERENCE NOISE CONTRIBUTION

- If reference noise is given as a spectral density, V_n (nV/√Hz) rms,
  - $V_{refnoise} = V_n \sqrt{BW}$, where BW = 0.262 f_{notch}

- If reference noise is given as a peak-to-peak value in the 0.1 to 10Hz bandwidth, V_{p-p}, then

$$V_n = \frac{V_{p-p}}{6.6\sqrt{10Hz}}$$

  - $V_{refnoise} = \frac{V_{p-p}}{6.6\sqrt{10Hz}} \cdot \sqrt{BW}$, BW = 0.262 f_{notch}

- Above are approximations, but are sufficiently accurate for estimation of reference noise

- Reference noise should be no greater than 50% of ADC noise

Figure 3.39

The AD780 is also a suitable reference for the AD7714 ADC, whose noise with a 10Hz update rate (G = 1, Vsupply = +5V) is 1.0µV rms. When the AD7714 is operating on a +3V supply, it requires a low-noise 1.25V reference, and the AD589 is a suitable choice.

When the AD771X-series ADCs are operated at higher output rates and higher input bandwidths, the ADC noise is significantly higher, and less effective resolution is required. This allows the use of higher noise, lower power references such as the REF192 (25µV p-p noise, 0.1 to 10Hz) which only requires 45µA of quiescent current, compared to 0.75mA for the AD780.

Regardless of the reference selected, it should be properly decoupled in order to act as a charge reservoir to transient load currents as well as a filter for wideband noise. This implies that the reference must be stable under capacitive loads, which is not necessarily the case in all references. In fact, some references actually *require* an external decoupling capacitor in order to maintain stability. Regardless of the reference selected, the data sheet should be carefully examined with respect to output capacitive loading. Further information on applying voltage references can be obtained in References 5, 6, and 7.

DC errors also affect conversion accuracy, but AD771X devices can calibrate themselves to correct dc errors. The AD7710, AD7711, AD7712, and AD7713 have four different calibration modes. These are summarized in Figure 3.41 and comprise Self Calibration, System Calibration, System-Offset Calibration, and Background Calibration. Each calibration cycle contains two conversions, one each for zero-scale and for full-scale calibration. The calibration modes for the AD7714 are similar, except that in the Background-Calibration Mode, only zero-scale is calibrated.

LOW VOLTAGE REFERENCE SUMMARY

REFERENCE PART #	OUTPUT (V)	TOLERANCE (mV) (max)	DRIFT ppm/°C (max)	NOISE (µV p-p, typ) 0.1 to 10Hz	SUPPLY CURRENT (mA) typ
AD780	+2.5 / +3.0	1 - 5	3 - 20	4	0.75
REF43	+2.5	15 - 50	10 - 25	5	0.45
REF192	+2.5	2 - 10	5 - 25	25	0.045
AD589*	+1.235	35	25 - 100	4	0.050 - 5

* Two Terminal

Figure 3.40

AD771X OFFERS 4 CALIBRATION OPTIONS

	SELF-CALIBRATION	SYSTEM CALIBRATION	SYSTEM OFFSET CALIBRATION	BACKGROUND CALIBRATION*
1st Cycle	Internally Short Inputs to Ground	Externally Short Inputs to Zero-Scale	Externally Short Inputs to Zero-Scale	Internally Short Inputs to Ground*
2nd Cycle	Internally Short Input to V_{REF}	Externally Short Input to Fullscale	Calibrate for Span from AV_{IN} to V_{REF}	Internally Short Inputs to V_{REF}*
Duration	9 ÷ Output Rate	4 ÷ Output Rate Each Step	9 ÷ Output Rate	6 ÷ Output Rate*

* AD7714 Background Calibration consists of zero-scale (Ground) calibration only.

Figure 3.41

To initiate a calibration cycle, the appropriate code must be sent to the control register. After the code is sent, the AD771X automatically conducts the entire operation, and clears the control register of the calibration command so that a separate command to stop calibration is not necessary. Since the filter in the sigma-delta converter must purge itself of its previous result for four output update cycles whenever the input sees a full-scale step the total calibration operation takes nine such cycles.

Self Calibration removes errors in an AD771X by connecting the input to ground and performing a conversion, and then connecting the input to V_{ref} and performing another. The results of these conversions are used to calibrate the device.

Background Calibration is a variation of Self Calibration. The only difference is that when an AD771X is placed in Background Calibration mode, it continually calibrates itself at regular intervals without further instructions. This ensures that the AD771X remains calibrated regardless of drift. The Background Calibration cycle alternates calibration conversions with signal conversions: zero calibrate/convert signal/full-scale calibrate/convert signal/zero calibrate/etc. This provides continuous calibration but reduces the output data rate by a factor of six. Background Calibration for the AD7714 only calibrates zero-scale.

System Calibration is intended to calibrate all the elements prior to the ADC which may contribute to system errors, as well as the ADC itself. (For example an instrumentation amplifier introduces errors into a system due to its own offset, drift and gain error. These errors can be removed by System Calibration.) However, System Calibration requires additional analog switches to connect *system* inputs to ground and a reference as well as to the original signal source. The first step in System Calibration requires external grounding of the system input terminal to calibrate out offsets. The second step requires that the input be connected to a reference, which calibrates gain error at full-scale. The System Calibration cycle requires the sending of two separate instructions to the control register as well as control of the analog switches at the system input. It must be repeated regularly to correct for drift with time and temperature.

The final calibration mode is System-Offset Calibration. This calibrates *system* offsets, and *the AD771X* gain. Again, it requires external analog switches at the system input, and separate instructions for zero and gain calibration. For the first cycle, the system input is connected to ground and the AD771X calibrates for system offsets. During the second cycle, the ADC input is connected to the reference for ADC gain calibration.

When calibration is complete, $\overline{DRDY}$ goes low - but it does not necessarily go high as soon as the calibration command is sent to the ADC, there may be a delay of up to one output data cycle before it does so. Controllers should therefore look for a 0→1 transition, rather than the presence of a 0, on $\overline{DRDY}$ to signal the completion of a calibration after it has been commanded.

Calibration is crucial to achieving the rated accuracy of AD771X devices and should be performed immediately after power-up and repeated regularly. A 1.25μV/°C temperature coefficient of input offset and a 2°C temperature change causes an LSB of error in a 20-bit 2.5V system. Any reference drift adds to the error. Frequent calibration ensures that temperature changes do not degrade the accuracy of conversions.

CALIBRATION ISSUES

- Always calibrate on power-up!

- Background calibration sequence: (Zero-Scale, Convert, Fullscale, Convert, Zero-Scale, Convert, . . .) AD7710, AD7711, AD7712, AD7713
    This reduces the data rate by a factor of 6.

- $\overline{DRDY}$ signals when calibration cycle is complete by going low.

- $\overline{DRDY}$ may already be low if a conversion is taking place.

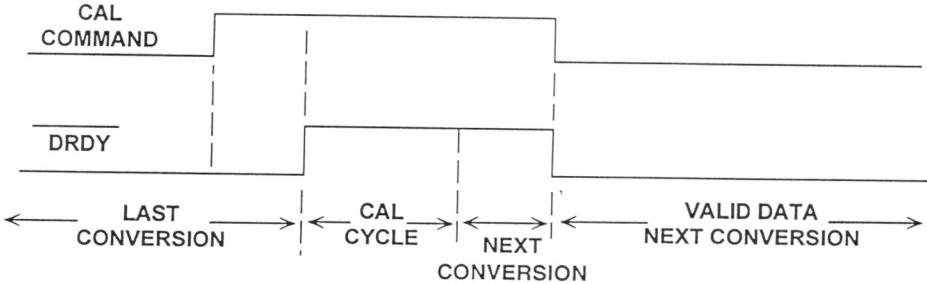

Figure 3.42

CALIBRATE OFTEN TO MAINTAIN ACCURACY

- Temperature Drift can cause errors

- Unipolar offset drift of 2.5µV is

    1 LSB in a 20 bit system (2.5V fullscale)

- Reference voltage drift adds to this error

- Calibration can remove gain errors created by input filters

- Therefore, to minimize errors, calibrate often.

- Always calibrate on power-up!

- Calibration coefficients can be manually adjusted

Figure 3.43

When the AD771X executes a calibration cycle, it saves two coefficients in internal registers. One register stores the full scale calibration coefficient, FSC, and the other stores the zero scale calibration coefficient, ZSC. Adjusting the calibration coefficients manually may be useful in some applications. For example, in a weigh scale application it may be necessary to insert an offset to account for a fixed weight. It is possible to read from and write to the calibration registers of members of the AD771X family, making adjustment of calibration coefficients a straightforward task. Details of this procedure are given in References 4 and 8.

A typical application of the AD7710 is in a weigh scale (Figure 3.44). These generally use a resistive bridge as their sensing element and require resolution of at least 16-bits and often more. The AD7710 dramatically simplifies the design of such a system: the bridge is connected directly to its differential inputs, making an external instrumentation amplifier unnecessary. The excitation for the bridge, and the reference for the AD7710, are provided by an AD780, whose low noise helps to preserve the system ENOB. Because the system bandwidth is limited (both by the conversion rate selected and the filter capacitors on the bridge) the ENOB achievable is quite high ($\approx$20-bits) but the conversion (and output data) rate is rather low at 10Hz.

WEIGH SCALE APPLICATION USING THE AD7710

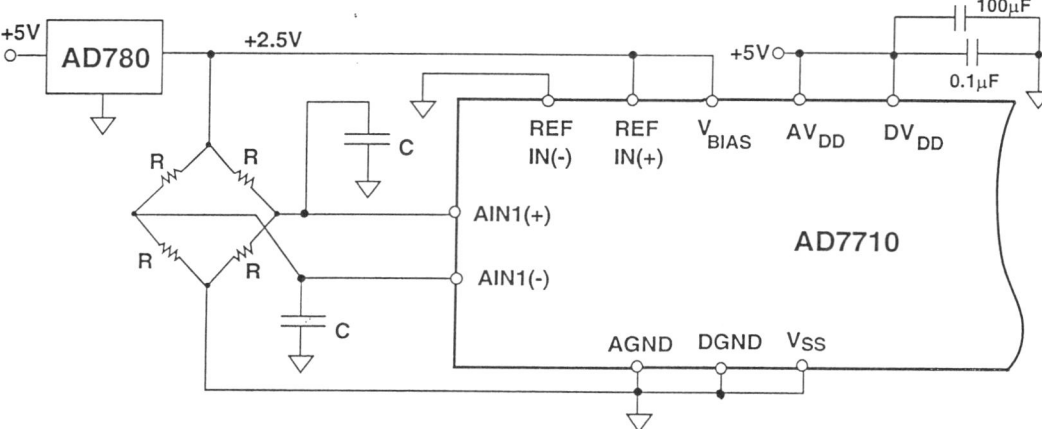

- High impedance differential input interfaces directly to bridge
- External reference used for accuracy
- Wide range of impedances can be used for bridge
- Add capacitors to input to filter noise

Figure 3.44

The converters in the AD771X family all have serial interfaces, which are described in greater detail in the data sheet. They have control registers that control all their operations. Changing the PGA gain, starting a calibration, and changing the filter parameters are all accomplished by writing to the appropriate register. On the other hand, data can be read either as a 16-bit or a 24-bit operation - one of the bits in the control register controls the size of the data word. The $\overline{\text{DRDY}}$ output indicates when a conversion is complete and valid data is available in the output register.

Figure 3.45 shows an isolated 4-wire interface to the AD7713 using common opto-isolators. Over 6kV of isolation is possible. The $\overline{\text{TFS}}$, A0, and $\overline{\text{SYNC}}$ lines are tied together at the converter to minimize the number of control lines. Tying $\overline{\text{TFS}}$ to A0 causes a write to the device to load data to the control register, and any read accesses the data register. The only restrictions of this method of control is that the controller cannot write to the calibration registers and cannot read from the control register. In many applications these capabilities are unnecessary. Four opto-isolators carry data and instructions from the controller to the ADC and a fifth, with a 74HC125 on each side of the isolation barrier, carries data to the controller. The AD7713 is ideal for this particular application because its low supply current minimizes the load on the isolated power supply.

HIGH RESOLUTION SIGNAL CONDITIONING ADCs

ISOLATED 4-WIRE INTERFACE USING AD7713

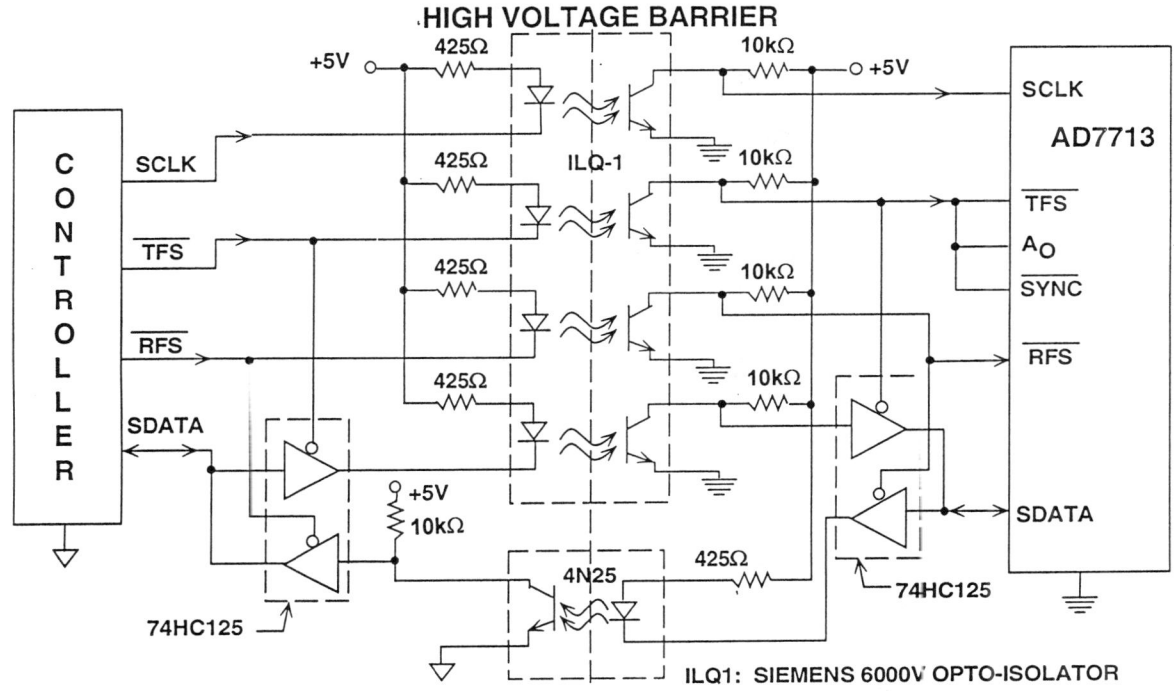

Figure 3.45

The AD771X family generally interfaces with some type of microprocessor. Their data sheet includes circuits and microcode for interfacing to the 8051 and 68HC11 microcontroller and the ADSP-2103/2105 DSP processor. Figure 3.46 shows how the AD7714 may be interfaced to the 68HC11 microcontroller. The diagram shows the minimum (three-wire) interface with $\overline{CS}$ on the AD7714 hard-wired low. In this scheme, the $\overline{DRDY}$ bit of the AD7714 Communications Register is monitored to determine when the Data Register is updated. Other schemes are described in the AD7714 data sheet.

HIGH RESOLUTION SIGNAL CONDITIONING ADCs

AD7714 TO 68HC11 MICROCONTROLLER INTERFACE

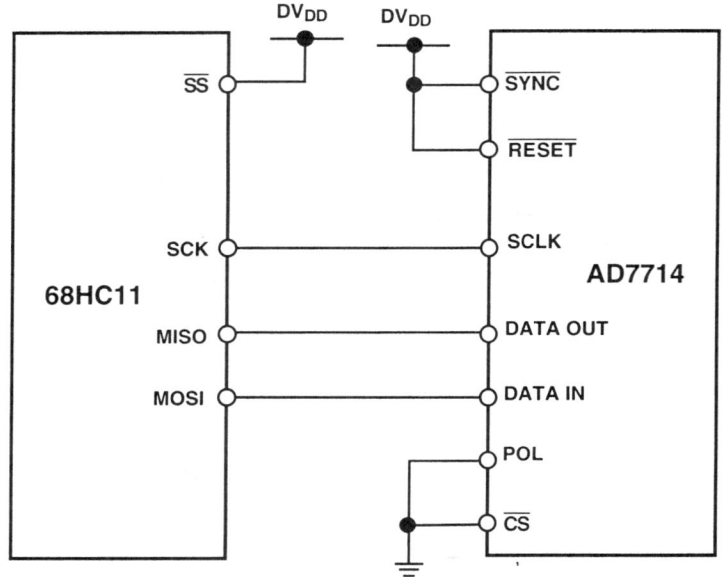

Figure 3.46

Interfacing to the ADSP-2103/2105 is also relatively straightforward. The $\overline{\text{DRDY}}$ bit of the Communications Register is again monitored to determine when the Data Register in the AD7714 is updated.

AD7714 TO ADSP-2103/2105 DSP INTERFACE

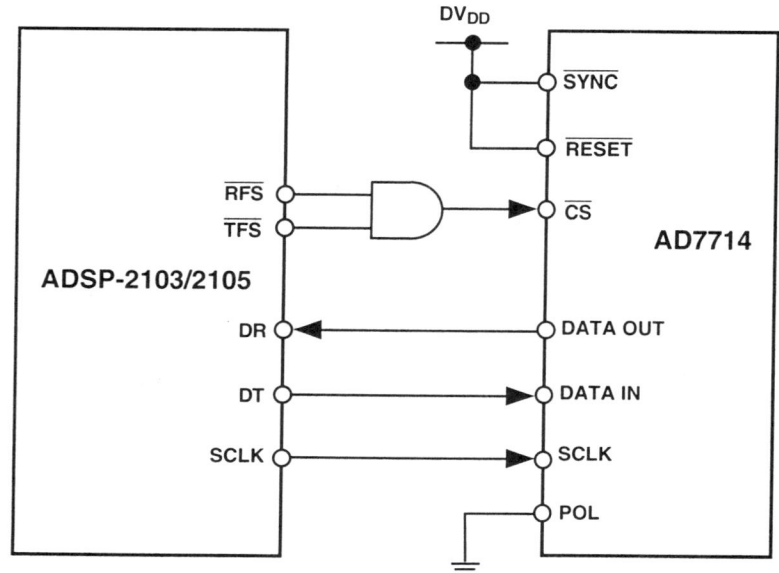

Figure 3.47

HIGH RESOLUTION SIGNAL CONDITIONING ADCs

The AD771X sigma-delta converters are powerful tools for building high accuracy systems. Every one of them combines high resolution, system calibration, a programmable gain amplifier, and high impedance differential inputs with great ease of design. Their adjustable digital filters provides flexibility in the choice of data rates and resolution and their serial interface minimizes their pin count, so that they fit in a 24-pin skinny DIP package, providing a high degree of functionality in a small space.

The AD7714 is especially suitable for low power applications. Figure 3.48 shows the total supply current required as a function of supply voltage for two clock frequencies: 2.4576MHz and 1MHz. These data are for an external clock with the AD7714 operating in the unbuffered mode. Figure 3.48 illustrates an important point which is applicable to a large number of low power data converters - the total power dissipation is a function of the clock frequency! Make sure to check the data sheet carefully for this dependency when estimating the total power requirement. The power dissipation of older, higher-power, bipolar data converters was generally much less sensitive to clock frequency than the modern low-power CMOS designs.

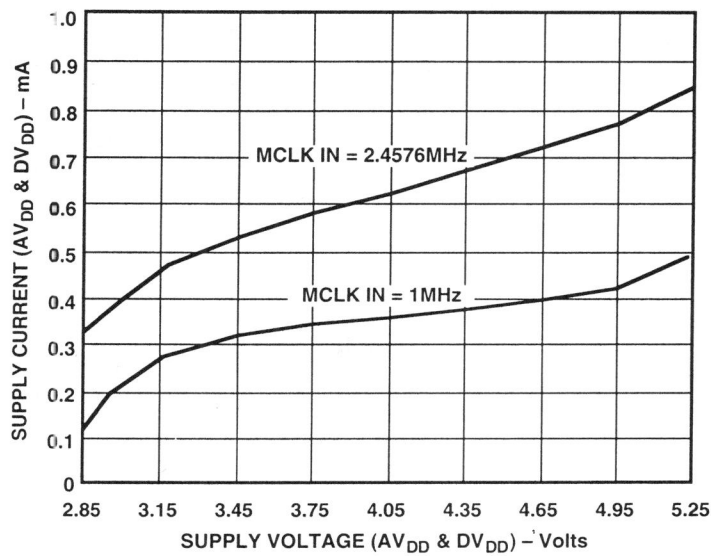

Figure 3.48

HIGH RESOLUTION SIGNAL CONDITIONING ADCs

An area where the low power, single supply, three wire interface capabilities of the AD7714 is of benefit is in smart transmitters (Figure 3.49). The entire smart transmitter must operate from the 4mA to 20mA loop. Tolerances in the loop mean that the amount of current available to power the entire transmitter is as low as 3.5mA. The AD7714 consumes only 500µA, leaving 3mA available for the rest of the transmitter. Not shown in Figure 3.49 is the isolated power source required to power the front end circuits, including the AD7714.

SMART TRANSMITTER USING AD7714 OPERATES ON 4mA TO 20mA LOOP CURRENT

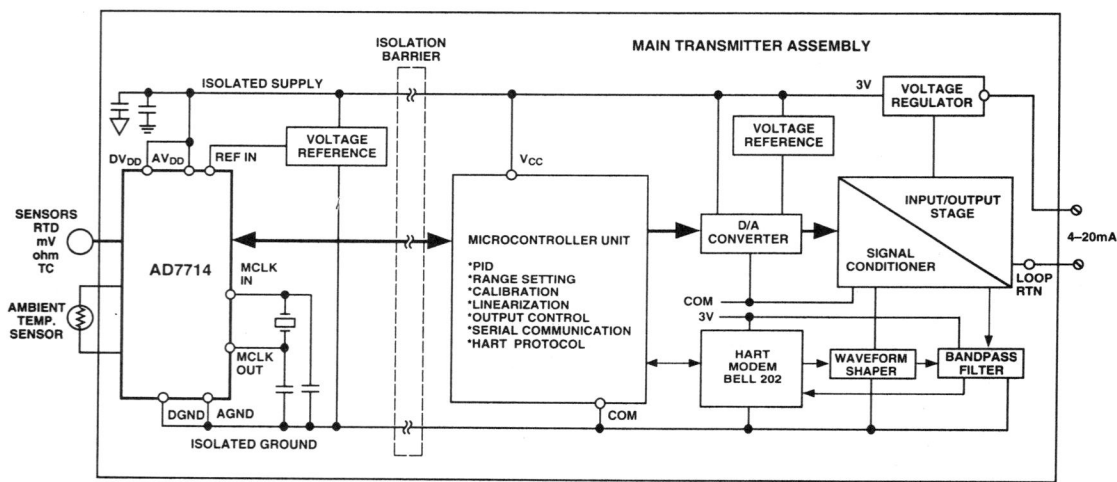

Figure 3.49

REFERENCES

1. S.A.Jantzi, M.Snelgrove & P.F.Ferguson Jr., *A 4th-Order Bandpass Sigma-Delta Modulator*, **IEEE Journal of Solid State Circuits**, Vol. 38, No. 3, March 1993, pp.282-291.

2. **System Applications Guide**, Analog Devices, Inc., 1993, Section 14.

3. **Mixed Signal Design Seminar**, Analog Devices, Inc., 1991, Section 6.

4. **AD7710, AD7711, AD7712, AD7713, AD7714 Data Sheets**, Analog Devices.

5. Walt Jung, *Getting the Most from IC Voltage References*, **Analog Dialogue**, 28-1, 1994, pp. 13-21.

6. Walt Jung, *Build an Ultra-Low-Noise Voltage Reference*, **Electronic Design Analog Applications Issue**, June 24, 1993.

7. **Linear Design Seminar**, Analog Devices, Inc., 1995, Section 8.

8. **System Applications Guide**, Analog Devices, Inc., 1993, Section 6.

High Resolution Signal Conditioning ADCs

SECTION 4

HIGH SPEED SAMPLING ADCs

- ADC Dynamic Considerations
- Selecting the Drive Amplifier Based on ADC Dynamic Performance
- Driving Flash Converters
- Driving the AD9050 Single-Supply ADC
- Driving ADCs with Switched Capacitor Inputs
- Gain Setting and Level Shifting
- External Reference Voltage Generation
- ADC Input Protection and Clamping
- Applications for Clamping Amplifiers
- Noise Considerations in High Speed Sampling ADC Applications

HIGH SPEED SAMPLING ADCs

SECTION 4

HIGH SPEED SAMPLING ADCs
Walt Kester

Modern high speed sampling ADCs are designed to give low distortion and wide dynamic range in signal processing systems. Realization of specified performance levels depends upon a number of factors external to the ADC itself, including proper design of any necessary support circuitry. The analog input drive circuitry is especially critical, because it can degrade the inherent ADC dynamic performance if not designed properly.

Because of various process and design-related constraints, it is generally not possible to make the input of a high speed sampling ADC totally well-behaved, i.e., high impedance, low capacitance, ground-referenced, free from glitches, impervious to overdrive, etc. Therefore, the ADC drive amplifier must provide excellent ac performance while driving what may be a somewhat hostile load (depending upon the particular ADC selected).

The trend toward single-supply high speed designs adds additional constraints. The input voltage range of high speed single-supply ADCs may not be ground referenced (for valid design reasons), therefore level shifting with single-supply op amps (which may have limited common-mode input and output ranges) is usually required, unless the application allows the signal to be ac coupled.

Although there is no *standard* high speed ADC input structure, this section addresses the most common ones and provides guidelines for properly designing the appropriate input drive circuitry.

Some sampling ADCs also require external reference voltages. In other cases, performance improvements can be realized by using an external reference in lieu of an internal one. It is equally important that these reference circuits be designed with utmost care, since they too affect the overall ADC performance.

HIGH SPEED, LOW VOLTAGE SAMPLING ADCs

- Key specifications for sampling ADCs:
    - Distortion
    - Noise
    - Distortion Plus Noise
    - Effective Number of Bits
    - Bandwidth (Full Power and Small Signal)
    - Sampling Rate

- Modern Trends:
    - Low Power: CMOS, BiCMOS, or XFCB Processes
    - Low Voltage: ±5V, +5V, +5V (Analog) / +3V (Digital)

    - Input Voltage Ranges not always Ground-Referenced
    - Analog Input Can Generate Transient Currents

Figure 4.1

ADC DYNAMIC PERFORMANCE SPECIFICATIONS

- Distortion Specifications: (Narrowband)
    - Harmonic Distortion
    - Total Harmonic Distortion (THD)
    - Spurious Free Dynamic Range (SFDR)
    - Intermodulation Distortion (IMD), Two-Tone Input

- Noise Specifications: dc to $f_S / 2$
    - Signal-to-Noise Ratio without Harmonics (often called SNR, or S/N)

- Noise Plus Distortion Specifications: dc to $f_S / 2$
    - Signal-to-Noise and Distortion (S/N+D, SINAD), but often referred to as SNR (check definition carefully when evaluating ADCs), often converted to Effective Bits (ENOB)
    - Total Harmonic Distortion Plus Noise (THD + N)

- Broadband Noise can be reduced by filtering or averaging

Figure 4.2

ADC Dynamic Considerations

In order to make intelligent decisions regarding the input drive circuitry, it is necessary to understand first exactly how the dynamic performance of the ADC is characterized. Modern signal processing applications require ADCs with wide dynamic range, high bandwidth, low distortion, and low noise. As well as having traditional dc specifications (offset error, gain error, differential linearity error, and integral linearity error), *sampling* ADCs (ADCs with an internal sample-and-hold function) are generally specified in terms of Signal-to-Noise Ratio (SNR, or S/N), Signal-to-Noise-Plus Distortion Ratio [S/(N+D), or SINAD], Effective Number of Bits (ENOB), Harmonic Distortion, Total Harmonic Distortion (THD), Total Harmonic Distortion Plus Noise (THD+N), Intermodulation Distortion (IMD), and Spurious Free Dynamic Range (SFDR). Sampling ADC data sheets may provide some, but not all of these ac specifications. The ac specifications are usually tested by applying spectrally pure sinewaves to the ADC and analyzing its output in the frequency domain with a Fast Fourier Transform (FFT). The process is similar to using an analog spectrum analyzer to measure the ac performance of an amplifier. Because of the quantization process, however, an ADC produces some errors not found in amplifiers.

An ideal N-bit ADC, sampling at a rate f_s, produces quantization noise having an rms value of $q/\sqrt{12}$ measured in the Nyquist bandwidth dc to $f_s/2$, where q is the weight of the Least Significant Bit (LSB). The value of q is obtained by dividing the full scale input range of the ADC by the number of quantization levels, 2^N. For example, an ideal 10-bit ADC with a 2.048V peak-to-peak input range has $2^{10} = 1024$ quantization levels, an LSB of 2mV, and an rms quantization noise of $2mV/\sqrt{12} = 577\mu V$ rms. The derivation of the theoretical value of quantization noise, $q/\sqrt{12}$, makes the assumption that the quantization noise is not correlated in any fashion to the input signal, and may therefore be treated as Gaussian noise. This is normally true, but in certain cases where the input sinewave frequency happens to be an exact submultiple of the sampling rate, the quantization noise may tend to be concentrated at the harmonics of the input signal, even though the rms value is still approximately $q/\sqrt{12}$.

Another way to express quantization noise is to convert it into a Signal-to-Noise ratio by dividing the rms value of the input sinewave by the rms value of the quantization noise. Normally, this is measured with a full scale input sinewave, and the expression relating the two is given by the well-known equation,

$$SNR = 6.02N + 1.76 dB.$$

An actual ADC will produce noise in excess of the theoretical quantization noise, as well as distortion products caused by a non-linear transfer function. An FFT is used to calculate the rms value of all the distortion and noise products, and the actual signal-to-noise-plus-distortion, S/(N+D), is computed. The above equation is solved for N, yielding the well-known expression for Effective Number of Bits, ENOB:

$$ENOB = \frac{S/(N+D)_{ACTUAL} - 1.76 dB}{6.02}.$$

For example, if a 10-bit ADC has an actual measured S/(N+D) of 56dB (theoretical would be 61.96dB), then it will have 9 effective bits, i.e., the non-ideal 10-bit ADC yields the same performance as an ideal 9-bit one.

Even well-designed sampling ADCs have non-linearities which contribute to non-ideal low frequency performance, and additionally, performance degrades as the input frequency is increased. A useful way to evaluate the ac performance of ADCs is to plot Signal-to-Noise Plus Distortion, S/(N+D), (or convert it to ENOB) as a function of input frequency. This measurement is somewhat all-inclusive and includes the effects of both noise and distortion products.

In some instances, SNR may be specified both with and without the distortion products, and in other cases, distortion may be specified separately, either as individual harmonic components, or as total harmonic distortion (THD). Spurious Free Dynamic Range (SFDR) is simply another way of describing distortion products and is the ratio of the signal level to the worst frequency spur, under a given set of conditions. Intermodulation Distortion (IMD) is measured by applying two tones (F1 and F2) to the ADC and determining the ratio of the power in one of the tones to the various IMD as shown in Figure 4.4. Unless otherwise specified, the third-order products which occur at the frequencies 2F1 − F2 and 2F2 − F1 are the ones used in the measurement because they lie close to the original tones and are difficult to filter.

EFFECTIVE NUMBER OF BITS (ENOB)
INDICATES OVERALL DYNAMIC PERFORMANCE OF ADCs

- S/(N+D) = 6.02N + 1.76dB (Theoretical)

- ADC ACHIEVES S/(N+D) = XdB (Actual)

- $\text{ENOB} = \dfrac{\text{XdB} - 1.76\text{dB}}{6.02\text{dB}}$

- ENOB Includes Effects of All Noise and Distortion in the bandwidth DC to $f_S/2$

Figure 4.3

INTERMODULATION DISTORTION (IMD)

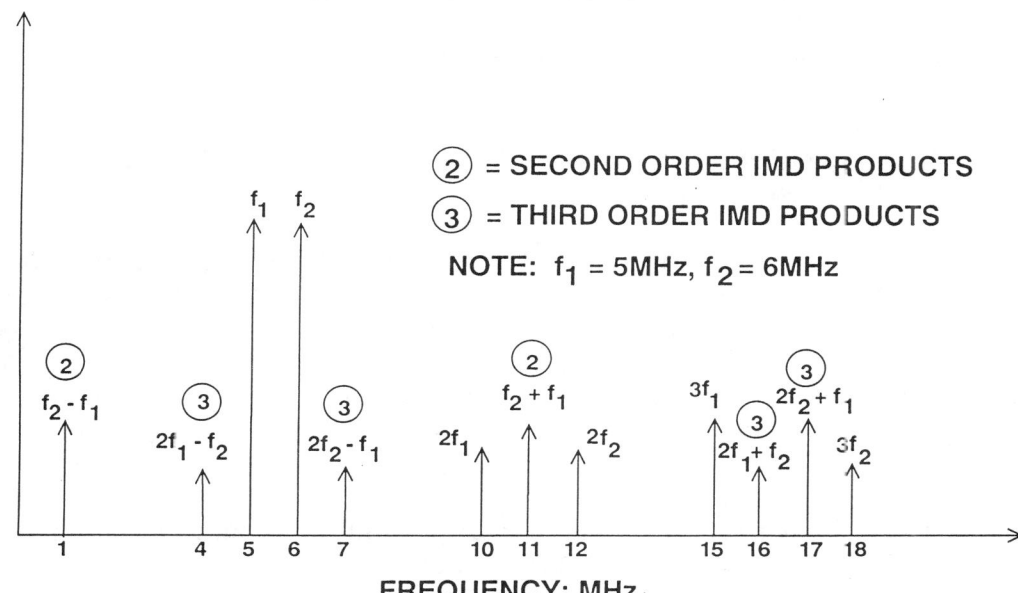

Figure 4.4

If we plot the gain of an amplifier with a small signal of a few millivolts or tens of millivolts, we find that as we increase the input frequency, there is a frequency at which the gain has dropped by 3 dB. This frequency is the upper limit of the *small signal bandwidth* of the amplifier and is set by the internal pole(s) in the amplifier response. If we drive the same amplifier with a large signal so that the output stage swings with its full rated peak-to-peak output voltage, we may find that the upper 3dB point is at a lower frequency, being limited by the slew rate of the amplifier output stage. This high-level 3dB point defines the *large signal bandwidth* of an amplifier. When defining the large signal bandwidth of an amplifier, a number of variables must be considered, including the power supply, the output amplitude (if slew rate is the only limiting factor, it is obvious that if the large signal amplitude is halved, the large signal bandwidth is doubled), and the load. Thus, large signal bandwidth is a rather uncertain parameter in an amplifier, since it depends on so many uncontrolled variables — in cases where the large signal bandwidth is less than the small signal bandwidth, it is better to define the output slew rate and calculate the maximum output swing at any particular frequency.

In an ADC, however, the maximum signal swing is always full scale, and the load seen by the signal is defined. It is therefore quite reasonable to define the large signal bandwidth (or full-power bandwidth) of an ADC and report it on the data sheet. In some cases, the small signal bandwidth may also be given.

ADC LARGE SIGNAL (OR FULL POWER) BANDWIDTH

- With Small Signal, the Bandwidth of a Circuit is limited by its Overall Frequency Response.

- At High Levels of Signal the Slew Rate of Some Stage May Control the Upper Frequency Limit.

- In Amplifiers There are so many Variables that *Large Signal Bandwidth* needs to be Redefined in every Individual Case, and *Slew Rate* is a more Useful Parameter for a Data Sheet.

- In ADCs the Maximum Signal Swing is the ADC's Full Scale Span, and is therefore Defined, so *Full Power Bandwidth* (FPBW) may Appear on the Data Sheet.

- HOWEVER the FPBW Specification Says Nothing About Distortion Levels. Effective Number of Bits (ENOB) is Much More Useful in Practical Applications.

Figure 4.5

However, the large signal bandwidth tells us the frequency at which the amplitude response of the ADC drops by 3dB — it tells us nothing at all about the relationship between distortion and frequency. If we study the behavior of an ADC as its input frequency is increased, we discover that, in general, noise and distortion increase with increasing frequency. This reduces the resolution that we can obtain from the ADC.

If we draw a graph of the ratio of signal-to-noise plus distortion (S/N+D) against its input frequency, we find a much more discouraging graph than that of its frequency response. The ratio of S/N+D can be expressed in dB or as effective number of bits (ENOB) as discussed above. As we have seen, the SNR of a perfect N-bit ADC (with a full scale sinewave input) is (6.02N + 1.76)dB. A graph of ENOB against the variations of input amplitude can be depressing when we see just how little of the dc resolution of the ADC can actually be used, but can sometimes show interesting features: the ADC in Figure 4.6, for instance, has a larger ENOB for signals at 10% of FS at 1MHz than for FS signals of the same frequency. A simple frequency response curve cannot have plots crossing in this way.

ADC GAIN AND ENOB VERSUS FREQUENCY
SHOWS IMPORTANCE OF ENOB SPECIFICATION

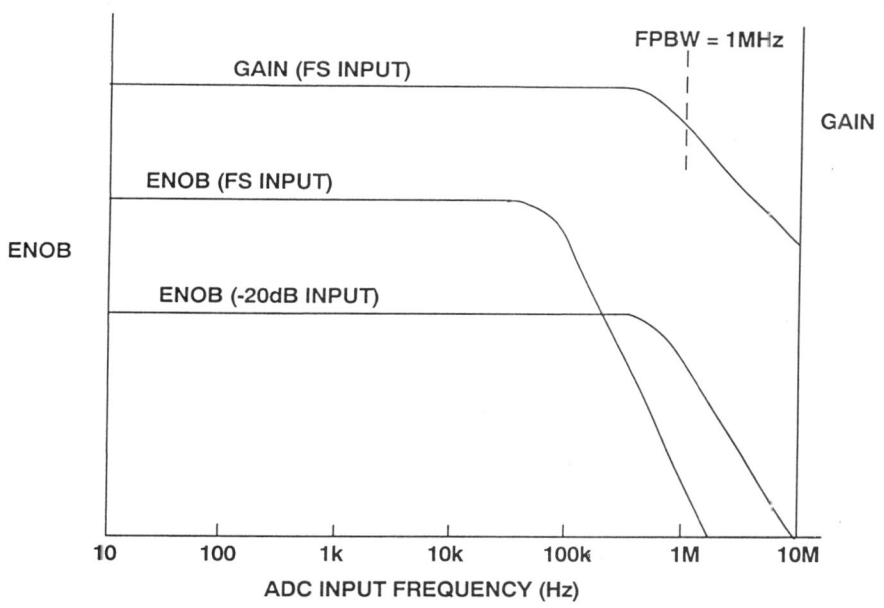

Figure 4.6

The causes of the loss of ENOB at higher input frequencies are varied. The linearity of the ADC transfer function degrades as the input frequency increases, thereby causing higher levels of distortion. Another reason that the SNR of an ADC decreases with input frequency may be deduced from Figure 4.7, which shows the effects of phase jitter on the sampling clock of an ADC. The phase jitter causes a voltage error which is a function of slew rate and results in an overall degradation in SNR as shown in Figure 4.8. This is quite serious, especially at higher input/output frequencies. Therefore, extreme care must be taken to minimize phase noise in the sampling/reconstruction clock of any sampled data system. This care must extend to all aspects of the clock signal: the oscillator itself (for example, a 555 timer is absolutely inadequate, but even a quartz crystal oscillator can give problems if it uses an active device which shares a chip with noisy logic); the transmission path (these clocks are very vulnerable to interference of all sorts), and phase noise introduced in the ADC or DAC. A very common source of phase noise in converter circuitry is aperture jitter in the integral sample-and-hold (SHA) circuitry.

HIGH SPEED SAMPLING ADCs

EFFECTS OF APERTURE AND SAMPLING CLOCK JITTER

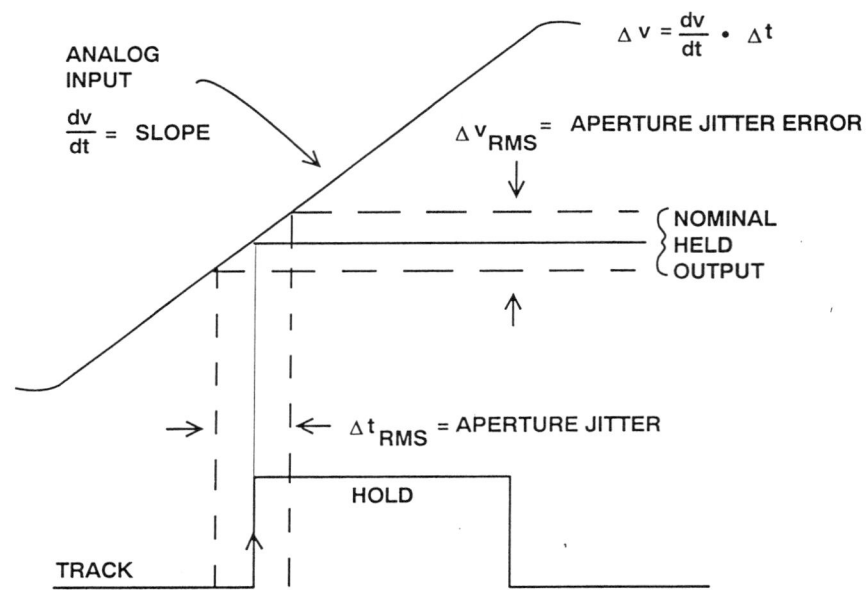

Figure 4.7

SNR DUE TO SAMPLING CLOCK JITTER (t_j)

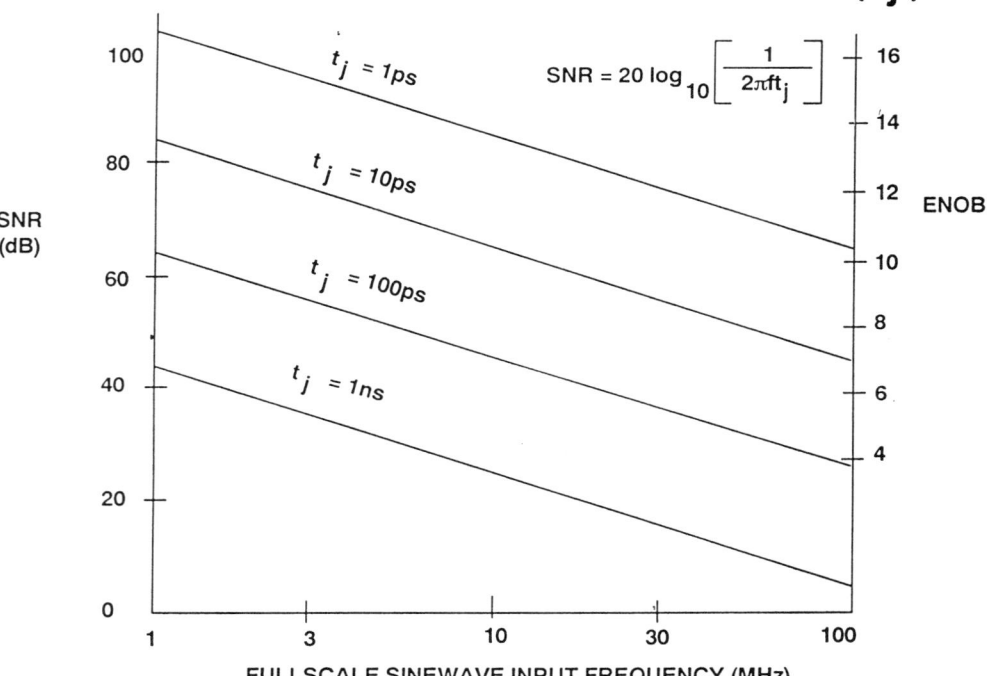

Figure 4.8

HIGH SPEED SAMPLING ADCs

A decade or so ago, sampling ADCs were built up from a separate SHA and ADC. Interface design was difficult, and a key parameter was aperture jitter in the SHA. Today, most sampled data systems use *sampling* ADCs which contain an integral SHA. The aperture jitter of the SHA may not be specified as such, but this is not a cause of concern if the SNR or ENOB is clearly specified, since a guarantee of a specific SNR is an implicit guarantee of an adequate aperture jitter specification. However, the use of an additional high-performance SHA will sometimes improve the high-frequency ENOB of a sampling ADC, and may be more cost-effective than replacing the ADC with a more expensive one.

It should be noted that there is also a fixed component which makes up the ADC aperture time. This component, usually called *effective aperture delay time*, does not produce an error. It simply results in a time offset between the time the ADC is asked to sample and when the actual sample takes place (see Figure 4.9). The variation or tolerance placed on this parameter from part to part is important in simultaneous sampling applications or other applications such as I and Q demodulation where several ADCs are required to track each other.

EFFECTIVE APERTURE DELAY TIME

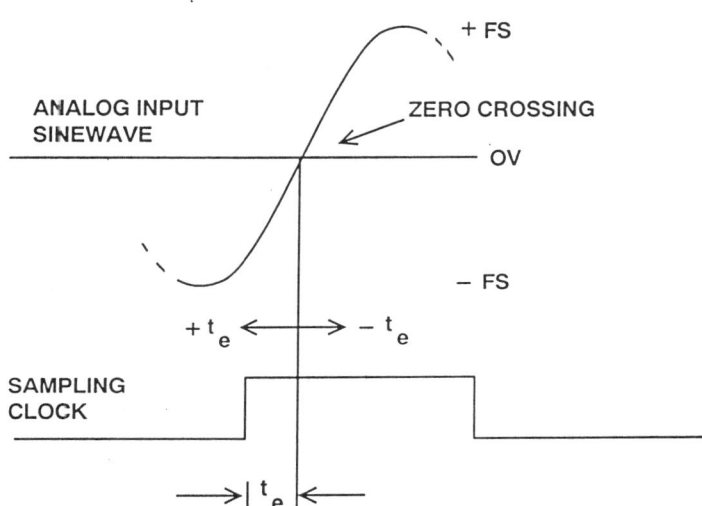

Figure 4.9

The distortion produced by an ADC or DAC cannot be analyzed in terms of second and third-order intercepts, as in the case of an amplifier. This is because there are two components of distortion in a high performance data converter. One component is due to the non-linearity associated with the analog circuits within the converter. This non-linearity has the familiar "bow" or "s"-shaped curve shown in Figure 4.10. (It may be polynomial or logarithmic in form). The distortion associated with this type of non-linearity is sometimes referred to as *soft* distortion and produces low-order distortion products. This component of distortion behaves in the traditional manner, and is a function of signal level. In a practical data converter, however, the soft distortion is usually much less than the other component of distortion, which is due to the differential nonlinearity of the transfer function. The converter transfer function is more likely to have discrete points of discontinuity across the signal range as shown in Figure 4.10.

The actual location of the points of discontinuity depends on the particular data converter architecture, but nevertheless, such discontinuities occur in practically all converters. Non-linearity of this type produces high-order distortion products which are relatively unpredictable with respect to input signal level, and therefore such specifications as *third order intercept point* may be less relevant to converters than to amplifiers and mixers. For lower-amplitude signals, this constant level *hard* distortion causes the SFDR of the converter to *decrease* as input amplitude decreases. The soft distortion in a well-designed converter is only significant for high frequency large-amplitude signals where it may rise above the hard distortion floor.

In a practical system design, the ADC is usually selected based primarily on the required dynamic performance at the required sampling rate and input signal frequency, using one or more of the above specifications. DC performance may be important also, but is generally of less concern in signal processing applications. Once the ADC is selected, the appropriate interface circuitry must be designed to preserve these levels of ac and dc performance.

Selecting the Drive Amplifier Based on ADC Dynamic Performance

The ADC drive amplifier performs several important functions in a system. First, it isolates the signal source and provides a low-impedance drive to the ADC input. A low-impedance dc and ac drive source is important because the input impedance of the ADC may be signal-dependent, and the input may also generate transient load currents during the actual conversion process. A low source impedance at high frequencies minimizes the errors produced by these effects. Second, the drive amplifier provides the necessary gain and level shifting to match the signal to the ADC input voltage range.

HIGH SPEED SAMPLING ADCs

TRANSFER CHARACTERISTICS FOR "SOFT" AND "HARD" DISTORTION IN ADCs

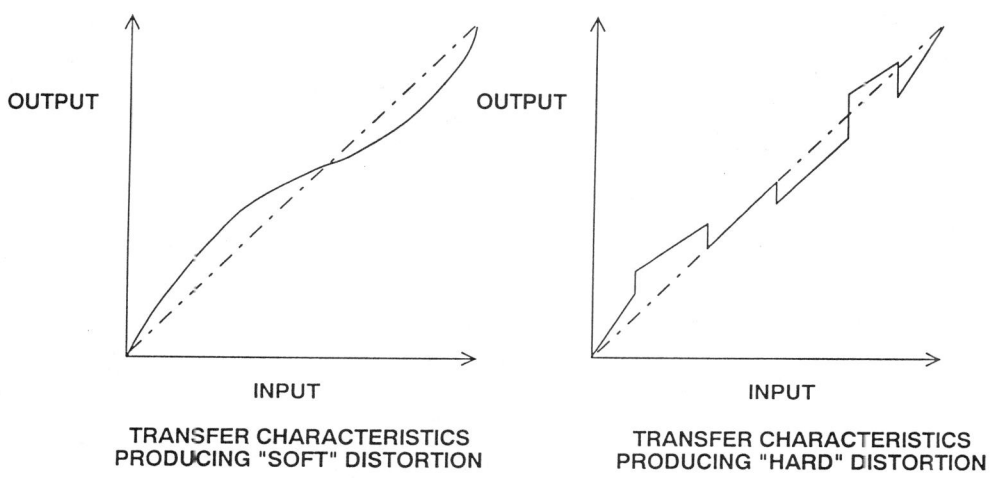

TRANSFER CHARACTERISTICS PRODUCING "SOFT" DISTORTION

TRANSFER CHARACTERISTICS PRODUCING "HARD" DISTORTION

Figure 4.10

FUNCTIONS OF THE ADC DRIVE AMPLIFIER

- Buffer the analog signal from the ADC input:
    - ADC input may not be a constant high impedance
    - ADC input may generate transient loads

- Provide other functions:
    - Gain
    - Level Shifting

- If the ADC input is constant high impedance with no transient loading, do not use a buffer amplifier unless required for gain or level shifting!!

Figure 4.11

HIGH SPEED SAMPLING ADCS

The S/(N+D) plot of the ADC should generally be used as the first selection criterion for the drive amplifier. If the Total Harmonic Distortion Plus Noise (THD+N) of the drive amplifier is always 6 to 10dB better than the S/(N+D) of the ADC over the frequency range of interest, then the overall degradation in S/(N+D) caused by the amplifier will be limited to between approximately 0.5dB and 1db, respectively. This will be illustrated using two state of the art components: the AD9022 12 bit, 20MSPS ADC and the AD9631 op amp. A block diagram of the AD9022 is shown in Figure 4.12, and key specifications in Figure 4.13. AD9631/AD9632 key specifications are given in Figure 4.14.

The AD9022 employs a three-pass subranging architecture and digital error correction. The analog input is applied to a 300Ω attenuator and passed to the sampling bridge of the first internal track-and-hold amplifier (T/H). The held value of the first T/H is applied to a 5-bit flash converter and a second T/H. The 5-bit flash converter resolves the most significant bits (MSBs) of the held analog voltage. These 5 bits are reconstructed via a 5-bit DAC and subtracted from the original T/H output signal to form a residue signal. A second T/H holds the amplified residue signal while it is encoded with a second 5-bit flash ADC. Again, the 5-bits are reconstructed and subtracted from the second T/H output to form a residue signal. This residue is amplified and encoded with a 4-bit flash ADC to provide the 3 least significant bits (LSBs) of the digital output and one bit of error correction. The digital error correction logic combines the data from the three flash converters and presents the result as a 12-bit parallel digital word. The output stage is TTL (AD9022), or ECL (AD9023). Output data can be strobed on the rising edge of the ENCODE command.

AD9022 12-BIT, 20MSPS SAMPLING ADC

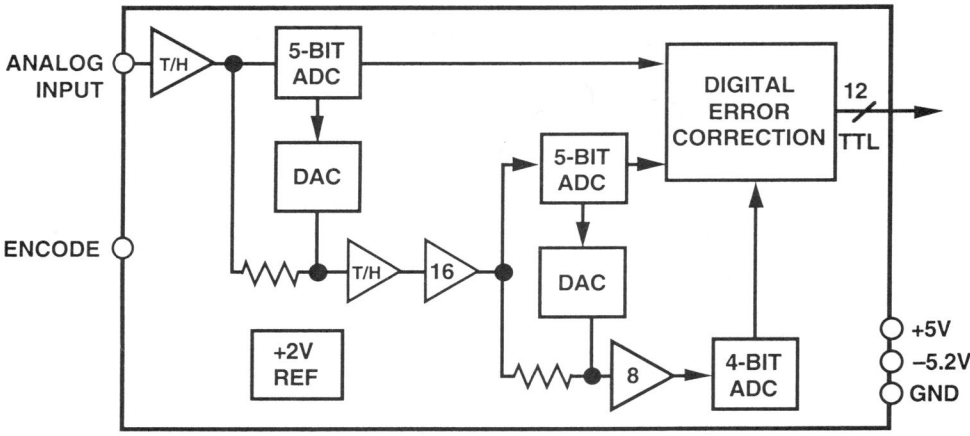

Figure 4.12

AD9022 ADC KEY SPECIFICATIONS

- 12-bit, 20MSPS Sampling ADC
- TTL Outputs (AD9023 has ECL outputs)
- On-Chip reference and SHA
- High Spurious Free Dynamic Range (SFDR):
  - 76dB @ 1MHz Input f_S = 20MSPS
  - 74dB @ 9.6MHz Input f_S = 20MSPS
- Analog Input Bandwidth: 110MHz
- Well-Behaved analog input with no transients
- Input Range: ±1.024V, Input Impedance: 300Ω, 5pF
- Dual Supplies (+5, -5.2V), 1.4W Power Dissipation

Figure 4.13

AD9631/AD9632 OP AMP KEY SPECIFICATIONS

- Current-Feedback performance with voltage-feedback amps
- Small Signal Bandwidth: 320MHz (AD9631, G = +1)
                             250MHz (AD9632, G = +2)
- Low Distortion: -113dBc   @ 1MHz
                     - 95dBc   @ 5Mhz
                     - 72dBc   @ 20MHz
- Slew Rate: 1300V / µs
- Settling Time: 16ns to 0.01%, 2V step
- Low Noise: Voltage: 7nV/√Hz, Current: 2pA/√Hz
- ±3V to ±5V Supply Operation, 17mA Supply Current

Figure 4.14

HIGH SPEED SAMPLING ADCS

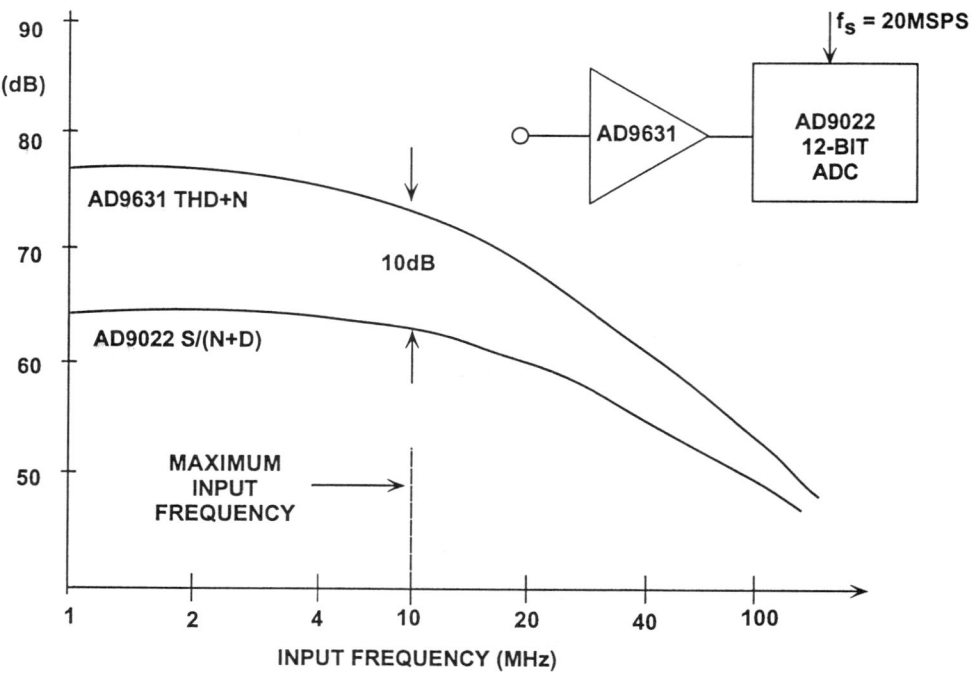

Figure 4.15

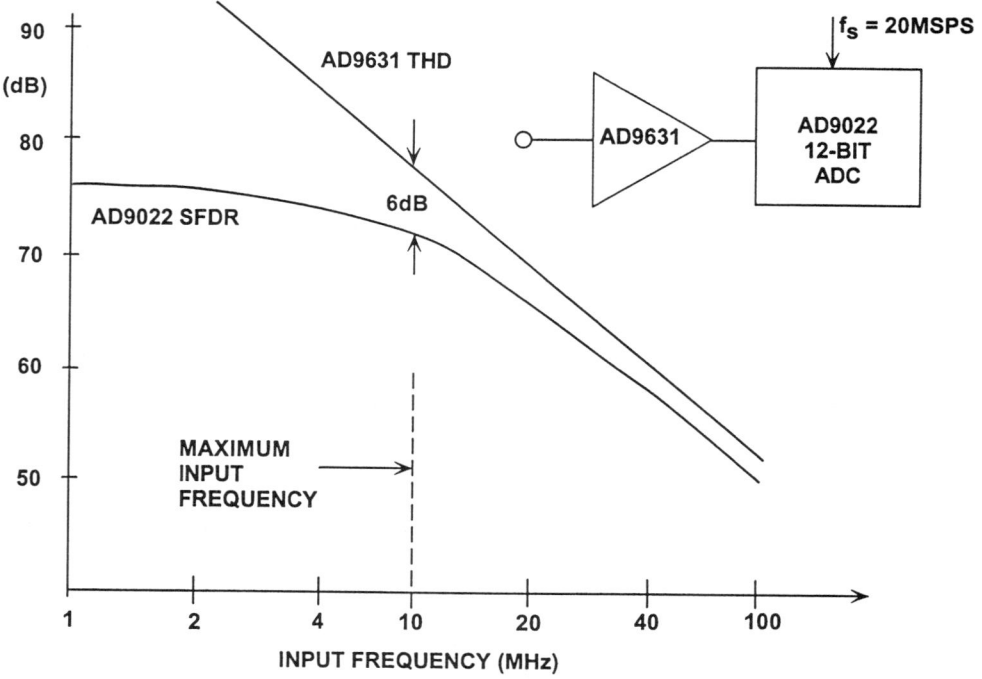

Figure 4.16

Figure 4.15 shows the THD+N of the AD9631 drive amplifier superimposed on the S/(N+D) plot for the AD9022 ADC (12-bits, 20MSPS). Notice that the amplifier THD+ N is at least 10dB better than the ADC S/(N+D) for input frequencies up to about 10MHz (the Nyquist frequency). In performing this comparison, it is important that the data for the op amp be obtained under the final operating conditions encountered in the actual circuit, i.e., gain, signal level, power supply voltage, etc.

While S/(N+D) and THD+N are useful ac performance indicators, there are a number of applications where low distortion is more important than low noise. In spectral analysis using FFTs, or other applications where averaging techniques can be used to reduce the effects of noise, the amplifier THD and the ADC distortion (generally SFDR) should be used as the selection criteria.

These characteristics should be plotted on the same scale, and the drive amplifier THD should be at least 6 to 10dB better than the ADC SFDR over the frequency range of interest. Such a plot for the AD9631 op amp and the AD9022 ADC is shown in Figure 4.16.

The above ac selection criterion works well if the ADC input is relatively benign, but may give overly optimistic results if the input impedance is signal dependent, or the input produces transient currents. The existence of either of these two conditions requires further investigation. The implications of signal-dependent input impedance will be demonstrated using a flash converter. Dealing with ADC input transient currents will be illustrated by examining a fast single-supply sampling ADC with a CMOS switched capacitor input stage.

DRIVING FLASH CONVERTERS

A typical flash converter (Figure 4.17) generally exhibits a signal-dependent input impedance (often referred to as *non-linear* input impedance), where the effective input capacitance is a function of signal level. The signal-dependent capacitance can be modeled as the junction capacitance of a diode, C_j. At the negative end of the input range, all the parallel comparators in the flash converter are "off", and the capacitance is low (modeled by a reverse-biased diode). At the positive end of the input range, all comparators are "on", thereby increasing the effective input capacitance (modeled by a zero-biased diode). For the example in the diagram, the Spice model (Figure 4.18) for the flash converter input under consideration is a 7.5nH inductor (simulating package pin and wirebond inductance), a 10pF fixed capacitor, and a diode having a 6pF zero-bias junction capacitance (C_{JO}). The total input capacitance changes from 16pF (0V input) to 12.5pF (−2V input) as shown in Figure 4.18. The 50Ω series resistor, R_s, is required to isolate the wideband op amp output from the flash converter input capacitance. Selecting the correct value for the series resistance is critical. If it is too low, the wideband, low-distortion op amp may be unstable because of the flash converter capacitive load. If it is too large, the distortion due to the non-linear input impedance may become significant, and bandwidth will be reduced because of the lowpass filter formed by the series resistor and the input capacitance. Data sheets for wideband low distortion amplifiers generally have curves showing the

HIGH SPEED SAMPLING ADCs

optimum value of series resistance as a function of the load capacitance. Typical recommended resistor values range from about 10Ω to 100Ω, depending on the amplifier and the load capacitance.

In the model, a value of 50Ω was chosen. Figure 4.19 shows the simulated THD produced by the equivalent circuit (assuming an ideal op amp, of course).

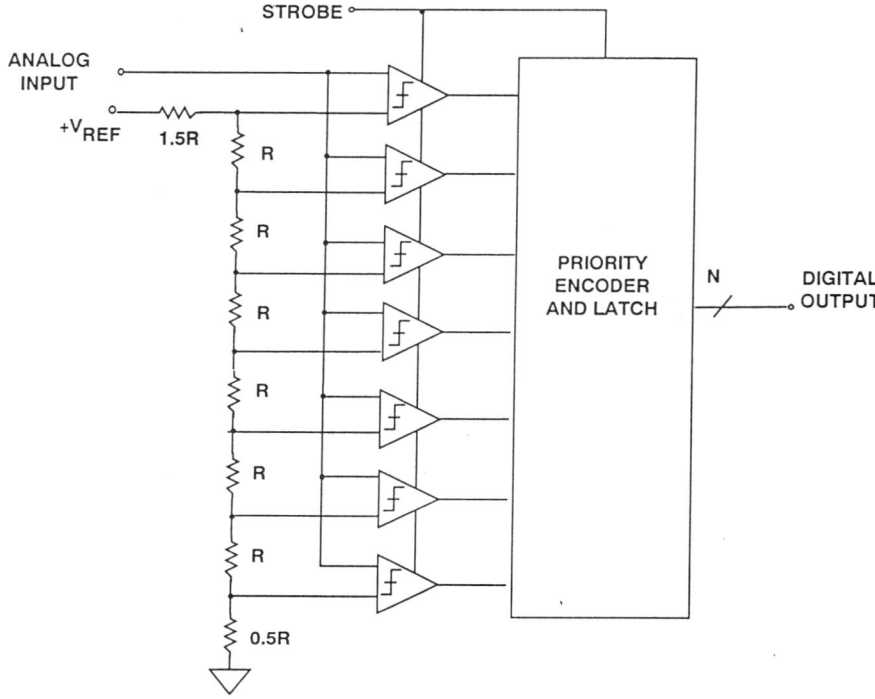

Figure 4.17

FLASH ADC INPUT MODEL SHOWS CAPACITANCE IS A FUNCTION OF INPUT SIGNAL

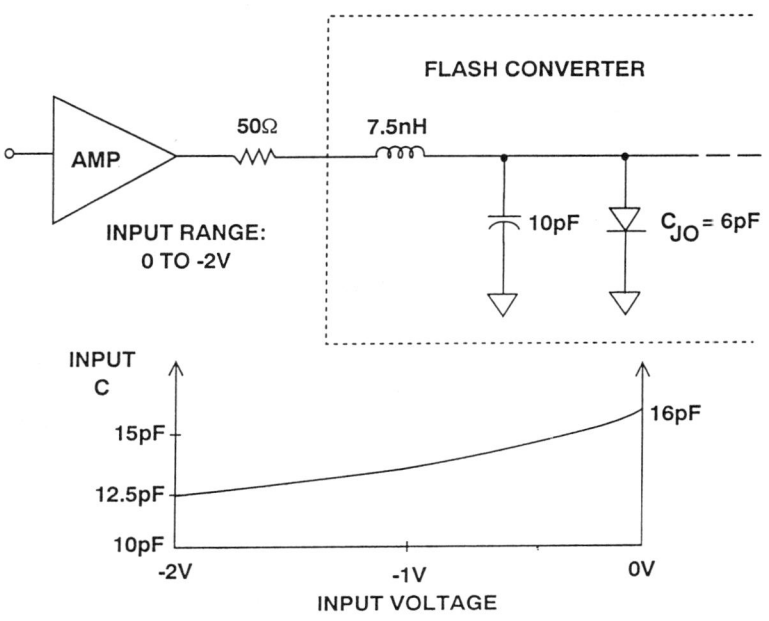

Figure 4.18

ADC TOTAL HARMONIC DISTORTION VERSUS INPUT FREQUENCY AS PREDICTED BY MODEL

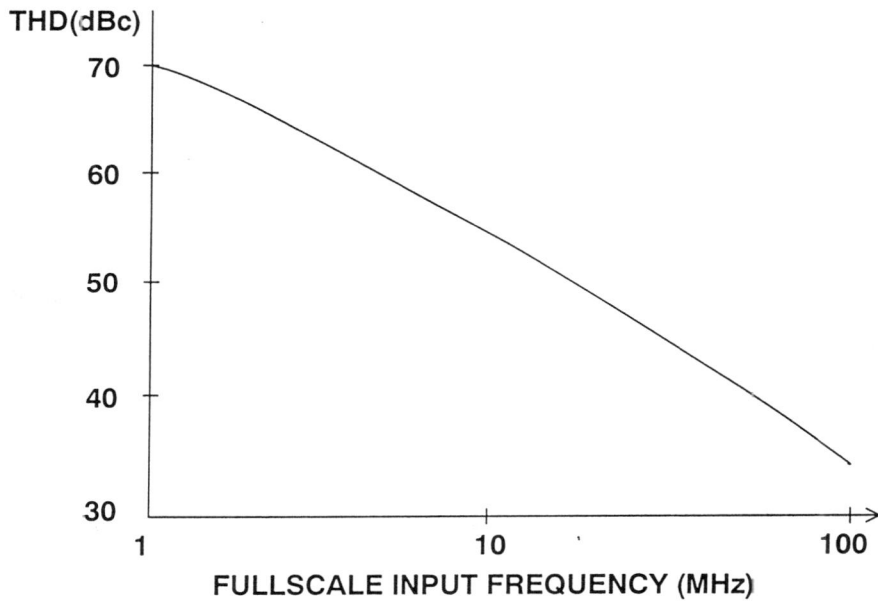

Figure 4.19

Driving the AD9050 Single-Supply 10-Bit, 40MSPS ADC

The AD9050 is a 10-bit, 40MSPS single supply ADC designed for wide dynamic range applications such as ultrasound, instrumentation, digital communications, and professional video. A block diagram of the AD9050 (Figure 4.20) illustrates the two-step subranging architecture, and key specifications are summarized in Figure 4.21.

The analog input circuit of the AD9050 (see Figure 4.22) is differential, but can be driven either single-endedly or differentially with equal performance. The input signal range of the AD9050 is ±0.5V centered around a common-mode voltage of +3.3V.

The input circuit of the AD9050 is a relatively benign and constant 5kΩ in parallel with approximately 5pF. Because of its well-behaved input, the AD9050 can be driven directly from 50, 75, or 100Ω sources without the need for a low-distortion buffer amplifier. In ultrasound applications, it is normal to ac couple the signal (generally between 1MHz and 15MHz) into the AD9050 differential inputs using a wideband transformer as shown in Figure 4.23 (A). Signal-to-noise plus distortion (S/N+D) values of 57dB (9.2 ENOB) are typical for a 10MHz input signal. If the input signal comes directly from a 50, 75, or 100Ω single-ended source, capacitive coupling as shown in Figure 4.23 (B) can be used.

AD9050 10-BIT, 40MSPS SINGLE SUPPLY ADC

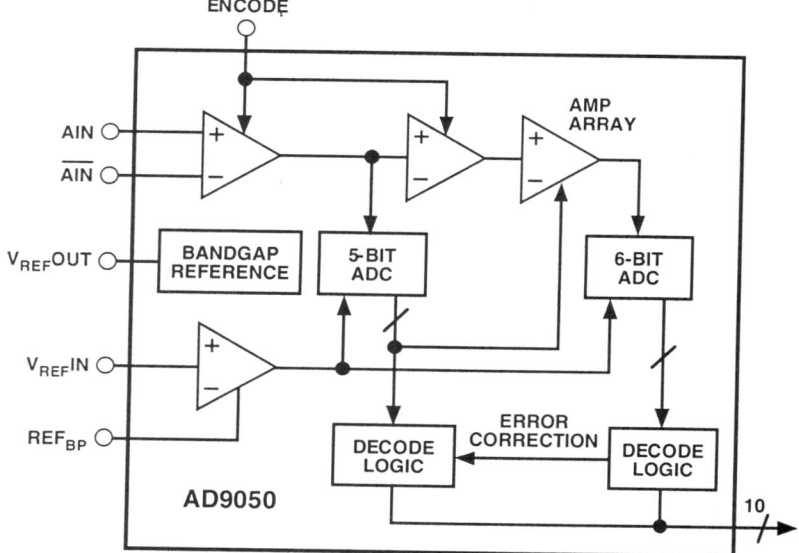

Figure 4.20

AD9050 10-BIT, 40MSPS ADC KEY SPECIFICATIONS

- 10-Bits, 40MSPS, Single +5V Supply
- Selectable Digital Supply: +5V, or +3V
- Low Power: 300mW on BiCMOS Process
- On-Chip SHA and +2.5V reference
- 56dB S/(N+D), 9 Effective Bits, with 10.3MHz Input Signal
- No input transients, Input Impedance 5kΩ, 5pF
- Input Range +3.3V ±0.5V Single-Ended or Differential
- 28-pin SOIC / SSOP Packages
- Ideal for Digital Beamforming Ultrasound Systems

Figure 4.21

AD9050 SIMPLIFIED INPUT CIRCUIT

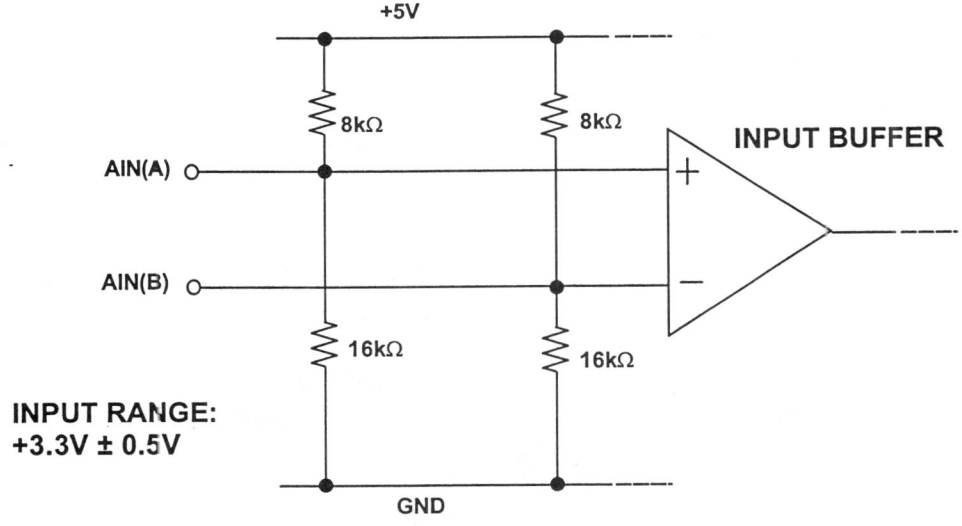

Figure 4.22

High Speed Sampling ADCs

AC COUPLING INTO THE INPUT OF THE AD9050 ADC

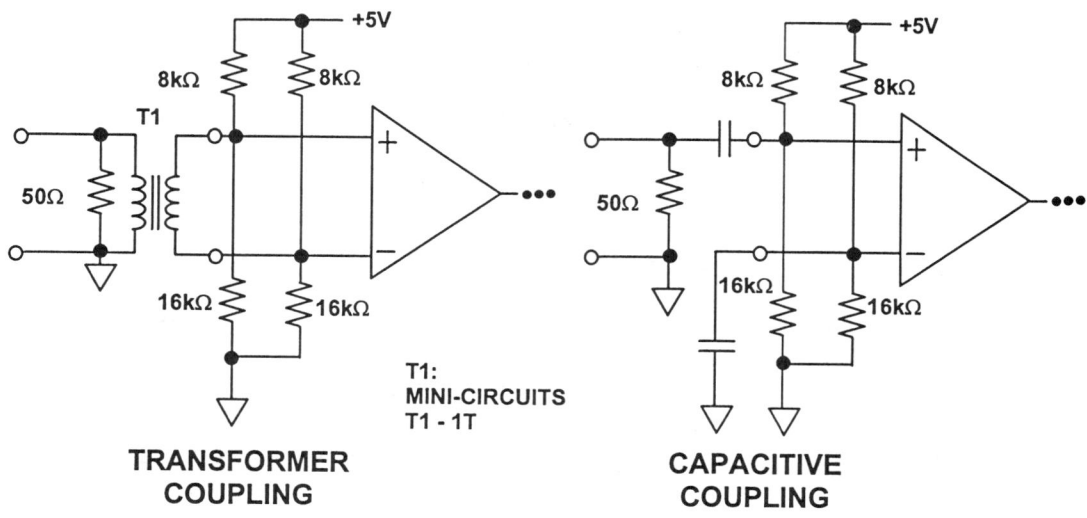

Figure 4.23

Driving ADCs with Switched Capacitor Inputs

Many ADCs, including fast sampling ones, have switched capacitor input circuits. Not only can the effective input impedance be a function of the sampling rate, but the switches (usually CMOS) may inject charge on the ADC's analog input. For instance, the internal track-and-hold amplifier (THA) may generate a current spike on the analog input when it switches from the *track* mode to the *hold* mode, and vice versa. Other spikes may be generated during the actual conversion. These fast current spikes appear on the output of the external ADC drive amplifier, producing corresponding voltage spikes (because of the closed-loop high frequency op amp output impedance), and conversion errors will result if the amplifier settling time to them is not adequate.

The AD876 is a 10-bit, 20MSPS, low power (150mW), CMOS ADC with a switched capacitor track-and-hold input circuit. The overall block diagram of the ADC is shown in Figure 4.24, and key specifications are given in Figure 4.25.

4 - 20

HIGH SPEED SAMPLING ADCs

AD876 10-BIT, 20MSPS LOW POWER SINGLE SUPPLY ADC SIMPLIFIED BLOCK DIAGRAM

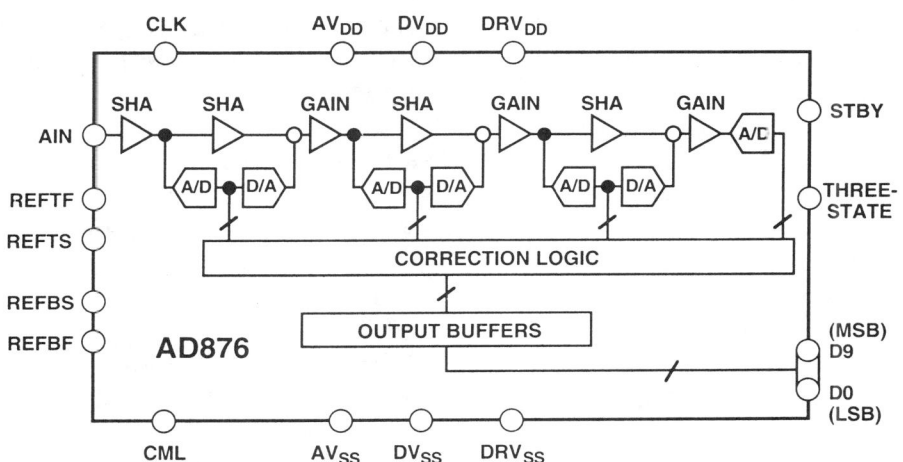

Figure 4.24

AD876 LOW POWER SINGLE SUPPLY ADC KEY SPECIFICATIONS

- 10-Bits, 20MSPS, Single-Supply

- Low Power CMOS Design: 140mW

- Standby Mode Power: < 50mW

- S/(N + D):   56dB   @   1MHz
              55dB   @   3.58MHz
              51dB   @   10MHz

- Input Bandwidth: 250MHz

- Input Range: 2V peak-to-peak

- Differential Gain: 1%, Differential Phase: 0.5°

Figure 4.25

4 - 21

Operation of the AD876 switched capacitor input circuit is illustrated in Figure 4.26, and the associated switching waveforms in Figure 4.27. The CMOS switches S1, S2, and S3 control the action of the internal sample and hold. They are shown in the track mode. Notice that in the track mode, the CMOS switch, S2, connects the input, V_{IN}, to the 3pF hold capacitor which must be charged by the drive amplifier.

When the circuit goes into the hold mode the following sequential switching occurs: S1 opens, S2 opens, and S3 closes (the entire sequence occurs in a few nanoseconds). The held voltage across C_H is thus transferred to the output of the internal op amp, A1. Opening CMOS switch S2 injects a small amount of charge into the ADC input (equivalent to approximately 1mA of current, having a duration of a few nanoseconds). This current produces a transient voltage across the op amp closed loop output impedance (approximately 10Ω at 10MHz, and 100Ω at 100MHz) in series with the 30Ω isolation resistor, R_s. The resulting voltage spike is approximately 100mV, corresponding to the product of the 1mA transient and the total ac source impedance of 100Ω (the op amp Z_o of about 70Ω plus the 30Ω isolation resistor). During the hold mode, the AD876 performs the conversion, while the input signal may continue to change. The ADC input signal in Figure 4.27 goes from negative full scale (+1.7V) to positive full scale (+3.7V) during the first hold interval shown in the timing diagram. This represents a worst case condition, and input slew rates (and the corresponding charging transient) are quite likely to be less in a practical application.

At the end of the conversion, the switches return to their track mode state in the following sequence: S3 opens, S1 closes, and S2 closes (the entire sequence occurs in a few nanoseconds). When S2 closes, a small amount of charge is injected into the op amp output, but the dominant current spike (5mA) is the instantaneous current required to charge the hold capacitor, C_H, to the new signal value. The external drive amplifier must therefore charge C_H to the new signal value and settle to the required accuracy (1/2 LSB, or 1mV) before the initiation of the next conversion (the settling time must be less than 25ns if the ADC is sampling at its maximum rate of 20MSPS as shown). The charging current dominates and is shown on the ADC input waveform as a negative-going 500mV spike. The addition of an external capacitor, C_P = 15pF, in parallel with the ADC input helps absorb some of the transient charge and is small enough so that bandwidth is not compromised. The optimum value of 15pF was determined empirically.

High Speed Sampling ADCs

AD876 ADC SWITCHED CAPACITOR INPUT CIRCUIT

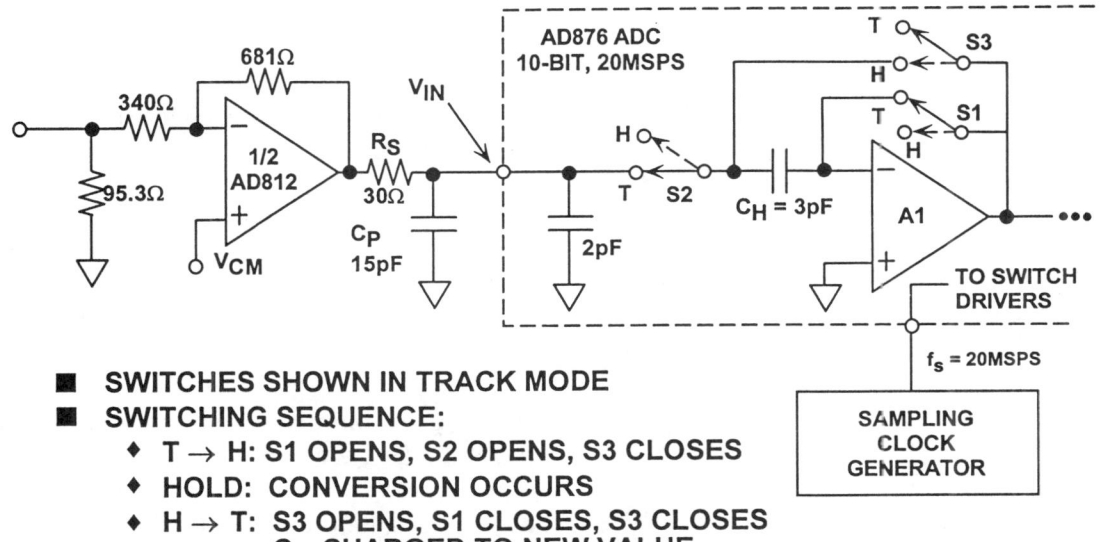

- SWITCHES SHOWN IN TRACK MODE
- SWITCHING SEQUENCE:
  - T → H: S1 OPENS, S2 OPENS, S3 CLOSES
  - HOLD: CONVERSION OCCURS
  - H → T: S3 OPENS, S1 CLOSES, S3 CLOSES
    C_H CHARGED TO NEW VALUE

Figure 4.26

AD876 ANALOG INPUT TRANSIENTS SHOWN FOR SAMPLING FREQUENCY OF 20MSPS

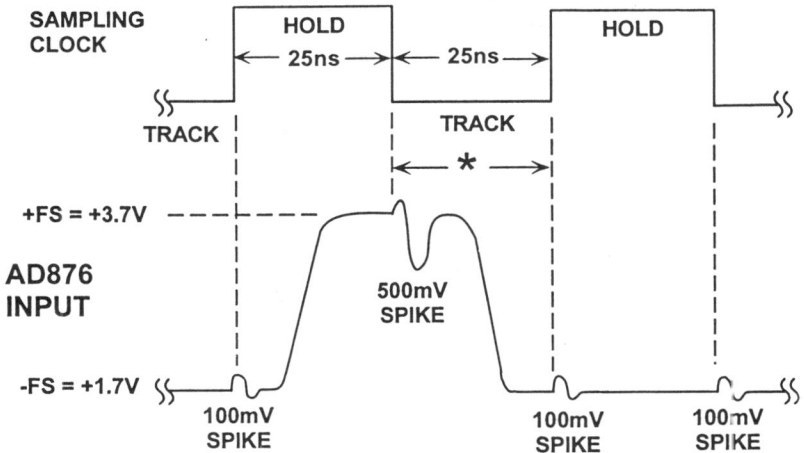

* INPUT MUST SETTLE AND CHARGE C_H DURING THIS INTERVAL

Figure 4.27

4 - 23

An empirical way to determine if the op amp transient load current settling time is adequate is to connect it to the ADC and observe the ADC input directly with a fast digital oscilloscope which is not sensitive to overdrive. If this is not possible, a good rule of thumb is to estimate the closed loop bandwidth, f_{cl}, required of the op amp to meet the required settling time. This is done as follows. The amplitude of the voltage spike, $V_{error}(t)$, at the ADC input is estimated by multiplying the input step current, ΔI, by the total driving impedance, Z_{out}, (composed of the closed loop output impedance plus the series isolation resistor, R_s). The output impedance of a typical high speed op amp (bandwidth of 50MHz or greater) is generally between 50Ω and 100Ω at the frequencies contained in the transient. If we assume a single-pole system with a bandwidth of f_{cl}, the transient exhibits an exponential decay described by the following equation:

$$V_{error}(t) = \Delta I \cdot Z_{out}\, e^{-t/\tau}.$$

Solving the equation for t:

$$t = -\tau \ln\left(\frac{V_{error}(t)}{\Delta I \cdot Z_{out}}\right).$$

Substituting $\tau = \dfrac{1}{2\pi f_{cl}}$, and solving for f_{cl}:

$$f_{cl} = -\frac{1}{2\pi t} \ln\left(\frac{V_{error}(t)}{\Delta I \cdot Z_{out}}\right)$$

Now, let $t = T_s/2 = 1/2f_s$ (where T_s = Sampling Period = $1/f_s$), and $V_{error}(t) = q/2$, q = weight of LSB:

$$f_{cl} = -\frac{f_s}{\pi} \ln\left(\frac{q/2}{\Delta I \cdot Z_{out}}\right),$$

the minimum required op amp closed loop bandwidth.

This equation determines the minimum closed-loop op amp bandwidth required based on the sampling rate, LSB weight, and the input voltage step ($\Delta I \cdot Z_{out}$).

In the example previously shown for the AD876 ADC, we can use the above equation to estimate the required op amp closed loop bandwidth by letting f_s = 20MSPS, q = 2mV, and $\Delta I \cdot Z_{out}$ = 500mV. Solving yields f_{cl} = 39.6MHz. The AD812 closed-loop bandwidth is approximately 60MHz in the configuration shown in Figure 4.28, which is more than sufficient to provide adequate settling time to the transient. Key specifications for the AD812 single-supply op amp are given in Figure 4.29.

SIMPLIFIED MODEL PREDICTS
INPUT TRANSIENT SETTLING TIME OF ADC DRIVE AMP

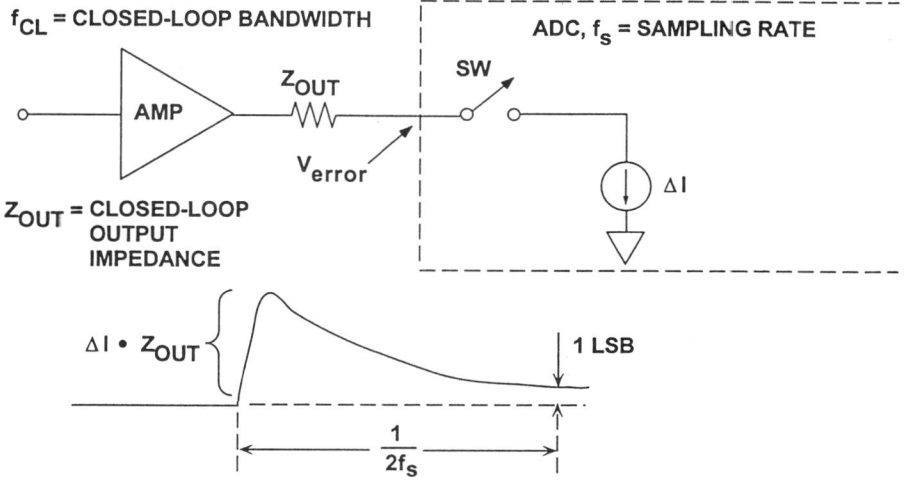

- LET $V_{error}(t) = \Delta I \cdot Z_{OUT} e^{-t/\tau} = 1$ LSB
  $t = 1/2f_s$
  $\tau = 1/2\pi f_{CL}$
- SOLVE FOR MINIMUM f_{CL}

Figure 4.28

KEY SPECIFICATIONS OF AD812 DUAL OP AMP

- Dual, Current Feedback, Low Current (11mA)
- Specified for ±15, ±5, +5, and +3V
- Input and Output CM Voltage Range (+1V to +4V), Vs = +5V
- Optimized for Video Applications:
  - Gain Flatness: 0.1dB to 40MHz
  - Differential Gain: 0.02%, Differential Phase: 0.02°
- 145MHz Bandwidth (3dB)
- 1600V / μs Slew Rate
- 50mA Output Current

Figure 4.29

It is generally true that if you select the op amp first based on the required distortion performance at the maximum input frequency of interest, then its bandwidth will be much greater than the ADC sampling rate, and the op amp will have adequate transient load current settling time.

The drive amplifier selection process can be summarized as follows. First, choose an op amp which provides the necessary bandwidth, distortion, and output voltage compatible with the ADC. A good ADC data sheet will recommend one or two op amps, generally selected to optimize ac performance at the higher frequencies. However, other choices may be better because of system considerations. For example, many tradeoffs are possible between ac and dc performance. If extremely low distortion at low frequencies is required (at the expense of high frequency performance), other low distortion amplifiers may provide optimum system performance. Other factors such a single-supply versus dual-supply may influence the decision.

The next step is to examine the ADC data sheet carefully to determine if the input structure presents any transient loads to the op amp. If transient loads are present, the op amp settling time is important, and the ADC data sheet should be consulted for specific requirements. If the data sheet does not specify the op amp settling time, a conservative approach is to choose an op amp with a settling time (to the required accuracy) of less than one-half the minimum sampling period. The required closed-loop bandwidth, f_{cl}, corresponding to the settling time can be estimated using the procedure and equations previously described. Finally, it is most important to construct a prototype of the system and perform an actual evaluation of the combined op amp and ADC performance. Manufacturer's evaluation boards are useful for this purpose.

DRIVE OP AMP SELECTION CRITERIA BASED ON ADC DYNAMIC PERFORMANCE: SUMMARY

- Select Op Amp with distortion and noise better than ADC
- Consider other factors also:
    - Output Voltage Swing must match ADC input
    - Single or Dual-Supply System?
    - DC Accuracy / Drift if DC coupled
- Examine ADC for transient load currents, if any
    - Op amp must have sufficient closed-loop bandwidth to settle to transient currents
    - Use exponential decay model to estimate required f_{cl}

Figure 4.30

GAIN SETTING AND LEVEL SHIFTING

In dc coupled applications, the drive amplifier must provide the required gain and offset to match the signal to the input voltage range of the ADC. Figure 4.31 summarizes various gain and level shifting options. The circuit of Figure 4.31A operates in the non-inverting mode and uses a reference voltage, V_{ref}, to offset the output. Gain and offset interact according to the equation:

$$V_{out} = \left(1 + \frac{R2}{R1}\right)V_{in} - \frac{R2}{R1}V_{ref}$$

The circuit in Figure 4.31B operates in the inverting mode, and the signal gain is independent of the offset. The disadvantage of this circuit is that the addition of R3 increases the noise gain, and hence the sensitivity to the op amp input offset voltage and noise. The input/output equation is given by:

$$V_{out} = -\frac{R2}{R1}V_{in} - \frac{R2}{R3}V_{ref}$$

The circuit in Figure 4.31C operates in the inverting mode, and the offset is applied to the non-inverting input, with no noise gain penalty. This circuit is also attractive for single-supply applications where $V_{ref} > 0$. The input/output equation is given by:

$$V_{out} = -\frac{R2}{R1}V_{in} + \left(\frac{R4}{R3+R4}\right)\left(1 + \frac{R2}{R1}\right)V_{ref}$$

OP AMP LEVEL SHIFTING CIRCUITS

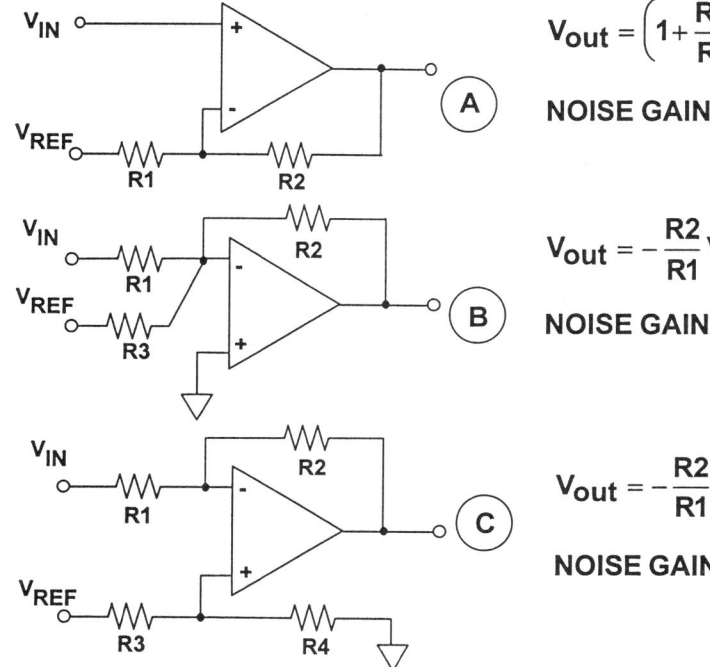

Figure 4.31

HIGH SPEED SAMPLING ADCs

A practical example of a single-supply video signal-processing digitizing circuit is shown in Figure 4.32. The AD876 (10-bit, 20MSPS) ADC operates on a single +5V supply, and its nominal input voltage range is 2V peak-to-peak centered around an allowable common-mode voltage between +2.7V and +3.1V. The input voltage range of the AD876 is set by external references. The AD812 drive amplifier is a fast video op amp with a common-mode input voltage range of +1V to +4V, and a +1V to +4V output voltage range. With this amplifier/ADC combination, optimum performance is obtained by setting the AD876 input common-mode voltage at +2.7V (corresponding to an upper and lower input range of +3.7 and +1.7V, respectively).

SINGLE-SUPPLY DC-COUPLED DRIVE CIRCUIT FOR AD876 10-BIT, 20MSPS ADC USING AD812 OP AMP

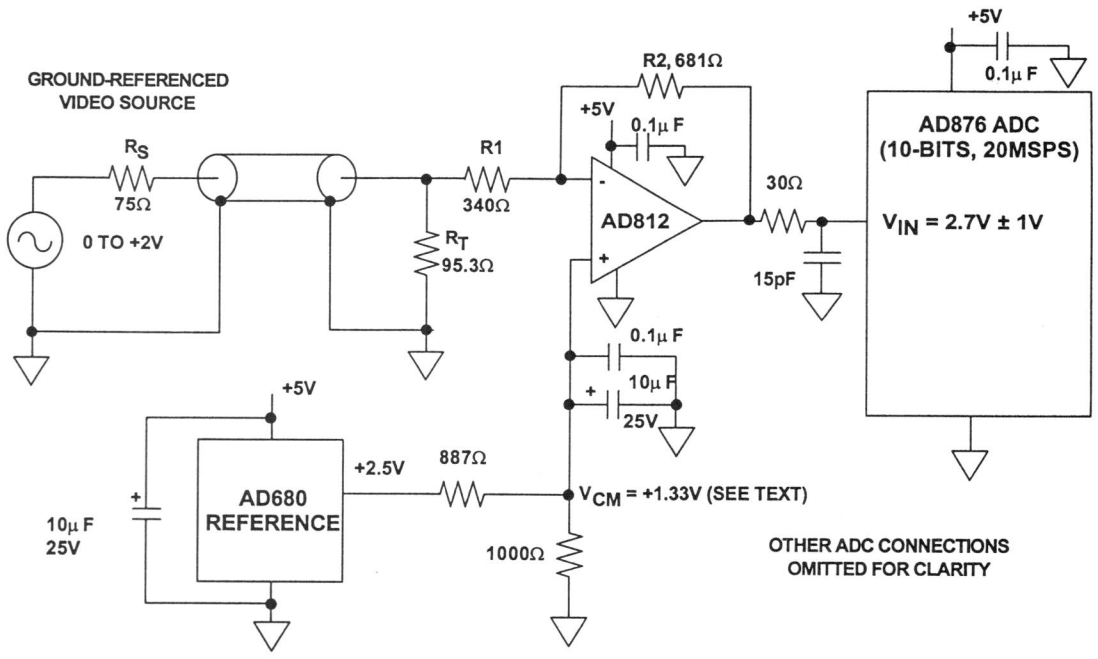

Figure 4.32

High Speed Sampling ADCs

The Thevenin equivalent circuit of the video signal is a ground-referenced, 0 to +2V source with a 75Ω source impedance designed to drive a 75Ω terminated coaxial cable, producing a standard 0 to +1V video signal level at the load termination, R_T. (The video source can also be modeled as a Norton equivalent circuit with a 0 to +26.7mA current source in parallel with the 75Ω source resistance.)

The AD812 op amp is operated as an inverting level shifter, similar to the circuit previously shown in Figure 4.31C. The feedback resistor, R2, is chosen to be 681Ω for optimum flatness over the video bandwidth per the AD812 data sheet. The feedforward resistor, R1, is selected to give a signal gain of −2. The termination resistor, R_T, is chosen so that the parallel combination of R_T and R1 is 75Ω. The common-mode voltage, V_{cm}, required on the non-inverting op amp input must now be determined. Assume that the video source is zero volts. The corresponding op amp output voltage should be +3.7V. The common-mode voltage is determined by the voltage divider formed by R2, R1, R_T, and the 75Ω source resistance:

$$V_{cm} = 3.7\left(\frac{R_S \| R_T + R1}{R_S \| R_T + R1 + R2}\right) = 3.7\left(\frac{42 + 340}{42 + 340 + 681}\right) = 1.33V$$

The +1.33V common-mode voltage is derived from the AD680 +2.5V reference using a resistor divider and is decoupled with a 10µF/25V tantalum capacitor in parallel with a 0.1µF low-inductance ceramic one.

The AD9050 10-bit, 40MSPS single-supply ADC input range of ±0.5V is centered around a common-mode voltage of +3.3V (corresponding to an upper and lower limit of +3.8V and +2.8V, respectively). An appropriate single-supply dc-coupled drive circuit based upon the AD8011 low power, low distortion op amp is shown in Figure 4.33. The source is a ground-referenced 0 to +2V signal which is series-terminated in 75Ω. The termination resistor, R_T, is chosen such that the parallel combination of R_T and R1 is 75Ω. The AD8011 current-feedback op amp is configured for a gain of −1. The feedback resistor, R2, is the value recommended for optimum bandwidth. Assume that the video source is at zero volts. The corresponding ADC input voltage should be +3.8V. The common-mode voltage, V_{cm}, is determined from the following equation:

$$V_{cm} = 3.8\left(\frac{R_S \| R_T + R1}{R_S \| R_T + R1 + R2}\right) = 3.8\left(\frac{38.8 + 1000}{38.8 + 1000 + 1000}\right) = 1.94V$$

The common-mode voltage, V_{cm}, is derived from the common-mode voltage at the inverting input of the AD9050. The +3.3V is buffered by the AD820 single-supply FET-input op amp. A divider network generates the required +1.94V for the AD8011, and a potentiometer provides offset adjustment capability.

The AD8011 current feedback op amp was chosen because of its low power (5mW), wide bandwidth (200MHz), and low distortion (–70dBc at 5MHz). It is fully specified for both ±5V and +5V operation. When operating on a single +5V supply, the input common-mode range is +1.5V to +3.5V, and the output swing is +1.2V to +3.5V. The high speed level-shifting PNP transistor at the output of the AD8011 allows the op amp to operate within in its recommended output range and ensures best distortion performance. Distortion performance of the entire circuit including the ADC is better than –60dBc for an input frequency of 10MHz and a sampling rate of 40MSPS.

DC-COUPLED SINGLE-SUPPLY DRIVE CIRCUIT FOR AD9050 10-BIT, 40MSPS ADC USING AD8011 OP AMP

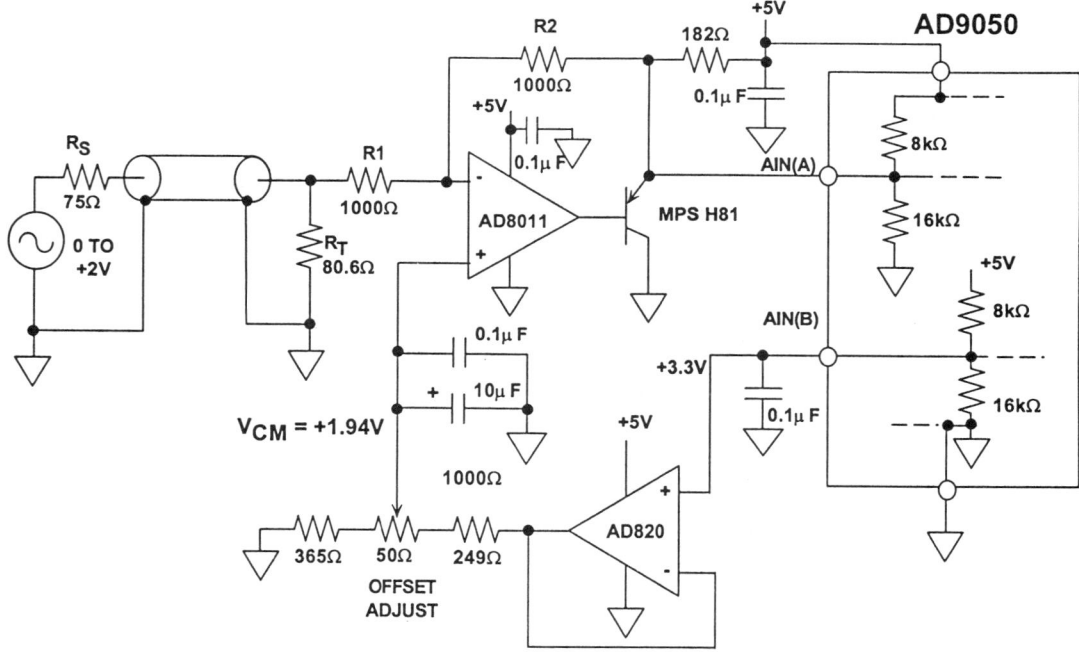

Figure 4.33

AD8011 OP AMP KEY SPECIFICATIONS

- Low Power: 1mA Current (5mW on +5V Supply)
- Bandwidth: 320MHz (G=+1), 180MHz (G=+2)
- Settling Time: 25ns to 0.1%
- Low Distortion: −76dBc at 5MHz
- Input Common Mode Voltage (+5V Supply):   +1.5V to +3.5V
- Output Voltage Swing (+5V Supply):   +1.2V to +3.8V
- Fully Specified for Single or Dual-Supply Operation

Figure 4.34

HIGH SPEED SAMPLING ADC EXTERNAL REFERENCE VOLTAGE GENERATION

Due to process and design-related constraints, it is not always possible to integrate the reference and the ADC on the same chip. In some ADCs which do have an internal reference, performance improvements (less noise and drift) may be obtained by using an external reference rather than the internal one. We saw in the case of the AD77XX series that this was the case.

There are a number of low-cost, low-noise, low-voltage references suitable for use with high performance sampling ADCs. Reference voltages of +1.25V, +2.048V, +2.5V, +3.0V, +3.3V, +4.096V, and +4.5V are ideal for single-supply (+3V or +5V) ADCs. (See Figure 4.35).

HIGH SPEED SAMPLING ADCs

LOW VOLTAGE REFERENCE SUMMARY

REFERENCE PART #	OUTPUT (V)	TOLERANCE (mV) (max)	DRIFT ppm/°C (max)	NOISE (µV p-p, typ) 0.1 to 10Hz	SUPPLY CURRENT (mA) typ
AD780	+2.5 / +3.0	1 - 5	3 - 20	4	0.75
AD680	+2.5	5 - 10	20 - 30	8	0.195
REF43	+2.5	15 - 50	10 - 25	5	0.45
REF191	+2.048	2 - 10	5 - 25	20	0.045
REF192	+2.5	2 - 10	5 - 25	25	0.045
REF193	+3.00	10	10 - 25	30	0.045
REF196	+3.3	10	20 - 25	33	0.045
REF198	+4.096	2 - 10	5 - 25	40	0.045
REF194	+4.5	2 - 10	5 - 25	45	0.045
AD589*	+1.235	35	25 - 100	4	0.050 - 5

* Two Terminal

Figure 4.35

The AD876 requires two references: one for each end of its input range which is nominally set for +1.7V and +3.7V. The output impedance of the drive sources must be low at high frequencies to absorb the transient currents generated at the ADC reference input terminals by the internal switched capacitor circuits.

The circuit shown in Figure 4.36 makes use of the REF198 (+4.096V) reference (see previous discussion regarding the analog input drive circuit for the AD876) and a dual FET-input single-supply op amp (AD822) to generate the two voltages. The AD876 reference inputs each have a FORCE (F) and SENSE (S) pin. The Kelvin connection compensates for the voltage drop in the internal parasitic resistances (approximately 5Ω). The internal ADC reference ladder impedance is approximately 300Ω, requiring the AD822s to source and sink approximately 6.7mA. There is an additional resistance of approximately 300Ω in series with each SENSE line. The two reference FORCE pins are decoupled at low and high frequencies using both a tantalum and a ceramic capacitor. The additional 20µF across the two SENSE pins adds additional decoupling for differential transients. Note that the AD822 must be properly compensated to drive the large capacitive load. Key specifications for the AD822 are summarized in Figure 4.37.

HIGH SPEED SAMPLING ADCs

REFERENCE VOLTAGE GENERATOR FOR AD876 10-BIT, 20MSPS SINGLE-SUPPLY ADC

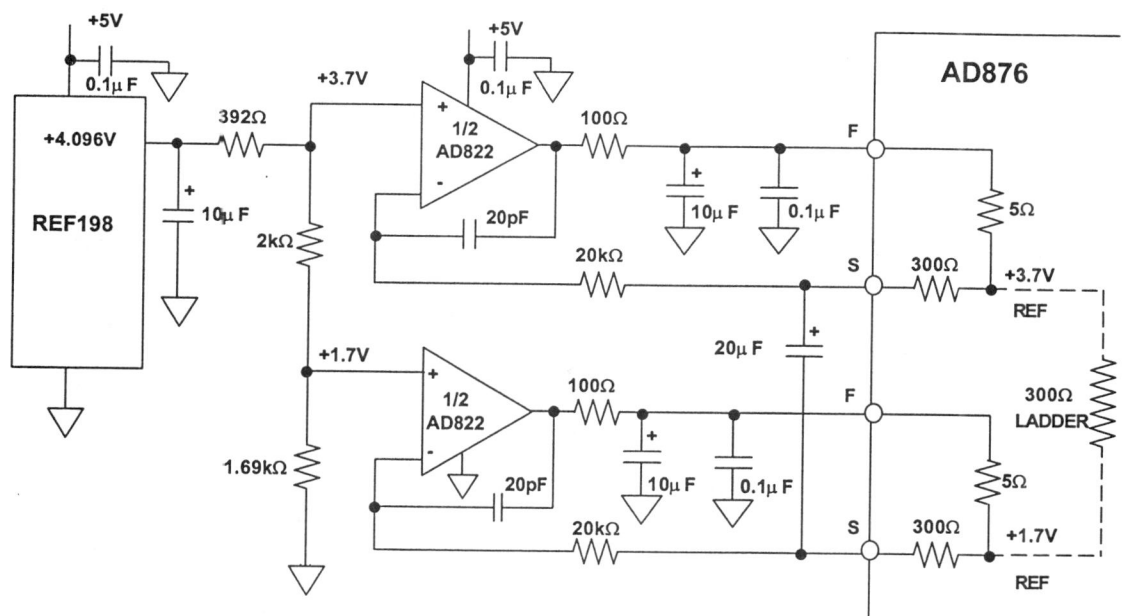

Figure 4.36

AD822 OP AMP KEY SPECIFICATIONS

- True Single-Supply FET-Input Dual Op Amp
- Complete Specifications for ±15V, ±5V, +5V, +3V
- Input Voltage Range Extends 200mV Below Ground and to within 1V of $+V_S$
- Output Goes to Within 5mV of Supplies (Open-Collector Complementary Output Stage Limited by V_{cesat})
- 1.8MHz Unity-Gain Bandwidth
- 800µV Offset Voltage, 2µV/°C Offset Drift
- Low Noise: 13nV/√Hz
- Low Power: 800µA / Amplifier
- Single Version Available (AD820)

Figure 4.37

ADC Input Protection and Clamping

The input to high speed ADCs should be protected from overdrive to prevent catastrophic damage or performance degradation. A good rule of thumb is never let the analog input exceed the supply voltage by more than 0.3V (this not only applies when the supply is on, but also when it is off, i.e., if the supply is off, the analog input should not exceed ±0.3V). In a dual supply system, this rule applies to both supplies. The rule of thumb protects most devices, but the data sheet Absolute Maximum specifications should always be consulted to determine possible exceptions. In some ADCs, the analog input is protected internally by diodes connected to the supplies. In these cases, an external resistor is required to limit the input current to 5mA or less under the overvoltage condition. Several overdrive protection schemes which use external diodes are shown in Figure 4.38.

In Figure 4.38A, the op amp is powered from ±15V, and the ADC from ±5V. The addition of the Schottky diodes on the input will prevent the analog input from exceeding the ADC supplies. Some ADCs have internal diodes, but the addition of the external ones ensures protection for higher currents. An alternative solution is to generate the ±5V for the ADC from the op amp ±15V supplies using three terminal regulators (such as the 78L05 and 79L05). This eliminates possible sequencing and overdrive problems, and power dissipation in the regulator is not excessive if low power CMOS ADCs are used (see Figure 4.38B).

With the proliferation of high speed op amps and ADCs, both of which operate on dual 5V supplies, the situation shown in Figure 4.38C is quite common, and there is no sequencing problem provided both the amplifier and the ADC are operated from the same supplies.

Figure 4.38D shows the case where a flash converter (powered from a single −5V supply) is driven from an amplifier powered from ±5 V. The series resistor and the Schottky diode provide protection from forward-biasing the flash substrate diode more than a few tenths of a volt, thereby preventing possible damage.

Specially designed high speed, fast recovery clamping amplifiers offer an attractive alternative to designing external clamping/protection circuits. The AD8036/AD8037 low distortion, wide bandwidth clamp amplifiers represent a significant breakthrough in this technology. These devices allow the designer to specify a high (V_H) and low (V_L) clamp voltage. The output of the device clamps when the input exceeds either of these two levels. The AD8036/AD8037 offer superior clamping performance compared to traditional output-clamping devices. Recovery time from overdrive is less than 5ns.

The key to the AD8036 and AD8037's fast, accurate clamp and amplifier performance is their proprietary input clamp architecture. this new design reduces clamp errors by more than 10x over previous output clamp based circuits, as well as substantially increasing the bandwidth, precision, and versatility of the clamp inputs.

Figure 4.39 is an idealized block diagram of the AD8036 connected as a unity gain voltage follower. The primary signal path comprises A1 (a 1200V/μs, 240MHz high voltage gain, differential to single-ended amplifier) and A2 (a G=+1 high current gain output buffer). The AD8037 differs from the AD8036 only in that A1 is optimized for closed-loop gains of two or greater.

ADC INPUT OVERVOLTAGE PROTECTION CIRCUITS

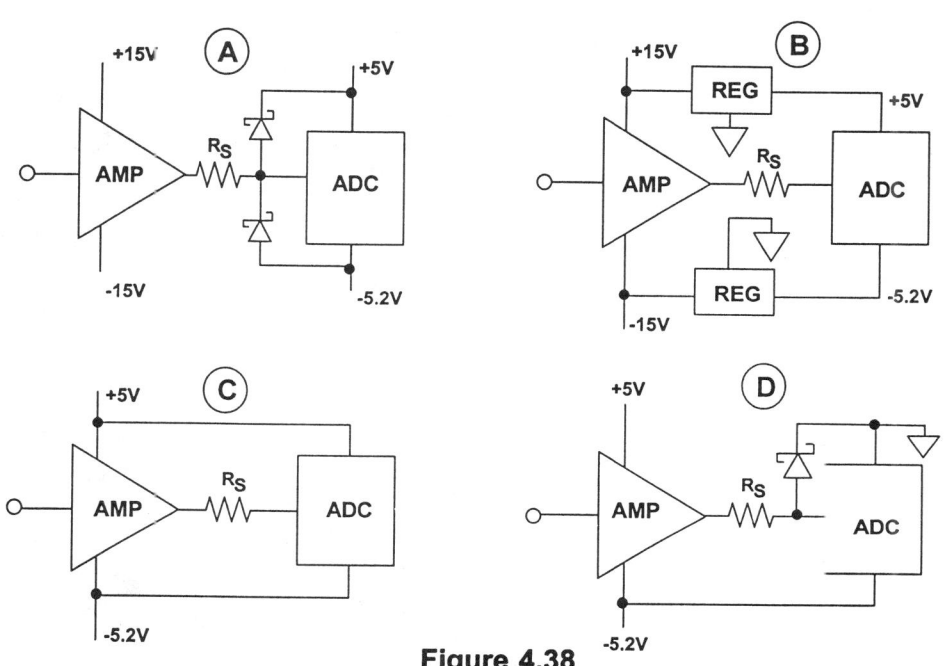

Figure 4.38

AD8036/AD8037 CLAMP AMPLIFIER EQUIVALENT CIRCUIT

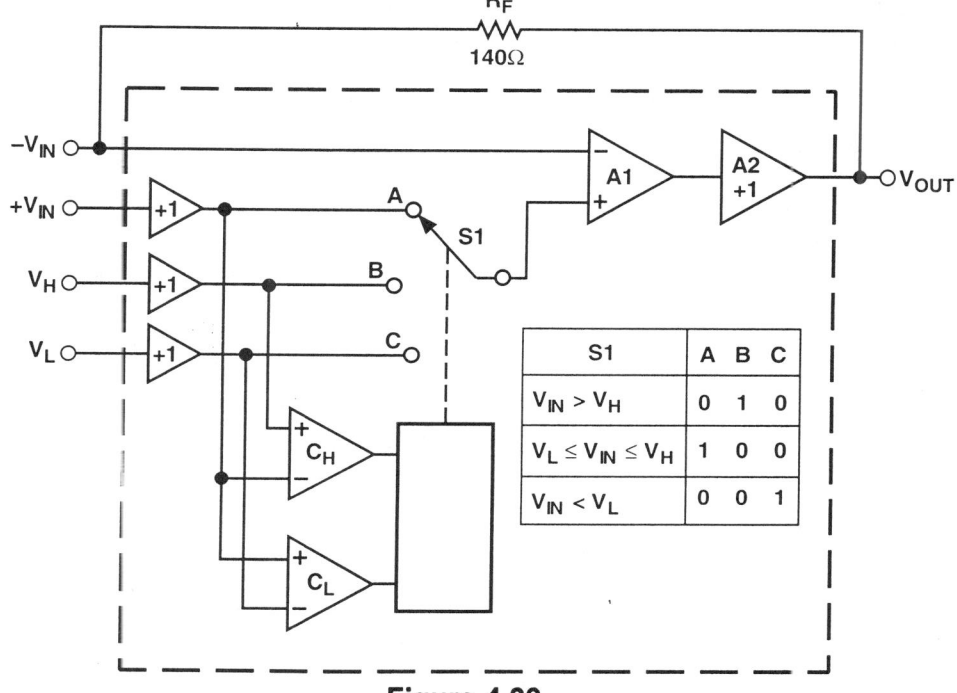

Figure 4.39

High Speed Sampling ADCs

The input clamp section is comprised of comparators C_H and C_L, which drive switch S1 through a decoder. The unity-gain buffers in series with the $+V_{IN}$, V_H, and V_L inputs isolate the input pins from the comparators and S1 without reducing bandwidth or precision.

The two comparators have about the same bandwidth as A1 (240MHz), so they can keep up with signals within the useful bandwidth of the AD8036. To illustrate the operation of the input clamp circuit, consider the case where V_H is referenced to +1V, V_L is open, and the AD8036 is set for a gain of +1 by connecting its output back to its inverting input through the recommended 140Ω feedback resistor. Note that the main signal path always operates closed loop, since the clamping circuit only affects A1's noninverting input.

If a 0V to +2V voltage ramp is applied to the AD8036's $+V_{IN}$ for the connection just described, V_{OUT} should track $+V_{IN}$ perfectly up to +1V, then should limit at exactly +1V as $+V_{IN}$ continues to +2V.

In practice, the AD8036 comes close to this ideal behavior. As the $+V_{IN}$ input voltage ramps from zero to 1V, the output of the high limit comparator C_H starts in the off state, as does the output of C_L. When $+V_{IN}$ just exceeds V_H (practically, by about 18mV), C_H changes state, switching S1 from "A" to "B" reference level. Since the + input of A1 is now connected to V_H, further increases in $+V_{IN}$ have no effect on the AD8036's output voltage. The AD8036 is now operating as a unity-gain buffer for the V_H input, as any variation in V_H, for $V_H > 1V$, will be faithfully produced at V_{OUT}.

Operation of the AD8036 for negative input voltages and negative clamp levels on V_L is similar, with comparator C_L controlling S1. Since the comparators see the voltage on the $+V_{IN}$ pin as their common reference level, the voltage V_H and V_L are defined as "High" or "Low" with respect to $+V_{IN}$. For example, if V_{IN} is set to zero volts, V_H is open, and V_L is +1V, comparator C_L will switch S1 to "C", so the AD8036 will buffer the voltage on V_L and ignore $+V_{IN}$.

The performance of the AD8036/AD8037 closely matches the ideal just described. The comparator's threshold extends from 60mV inside the clamp window defined by the voltages on V_L and V_H to 60mV beyond the window's edge. Switch S1 is implemented with current steering, so that A1's + input makes a continuous transition from say, V_{IN} to V_H as the input voltage traverses the comparator's input threshold from 0.9V to 1.0V for $V_H = 1.0V$.

The practical effect of the non-ideal operation is to soften the transition from amplification to clamping modes, without compromising the absolute clamp limit set by the input clamping circuit. Figure 4.40 is a graph of V_{OUT} versus V_{IN} for the AD8036 and a typical output clamp amplifier. Both amplifiers are set for G=+1 and $V_H = +1V$.

COMPARISON BETWEEN INPUT AND OUTPUT CLAMPING

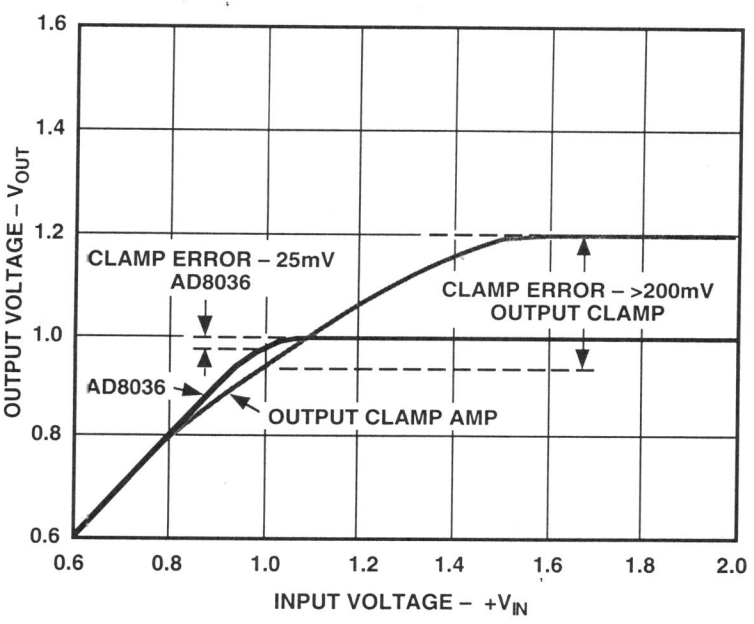

Figure 4.40

The worst case error between V_{OUT} (ideally clamped) and V_{OUT} (actual) is typically 18mV times the amplifier closed-loop gain. This occurs when V_{IN} equals V_H (or V_L). As V_{IN} goes above and/or below this limit, V_{OUT} will stay within 5mV of the ideal value.

In contrast, the output clamp amplifier's transfer curve typically will show some compression starting at an input of 0.8V, and can have an output voltage as far as 200mV over the clamp limit. In addition, since the output clamp causes the amplifier to operate open-loop in the clamp mode, the amplifier's output impedance will increase, potentially causing additional errors, and the recovery time is significantly longer.

It is important that a clamped amplifier such as the AD8036/AD8037 maintain low levels of distortion when the input signals are close the clamping voltages. Figure 4.41 shows the second and third harmonic distortion for the amplifiers as the output approaches the clamp voltages. The input signal is 20MHz, the output signal is 2V peak-to-peak, and the output load is 100Ω.

Recovery from step voltage which is two times over the clamping voltage is shown in Figure 4.42. The input step voltage starts at +2V and goes to 0V (left-hand traces on scope photo). The input clamp voltage (V_H) is set at +1V. The right-hand trace shows the output waveform. The key specifications for the AD8036/AD8037 clamped amplifiers are summarized in Figure 4.43.

AD8036/AD8037 DISTORTION NEAR CLAMPING REGION, OUTPUT = 2V p-p, LOAD = 100Ω, f = 20MHz

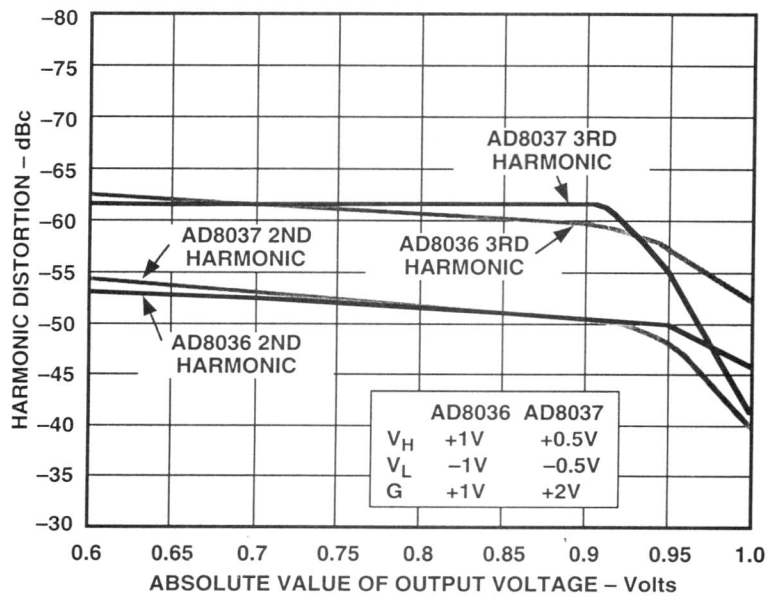

Figure 4.41

AD8036 / AD8037 OVERDRIVE (2x) RECOVERY

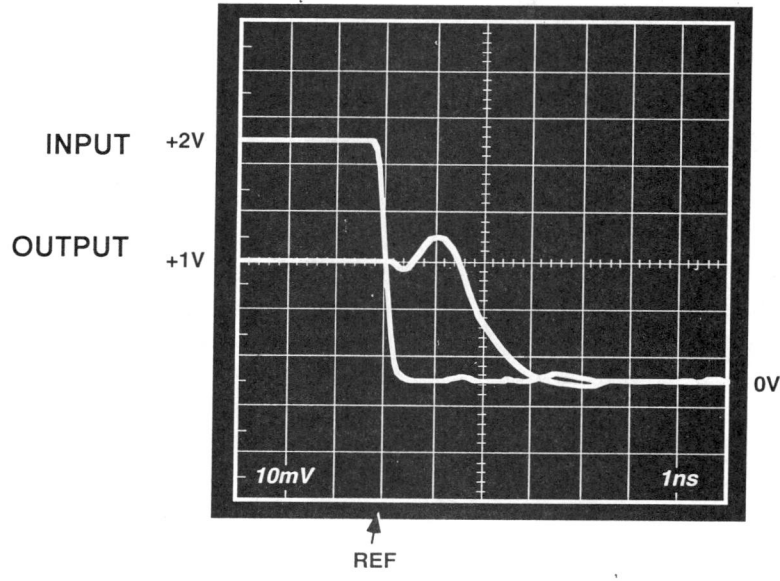

HORIZONTAL SCALE: 1ns/div

Figure 4.42

AD8036/AD8037 SUMMARY SPECIFICATIONS

- Proprietary Input Clamping Circuit with Minimized Nonlinear Clamping Region

- Small Signal Bandwidth: 240MHz (AD8036), 270MHz (AD8037)

- Slew Rate: 1500V/μs

- 1.5ns Overdrive Recovery

- Low Distortion: -72dBc @ 20MHz

- Low Noise: 4.5nv/√Hz, 2pA/√Hz

- 20mA Supply Current on ±5V

Figure 4.43

Figure 4.44 shows the AD9002 8-bit, 125MSPS flash converter driven by the AD8037 (240MHz bandwidth) clamping amplifier. In the circuit, the bandwidth of the AD8037 is 240MHz. The clamp voltages on the AD8037 are set to +0.55 and −0.55V, referenced to the ±0.5V input signal, with the external resistive dividers. The AD8037 also supplies a gain of two, and an offset of −1V (using the AD780 voltage reference), to match the 0 to −2V input range of the AD9002 flash converter. The output signal is clamped at +0.1V and −2.1V. This multi-function clamping circuit therefore performs several important functions as well as preventing damage to the flash converter which occurs if its input exceeds +0.5V, thereby forward biasing the substrate diode. The 1N5712 Schottky diode adds further protection during power-up.

AD9002 8-BIT, 125MSPS FLASH CONVERTER DRIVEN BY AD8037 CLAMP AMPLIFIER

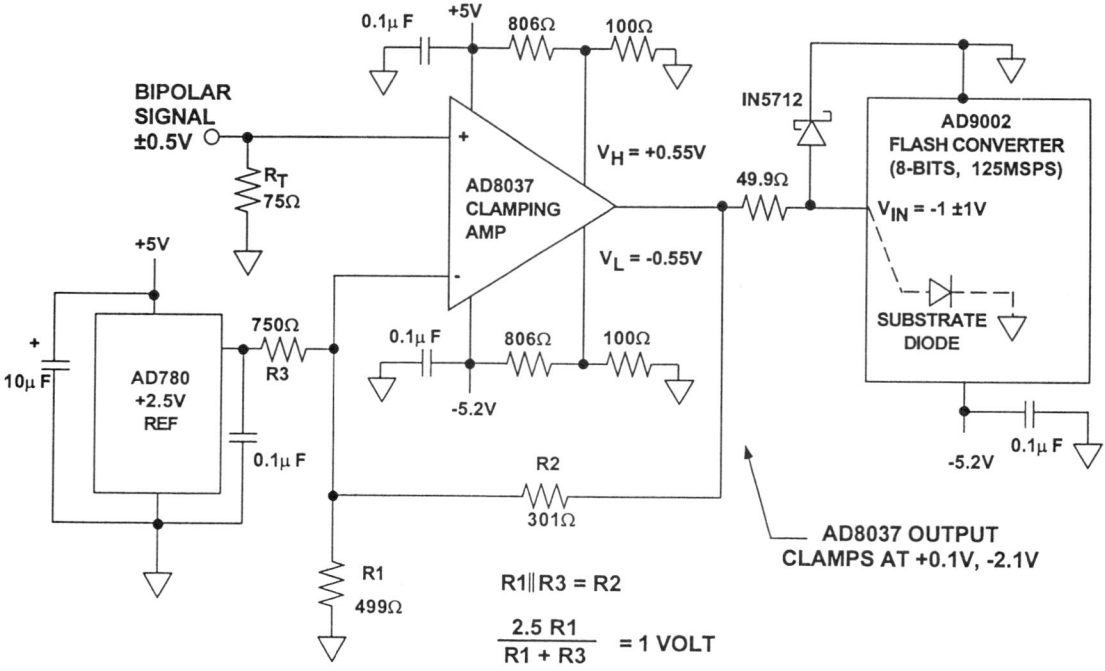

Figure 4.44

The feedback resistor, R2 = 301Ω, is selected for optimum bandwidth per the manufacturer's data sheet recommendation. In order to give a gain of two, the parallel combination of R1 and R3 must also equal R2:

$$\frac{R1 \cdot R3}{R1 + R3} = R2 = 301\Omega$$

(nearest 1% standard resistor value).

In addition, the Thevenin equivalent output voltage of the AD780 +2.5V reference and the R3/R1 divider must be +1V to provide the −1V offset at the output of the AD8037.

$$\frac{2.5 \cdot R1}{R1 + R3} = 1 \text{volt}$$

Solving the two simultaneous equations yields R1 = 499Ω, R3 = 750Ω (using the nearest 1% standard resistor values).

Other input and output voltages ranges can be accommodated by appropriate changes in the external resistors.

HIGH SPEED SAMPLING ADCs

OTHER APPLICATIONS FOR CLAMPING AMPLIFIERS

The AD8036/AD8037's clamp output can be set accurately and has a well controlled flat level. This, along with wide bandwidth and high slew rate make them well suited for a number of other applications. Figure 4.45 is a diagram of a programmable level pulse generator. The circuit accepts a TTL timing signal for its input and generates pulses at the output up to 24V p-p with 2500V/µs slew rate. The output levels can be programmed to anywhere in the range between –12V to +12V.

PROGRAMMABLE PULSE GENERATOR USING AD8037 CLAMP AMP AND AD811 OP AMP

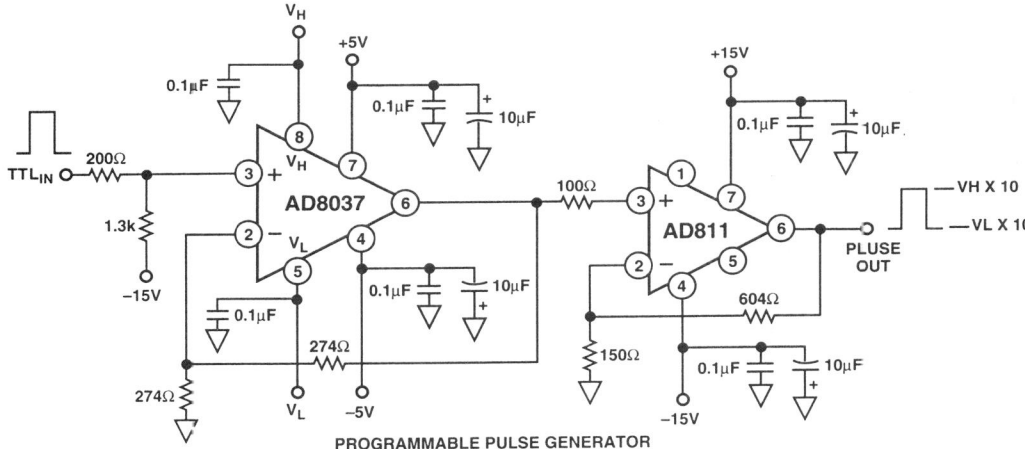

Figure 4.45

The circuit uses an AD8037 operating at a gain of two with an AD811 to boost the output to the ±12V range. The AD811 was chosen for its ability to operate with ±15V supplies and its high slew rate.

R1 and R2 act as a level shifter to make the TTL signal levels approximately symmetrical above and below ground. This ensures that both the high and low logic levels will be clamped by the AD8037. For well controlled signal levels in the output pulse, the high and low output levels result from the clamping action of the AD8037 and are not controlled by either the high or low logic levels passing through a linear amplifier. For good rise and fall times at the output pulse, a logic family with high speed edges should be used.

The high logic levels are clamped at 2 times the voltage at V_H, while the low logic levels are clamped at two times the voltage at V_L. The output of the AD8037 is amplified by the AD811 operating at a gain of 5. The overall gain of 10 will cause the high output level to be 10 times the voltage at V_H, and the low output level to be 10 times the voltage at V_L.

The clamping inputs are additional inputs to the input stage of the AD8036/AD8037. As such, they have an input bandwidth comparable to the amplifier inputs and lend themselves to some unique functions when they are driven dynamically.

Figure 4.46 is a schematic for a full wave rectifier, sometimes called an absolute value generator. It works well up to 20MHz and can operate at significantly higher frequencies with some degradation in performance. The distortion performance is significantly better than diode-based full-wave rectifiers, especially at high frequencies.

FULL-WAVE RECTIFIER USING AD8037 CLAMP AMP

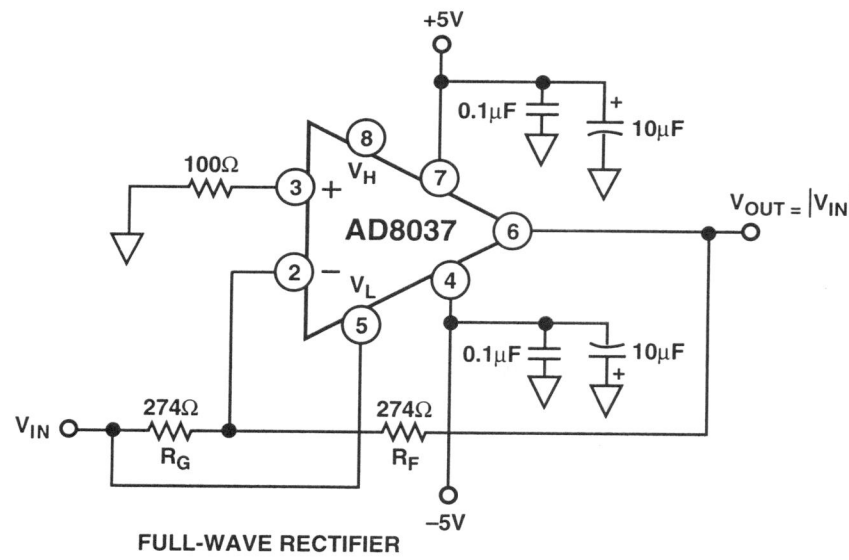

Figure 4.46

The AD8037 is configured as an inverting amplifier with a gain of unity. The input drives the inverting amplifier and also directly drives V_L, the lower level clamping input. The high level clamping input, V_H, is left floating and plays no role in the circuit.

When the input is negative, the amplifier acts as a unity-gain inverter and outputs a positive signal at the same amplitude as the input, with opposite polarity. V_L is driven negative by the input, so it performs no clamping action, because the positive output signal is always higher than the negative level driving V_L.

When the input is positive, the output result is the sum of two separate effects. First, the inverting amplifier multiplies the input by –1 because of the unity-gain inverting configuration. This effectively produces as offset as explained above, but with a dynamic level that is equal to –1 times the input. Second, although the positive input is grounded (through 100Ω), the output is clamped at two times the voltage applied to V_L (a positive, dynamic voltage in this case). The factor of two is because the noise gain of the amplifier is two.

The sum of these two actions results in an output that is equal to unity times the input signal for positive input signals as shown in Figure 4.47. An input/output scope photo with an input signal of 20MHz and an amplitude of ±1V is shown in Figure 4.48. Thus, for either positive or negative input signals, the output is unity times the absolute value of the input signal. The circuit can be easily configured to produce the negative absolute value of the input by applying the input to V_H rather than V_L.

FULL-WAVE RECTIFIER WAVEFORMS

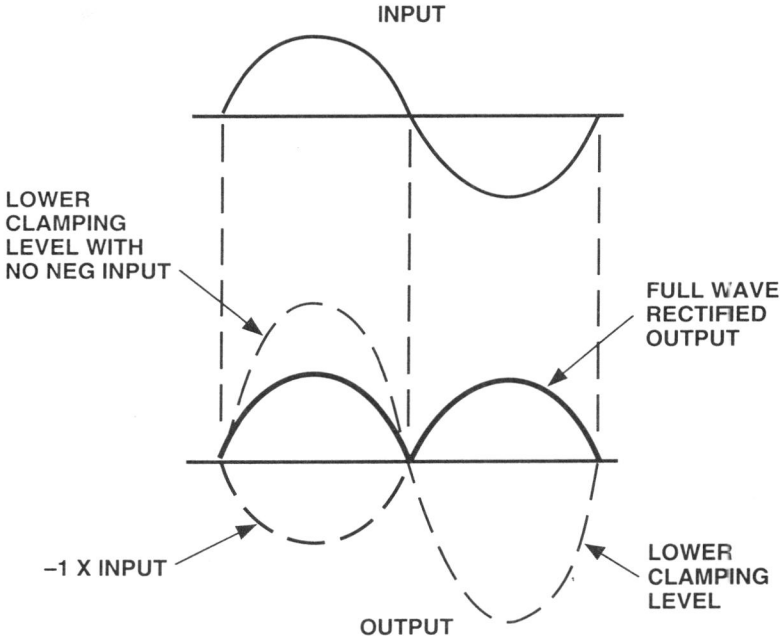

Figure 4.47

FULL-WAVE RECTIFIER INPUT/OUTPUT SCOPE WAVEFORM

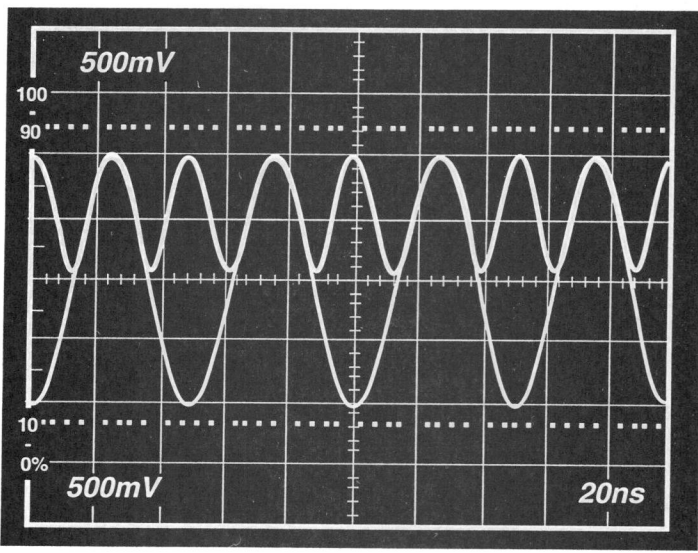

Figure 4.48

The circuit can get to within about 40mV of ground during the time when the input crosses zero. This voltage is fixed over a wide frequency range, and is a result of the switching between the conventional op amp input and the clamp input. But because there are no diodes to rapidly switch from forward to reverse bias, the performance far exceeds diode-based full wave rectifiers.

The 40mV offset can be removed by adding an offset to the circuit. A 27.4kΩ input resistor to the inverting input will have a gain of 0.01, while changing the gain of the circuit by only 1%. A plus or minus 4V dc level (depending on the polarity of the rectifier) into this resistor will compensate for the offset.

Full wave rectifiers are useful in many applications including AM signal detection, high frequency ac voltmeters, and various arithmetic operations.

The AD8037 can also be configured as an amplitude modulator as shown in Figure 4.49. The positive input of the AD8037 is driven with a square wave of sufficient amplitude to produce clamping action at both the high and low levels. This is the higher frequency carrier signal. The modulation signal is applied to both the input of a unity gain inverting amplifier and to V_L, the lower clamping input. V_H is biased at +0.5V.

AD8037 AMPLITUDE MODULATOR

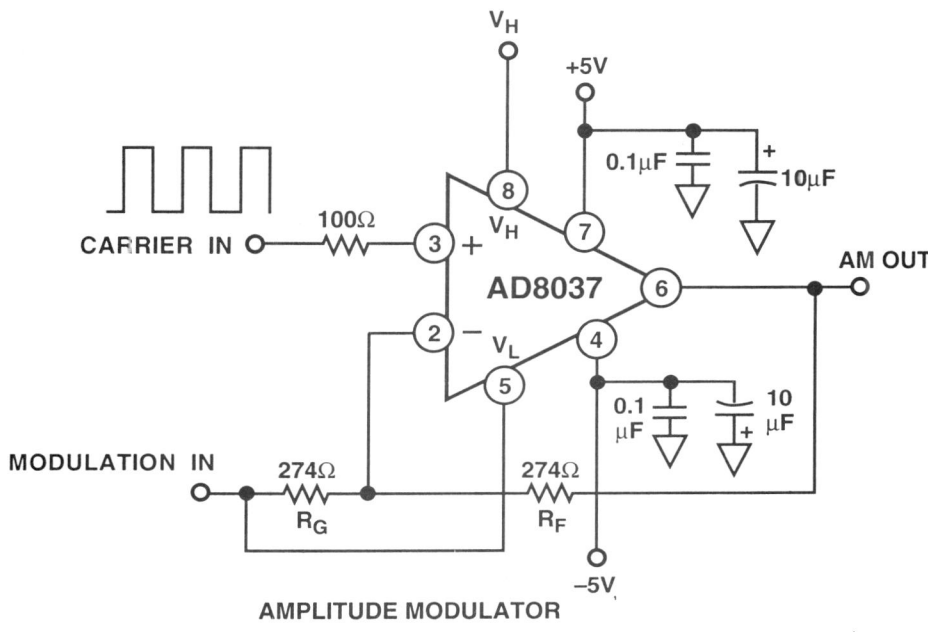

Figure 4.49

AMPLITUDE MODULATED WAVEFORM

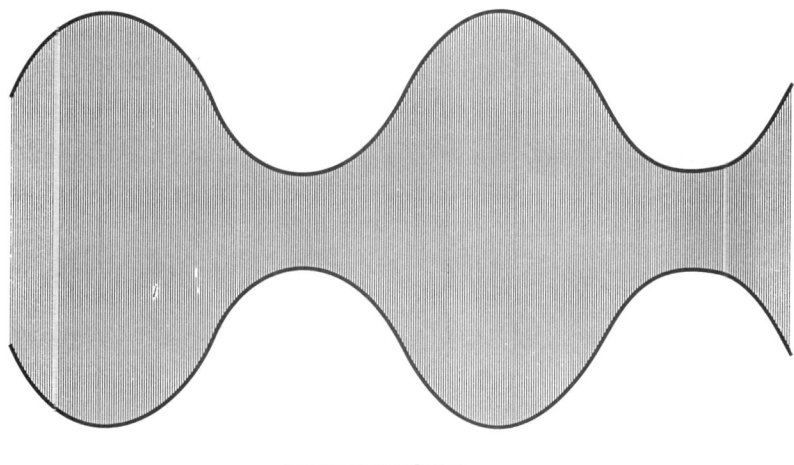

AM WAVEFORM

Figure 4.50

To understand the circuit operation, it is helpful to first consider a simpler circuit. If both V_L and V_H are dc biased at $-0.5V$ and the carrier and modulation inputs driven as above, the output would be a 2V p-p square wave at the carrier frequency riding on a waveform at the modulating frequency. The inverting input (modulation signal) is creating a varying offset to the 2V p-p square wave at the output. Both the high and low levels clamp at twice the input levels on the clamps because the noise gain of the circuit is two.

When V_L is driven by the modulation signal instead of being held at a dc level, a more complicated situation results. The resulting waveform is composed of an upper envelope and a lower envelope with the carrier square wave in between. The upper and lower envelopes are 180° out of phase as in a typical AM waveform.

The upper envelope is produced by the upper clamp level being offset by the waveform applied to the inverting input. This offset is the opposite polarity of the input waveform because of the inverting configuration.

The lower envelope is produced by the sum of two effects. First, it is offset by the waveform applied to the inverting input as in the case of the simpler circuit above. The polarity of this offset is in the same direction as the upper envelope. Second, the output is driven in the opposite direction of the offset at twice the offset voltage by the modulation signal being applied to V_L. This results from the noise gain being equal to two, and since there is no inversion in this connection, it is opposite polarity from the offset.

The result at the output for the lower envelope is the sum of these two effects, which produces the lower envelope of an AM waveform. The depth of modulation can be modified by changing the amplitude of the modulation signal. This changes the amplitude of the upper and lower envelope waveforms. The modulation depth can also be changed by changing the dc bias applied to V_H. In this case, the amplitudes of the upper and lower envelope waveforms stay constant, but the spacing between them changes. This alters the ratio of the envelope amplitude to the amplitude of the overall waveform.

Noise Considerations in High Speed Sampling ADC Applications

High speed, wide bandwidth sampling ADCs are optimized for dynamic performance over a wide range of analog input frequencies. Because of the wide bandwidth front ends coupled with internal device and resistor noise, DC inputs generally produce a range of output codes as shown in Figure 4.51.

The correct code appears most of the time, but adjacent codes appear with reduced probability. If a normal probability density curve is fitted to this distribution, the standard deviation will be equal to the equivalent rms input noise of the ADC.

OUTPUT HISTOGRAM OF AD9022 12-BIT, 20MSPS ADC SHOWS EFFECTIVE INPUT NOISE OF 0.57LSB FOR DC INPUT

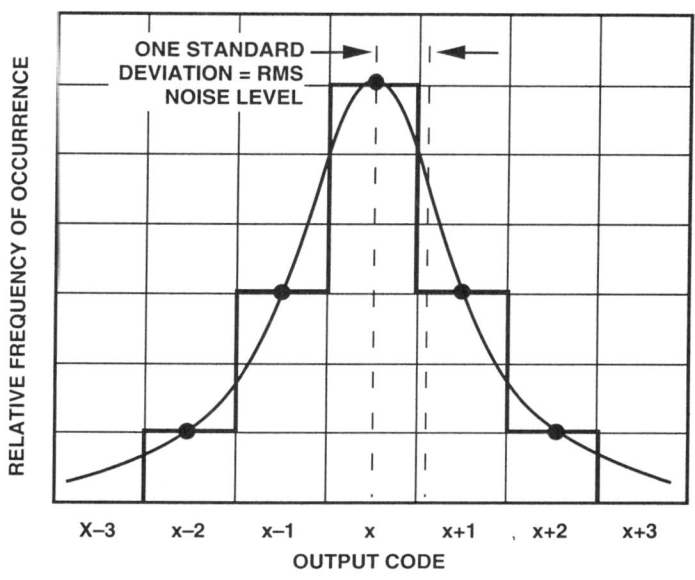

Figure 4.51

For instance, the equivalent input noise of the AD9022 12-bit, 20MSPS ADC is approximately 0.57LSB rms. This implies that the best full scale sinewave signal-to-noise ratio that can be obtained is approximately 68dB. (Full scale sinewave peak-to-peak amplitude = 4096LSBs, or 1448LSBs rms, from which the SNR is calculated as $20 \log_{10}[1448/0.57] = 68dB$). In fact, the SNR of the ADC is limited by other factors, such as quantization noise and distortion.

When driving sampling ADCs with wideband op amps, the output noise of the drive amplifier can contribute to the overall ADC noise floor. A few quick calculations should be made to estimate the total output noise of the op amp and see if it is significant with respect to the ADC noise.

The complete noise model for an op amp is shown in Figure 4.52. This model is accurate, provided there is less than 1 or 2dB peaking in the closed-loop output frequency response. Excessive peaking in the frequency response increases the effects of the wide band noise, and the simple approximation will give optimistic results.

It is rarely necessary to consider all noise sources, since sources which have noise contributions 50% smaller than the largest can be neglected. In high speed systems, where the resistors of the source and the op amp feedback network rarely exceed 1kΩ, the resistor Johnson noise can usually be neglected.

In the case of voltage feedback op amps, the input current noise can usually be neglected. For current feedback op amps operating at noise gains of approximately 4 or less, the inverting input current noise generally dominates. At higher noise gains, however, the effects of voltage noise become significant and should be included.

As an example to show the effects of noise, consider the circuit shown in Figure 4.54, where the AD9022 12-bit, 20MSPS ADC is driven by the AD9632 low-distortion op amp. Because the AD9632 is a voltage feedback op amp and the external resistor values are less than 1kΩ, only the voltage noise (4.3nV/√Hz) is significant to the calculation. Because the AD9632 is operated with a noise gain of 2, the output voltage noise is 8.6nV/√Hz (excluding any source noise).

GENERALIZED OP AMP NOISE MODEL

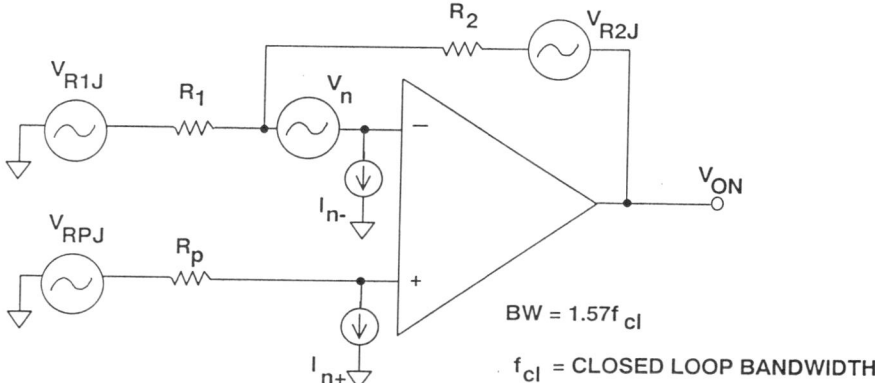

$$V_{ON} = \sqrt{BW} \sqrt{I_{n-}^2 R_2^2 + I_{n+}^2 R_p^2 \left[1 + \frac{R_2}{R_1}\right]^2 + V_n^2 \left[1 + \frac{R_2}{R_1}\right]^2 + 4kTR_2 + 4kTR_1 \left[\frac{R_2}{R_1}\right]^2 + 4kTR_p \left[1 + \frac{R_2}{R_1}\right]^2}$$

Figure 4.52

SIMPLIFICATIONS IN NOISE CALCULATIONS

- Voltage Feedback Op Amps:
    - Neglect Resistor Noise if Resistors < 1kΩ
    - Neglect Input Current Noise

- Current Feedback Op Amps:
    - Neglect Resistor Noise if Resistors < 1kΩ
    - Neglect Non-Inverting Input Current Noise
    - Evaluate Effects of Both Input Voltage Noise
      (Dominates at High Noise Gains)

  and

  Inverting Input Current Noise
  (Dominates at Low Noise Gains)

Figure 4.53

NOISE CALCULATIONS FOR AD9632 OP AMP DRIVING AD9022 12-BIT, 20MSPS ADC

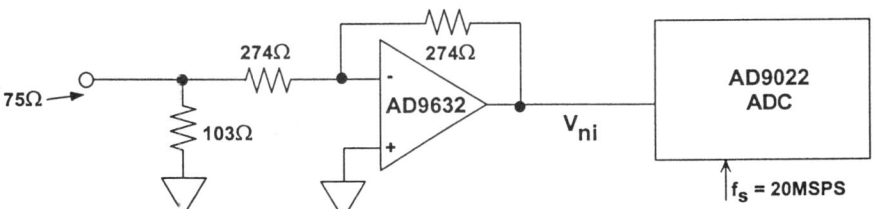

AD9632 OP AMP SPECIFICATIONS	AD9022 ADC SPECIFICATIONS
Input Voltage Noise = 4.3nV/√Hz	Effective Input Noise = 285μV rms
Closed-Loop Bandwidth = 250MHz	Input Bandwidth = 110Mhz

- AD9632 Output Noise Spectral Density = 2 × 4.3nV/√Hz = 8.6nV/√Hz

- Bandwidth for Integration = 110MHz (AD9022 Input Bandwidth)

- $V_{ni} = \dfrac{8.6nV}{\sqrt{Hz}} \cdot \sqrt{110 \times 10^6 \times 1.57 Hz} = 113 \mu V \text{ rms}$

- Less than 50% of AD9022 Effective Input Noise (285μV rms)

Figure 4.54

HIGH SPEED SAMPLING ADCs

The bandwidth for integration is either the op amp closed-loop bandwidth (250MHz), or the ADC input bandwidth (110MHz), whichever is less. This is an important point, because in most cases (especially when dealing with low-distortion, wide bandwidth amplifiers), the input of the ADC acts as the low pass filter to the op amp noise.

The calculation for the op amp noise contribution at the ADC input is simple:

$$V_{ni} = \frac{8.6nV}{\sqrt{Hz}} \cdot \sqrt{110 \times 10^6 \times 1.57 Hz} = 113 \mu V \text{ rms}$$

The factor of 1.57 is required to convert the single-pole 110MHz ADC input bandwidth into an equivalent noise bandwidth. The op amp contribution of 113µV rms is less than one-half the effective input noise of the AD9022 (0.57LSB = 285µV rms), and can therefore be neglected.

In most high speed system applications, a passive antialiasing filter (either lowpass for baseband sampling, or bandpass for harmonic or undersampling) is required. Placing this filter between the op amp and the ADC will further reduce the effects of the drive amplifier noise.

The op amp noise rarely limits the performance of high speed systems. The largest source of unwanted noise generally comes from improper attention to good high speed layout, grounding, and decoupling techniques.

PROPER POSITIONING OF THE ANTIALIASING FILTER WILL REDUCE THE EFFECTS OF THE OP AMP NOISE

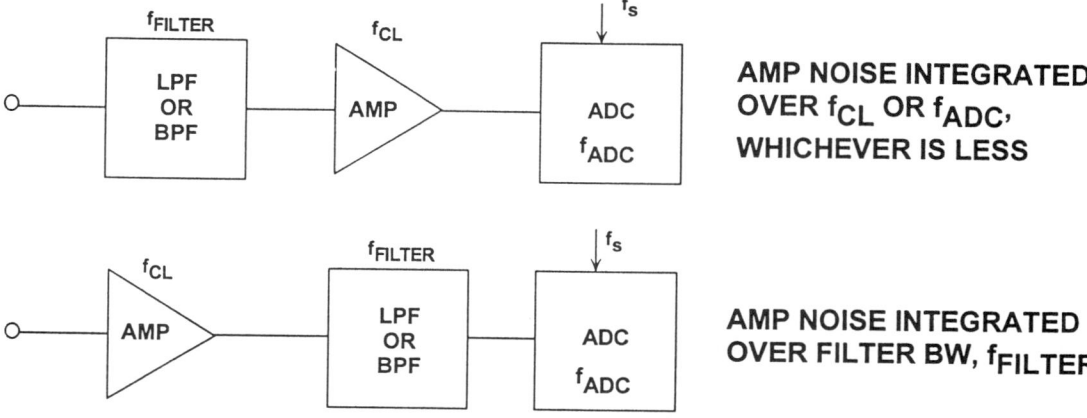

IN GENERAL, $f_{FILTER} < \frac{f_s}{2} \ll f_{ADC} < f_{CL}$

Figure 4.55

ADC NOISE SOURCES

- ADC Distortion and Quantization Noise
- ADC Equivalent Input Noise
- Internal SHA Aperture Jitter
- External Drive Amplifier
- Poor Grounding and Decoupling Techniques
- Poor Layout and Signal Routing Techniques
- Noisy Sampling Clock
- External Switching Power Supply

Figure 4.56

Proper power supply decoupling techniques must be used on each PC board in the system. Figure 4.57 shows an arrangement which will ensure minimum problems. The power supply input (usually brought into the PC board on multiple pins) is first decoupled to the large-area low-impedance ground plane with a good quality, low ESL and low ESR tantalum electrolytic capacitor. This capacitor bypasses low frequency noise to the ground plane. The ferrite bead reduces high frequency noise to the rest of the circuit. You should then place one low-inductance ceramic capacitor at each power pin on each IC. Ideally, you should use surface-mount chip capacitors for minimum inductance, but if you use leaded ceramics, be sure to minimize the lead lengths by mounting them flush on the PC board. Some ICs may require an additional small tantalum electrolytic capacitor (usually between 1 and 5µF). The data sheets for each IC should provide appropriate recommendations, but when in doubt, put them in!

PROPER POWER SUPPLY DECOUPLING AT EACH IC ON THE PC BOARD IS CRITICAL TO ACHIEVING GOOD HIGH SPEED SYSTEM PERFORMANCE

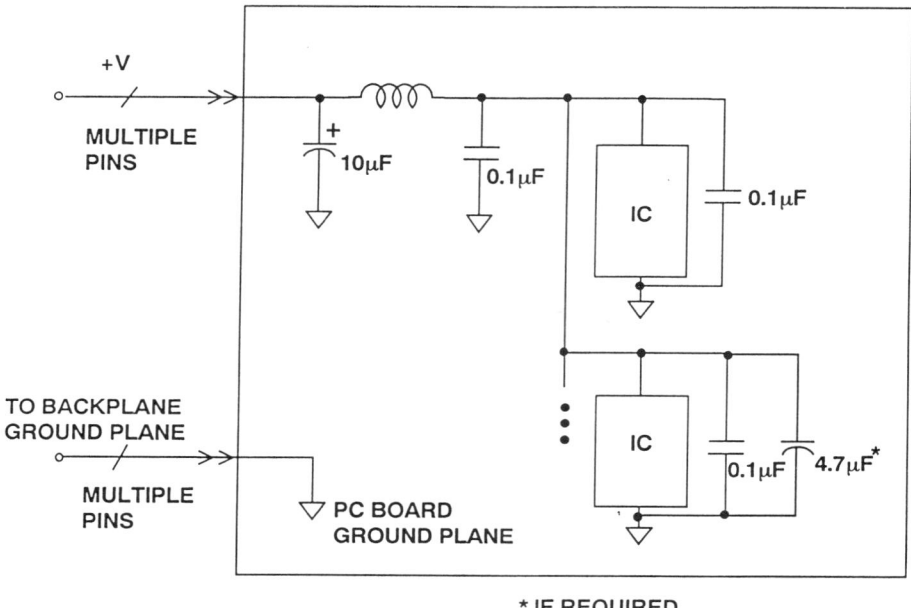

* IF REQUIRED

Figure 4.57

If a double-sided PC board is used, one side should be dedicated entirely (at least 75% of the total area) to the ground plane. The ICs are mounted on this side, and connections are made on the opposite side. Because of component interconnections, however, a few breaks in the ground plane are usually unavoidable. As more and more of the ground plane is eaten away for interconnections, its effectiveness diminishes. It is therefore recommended that multilayer PC boards be used where component packing density is high. Dedicate at least one entire layer to the ground plane.

When connecting to the backplane, use a number of pins (30 to 40%) on each PC board connector for ground. This will ensure that the low impedance ground plane is maintained between the various PC boards in a multicard system.

In practically all high speed systems, it is highly desirable to physically separate sensitive analog components from noisy digital components. It is usually a good idea to also establish separate analog and digital ground planes on each PC board as shown in Figure 4.58. The separate analog and digital ground planes are continued on the backplane using either motherboard ground planes or "ground screens" which are made up of a series of wired interconnections between the connector ground pins. The ground planes are joined together at the system *star ground*, or *single-point ground*, usually located at the common return point for the power supplies. The Schottky diodes are inserted to prevent accidental dc voltages from developing between the two ground systems.

HIGH SPEED SAMPLING ADCs

SEPARATING ANALOG AND DIGITAL GROUNDS IN A MULTICARD, STAR GROUND SYSTEM

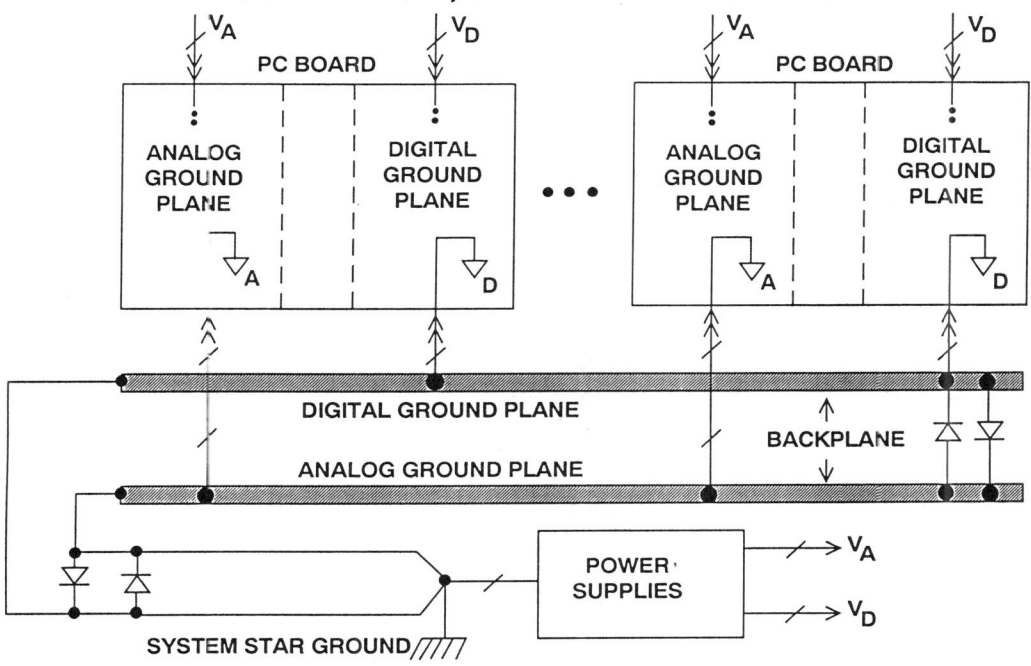

Figure 4.58

Sensitive analog components such as amplifiers and voltage references are referenced and decoupled to the analog ground plane. *The ADCs and DACs (and even some mixed-signal ICs) should be treated as analog circuits and also grounded and decoupled to the analog ground plane.* At first glance, this may seem somewhat contradictory, since a converter has an analog and digital interface and usually pins designated as analog ground (AGND) and digital ground (DGND). The diagram shown in Figure 4.59 will help to explain this seeming dilemma.

PROPER GROUNDING OF ADCs, DACs, AND OTHER MIXED-SIGNAL ICs

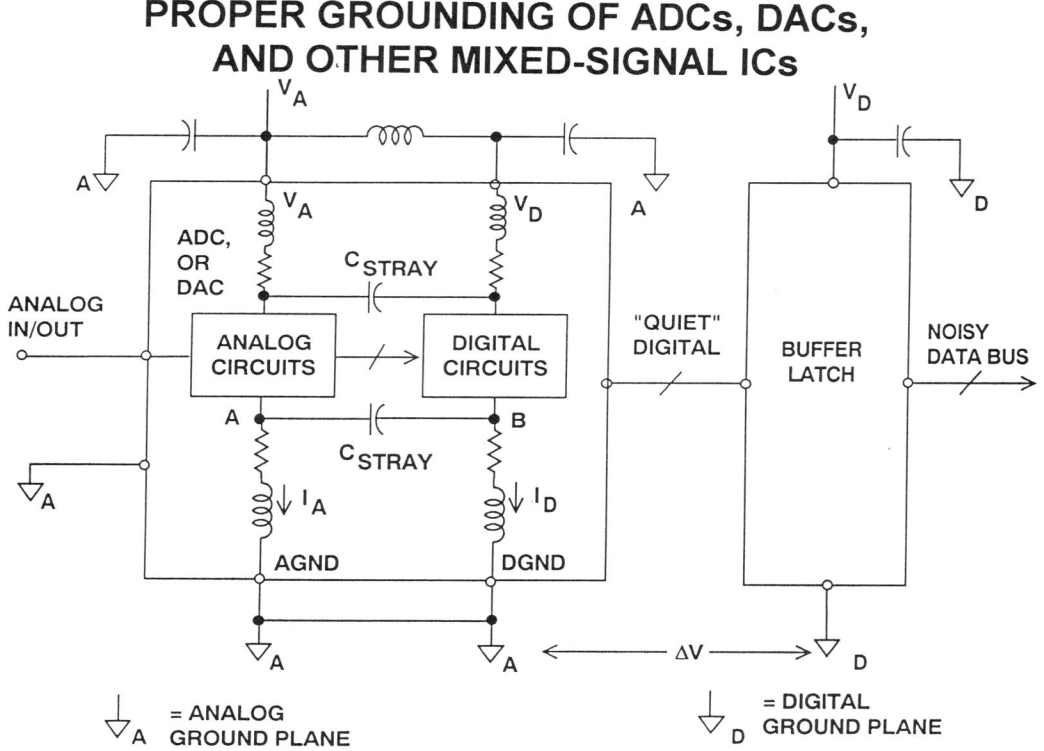

Figure 4.59

Inside an IC that has both analog and digital circuits, such as an ADC or a DAC, the grounds are usually kept separate to avoid coupling digital signals into the analog circuits. Figure 4.59 shows a simple model of a converter. There is nothing the IC designer can do about the wirebond inductance and resistance associated with connecting the pads on the chip to the package pins except to realize it's there. The rapidly changing digital currents produce a voltage at point B which will inevitably couple into point A of the analog circuits through the stray capacitance, C_{STRAY}. In addition, there is approximately 0.2pF unavoidable stray capacitance between every pin of the IC package! It's the IC designer's job to make the chip work in spite of this. However, in order to prevent further coupling, the AGND and DGND pins should be joined together externally to the *analog* ground plane with minimum lead lengths. Any extra impedance in the DGND connection will cause more digital noise to be developed at point B; it will, in turn, couple more digital noise into the analog circuit through the stray capacitance.

The name "DGND" on an IC tells us that this pin connects to the digital ground of the IC. It does not say that this pin must be connected to the digital ground of the system.

It is true that this arrangement will inject a small amount of digital noise on the analog ground plane. These currents should be quite small, and can be minimized by ensuring that the converter input/or output does not drive a large fanout. Minimizing the fanout on the converter's digital port will also keep the converter logic transitions relatively free from ringing, and thereby minimize any potential coupling into the analog port of the converter. The logic supply pin (V_D) can be further isolated from the analog supply by the insertion of a small ferrite bead as shown in Figure 4.60. The internal digital currents of the converter will return to ground through the V_D pin decoupling capacitor (mounted as close to the converter as possible) and will not appear in the external ground circuit. It is always a good idea (as shown in Figure 4.60) to place a buffer latch adjacent to the converter to isolate the converter's digital lines from any noise which may be on the data bus. Even though a few high speed converters have three-state outputs/inputs, this isolation latch represents good design practice.

The buffer latch and other digital circuits should be grounded and decoupled to the digital ground plane of the PC board. Notice that any noise between the analog and digital ground plane reduces the noise margin at the converter digital interface. Since digital noise immunity is of the orders of hundreds or thousands of millivolts, this is unlikely to matter.

POWER SUPPLY, GROUNDING, AND DECOUPLING POINTS

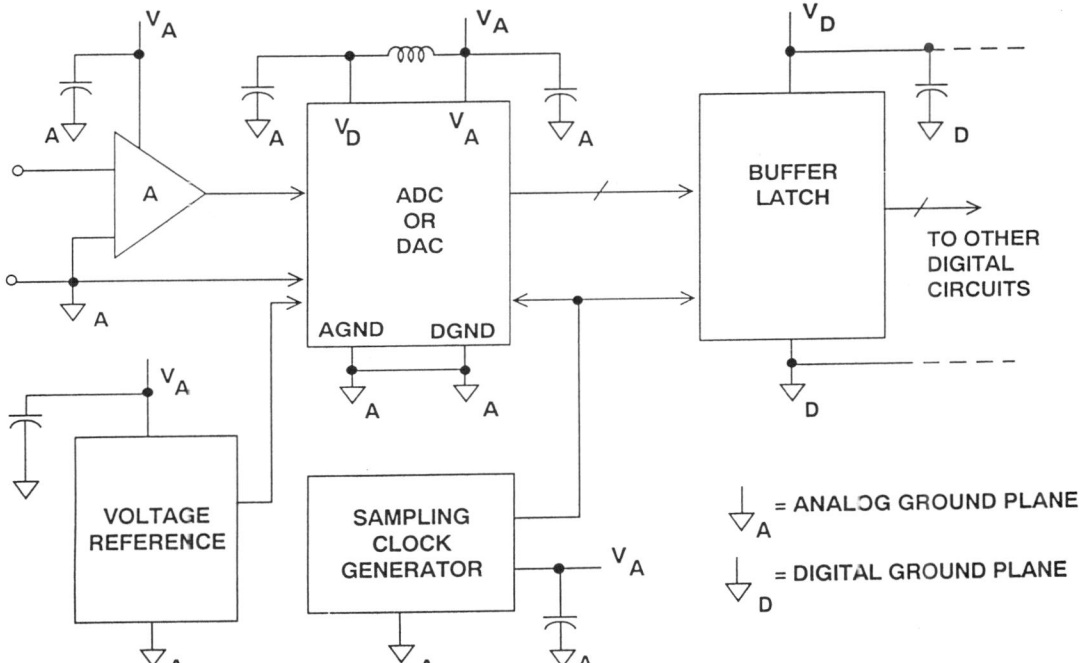

Figure 4.60

The sampling clock generation circuitry should also be grounded and heavily-decoupled to the analog ground plane. As previously discussed, phase noise on the sampling clock produces degradation in system SNR.

Separate power supplies for analog and digital circuits are also highly desirable. The analog supply should be used to power the converter. If the converter has a pin designated as a digital supply pin (V_D), it should either be powered from a separate analog supply, or filtered as shown in the diagram. All converter power pins should be decoupled to the analog ground plane, and all logic circuit power pins should be decoupled to the digital ground plane. If the digital power supply is relatively quiet, it may be possible to use it to supply analog circuits as well, but be very cautious.

A low phase-noise crystal oscillator should be used to generate the ADC sampling clock, because sampling clock jitter modulates the input signal and raises the noise and distortion floor. The sampling clock generator should be isolated from noisy digital circuits and grounded and decoupled to the analog ground plane, as is true for the op amp and the ADC.

It is evident that we can minimize noise by paying attention to the system layout and preventing different signals from interfering with each other. High level analog signals should be separated from low level analog signals, and both should be kept away from digital signals. We have seen elsewhere that in waveform sampling and reconstruction systems the sampling clock (which is a digital signal) is as vulnerable to noise as any analog signal, but is as liable to cause noise as any digital signal, and so must be kept isolated from both analog and digital systems.

If a ground plane is used, as it should in be most cases, it can act as a shield where sensitive signals cross. Figure 4.62 shows a good layout for a data acquisition system where all sensitive areas are isolated from each other and signal paths are kept as short as possible. While real life is rarely as tidy as this, the principle remains a valid one.

There are a number of important points to be considered when making signal and power connections. First of all a connector is one of the few places in the system where all signal conductors must run parallel — it is therefore a good idea to separate them with ground pins (creating a faraday shield) to reduce coupling between them.

SIGNAL ROUTING IN MIXED SIGNAL SYSTEMS

- Physically separate analog and digital signals.
- Avoid crossovers between analog and digital signals.
- Be careful with sampling clock and ADC/DAC analog runs.
- Use lots of ground plane.
- Use microstrip techniques at high frequencies for controlled impedances and controlled return current paths.
- Use surface mount components in high frequency systems to minimize parasitic capacitance and inductance.

Figure 4.61

A PC BOARD LAYOUT SHOWING GOOD SIGNAL ROUTING

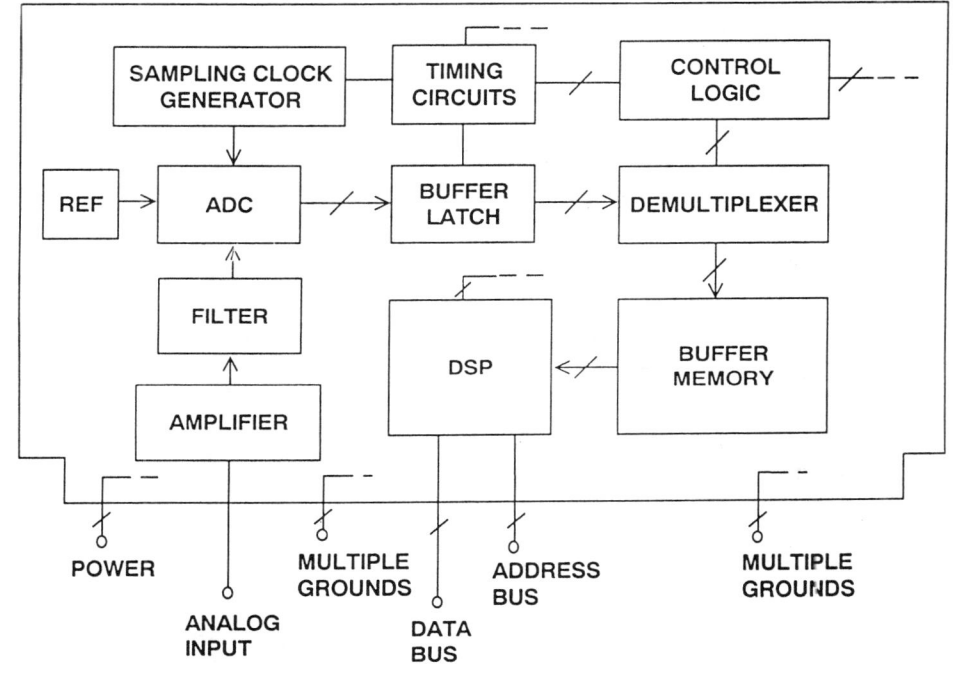

Figure 4.62

EDGE CONNECTIONS

- Separate sensitive signal by ground pins.

- Keep down ground impedances with multiple (30-40% of total) ground pins.

- Have several pins for each power line.

- Critical signals such as analog or sampling clocks may require a separate connector (possibly coax), or microstrip techniques.

Figure 4.63

USE OF SOCKETS WITH HIGH PERFORMANCE ANALOG CIRCUITS

- DON'T! (If at all possible)

- Use "Pin sockets" or "Cage jacks" such as Amp Part No: 5-330808-3 or 5-330808-6 (Capped & uncapped respectively).

- Always test the effect of sockets by comparing system performance with and without the use of sockets.

- Do not change the type of socket or manufacturer used without evaluating the effects of the change on performance.

Figure 4.64

Multiple ground pins are important for another reason: they keep down the ground impedance at the junction between the board and the backplane. The contact resistance of a single pin of a PCB connector is quite low (of the order of 10 mOhms) when the board is new - as the board gets older the contact resistance is likely to rise, and the board's performance may be compromised. It is therefore well worthwhile to afford extra PCB connector pins so that there are many ground connections (perhaps 30-40% of all the pins on the PCB connector should be ground pins). For similar reasons there should be several pins for each power connection, although there is no need to have as many as there are ground pins.

It is tempting to mount expensive ICs in sockets rather than soldering them in circuit — especially during circuit development. Engineers would do well not to succumb to this temptation.

Sockets add resistance, inductance and capacitance to the circuit and may degrade performance to quite unacceptable levels. When this occurs, though, it is always the IC manufacturer who is blamed - not the use of a socket. Even low profile, low insertion force sockets cannot be relied upon to ensure the performance of high performance (high speed or high precision or, worst of all, both) devices. As the socket ages and the board suffers vibration, the contact resistance of low insertion force sockets is very likely to rise. Where a socket must be used the best performance is achieved by using individual pin sockets (sometimes called "cage jacks") to make up a multi-pin socket in the PCB itself (see Section 9).

It really is best not to use IC sockets with high performance analog and mixed signal circuits. If their use can be avoided it should be. However at medium speeds and medium resolutions the trade-off between performance and convenience may fall on the side of convenience. It is very important, when sockets are used, to evaluate circuit performance with and without the socket chosen to ensure that the type of socket chosen really does have minimal effect on the way that the circuit behaves. The effects of a change of socket on the circuit should be evaluated as carefully as a change of IC would be, and the drawings should be prepared so that the change procedures for a socket are as rigorous as for an IC — in order to prevent a purchase clerk who knows nothing of electronics from devastating the system performance in order to save five cents on a socket.

The switching-mode power supply offers low cost, small size, high efficiency, high reliability and the possibility of operating from a wide range of input voltages without adjustment. Unfortunately, these supplies produce noise over a broad band of frequencies, and this noise occurs as conducted noise, radiated noise, and unwanted electric and magnetic fields. When used to supply logic circuits, even more noise is generated on the power supply bus. The noise transients on the output lines of switching supplies are short-duration voltage spikes. Although the actual switching frequencies may range from 10 to 100kHz, these spikes can contain frequency components that extend into the hundreds of megahertz.

Because of the wide variations in the noise characteristics of commercially available switching supplies, they should always be purchased in accordance with a specification-control drawing. Although specifying switching supplies in terms of rms noise is common practice, you should also specify

HIGH SPEED SAMPLING ADCs

the peak amplitudes of the switching spikes under the output loading conditions you expect in your system. You should also insist that the switching-supply manufacturer inform you of any internal supply design changes that may alter the spike amplitudes, duration, or switching frequency. These changes may require corresponding changes in the external power-supply filtering networks.

Filtering switching supply outputs that provide several amps and generate voltage spikes having high frequency components is a challenge. For this reason, you should place the initial filtering burden on the switching supply manufacturer. Even so, external filtering such as shown in Figure 4.66 should be added. The series inductors isolate both the output and common lines from the external circuits. Because the load currents may be large, make sure that the inductors selected do not saturate. Split-core inductors or large ferrite beads make a good choice. Because the switching power supplies generate high and low frequency electric and magnetic fields, they should be physically separated as far as possible from critical analog circuitry. This is especially important in preventing the inductive coupling of low frequency magnetic fields.

SWITCHING-MODE POWER SUPPLIES

- Generate Conducted and Radiated Noise as Well as Electric and Magnetic Fields (HF and LF)

- Outputs Must be Adequately Filtered if Powering Sensitive Analog Circuits

- Optimum Filter Design Depends on Power Supply Characteristics. Beware of Power Supply Design Changes.

- Use Faraday Shields to Reduce HF Electric and HF Magnetic Fields

- Physically Isolate Supply from Analog Circuits

- Temporarily Replace Switching Supply with Low-Noise Linear Supply or Battery when Suspicious of Switching Supply Noise

Figure 4.65

FILTERING A SWITCHING SUPPLY OUTPUT

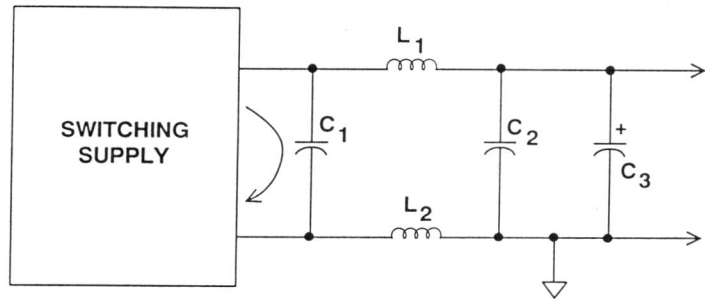

- C_1 MUST HAVE LOW INDUCTANCE AND BE CLOSE TO THE SUPPLY TO MINIMIZE HF CURRENT LOOPS AND RESULTANT HF MAGNETIC FIELDS

- C_2 IS ALSO LOW INDUCTANCE, C_3 IS ELECTROLYTIC

- IF THE SWITCHING SUPPLY IS INTERNALLY GROUNDED, L_2 SHOULD BE OMITTED

Figure 4.66

REFERENCES

1. **Linear Design Seminar**, Analog Devices, 1995, Chapter 4, 5.

2. **System Applications Guide**, Analog Devices, 1993, Chapter 12, 13.

3. **Amplifier Applications Guide**, Analog Devices, 1992, Chapter 7.

4. Walt Kester, *Drive Circuitry is Critical to High-Speed Sampling ADCs,* **Electronic Design Special Analog Issue**, Nov. 7, 1994, pp. 43-50.

5. Walt Kester, *Basic Characteristics Distinguish Sampling A/D Converters,* **EDN**, Sept. 3, 1992, pp. 135-144.

6. Walt Kester, *Peripheral Circuits Can Make or Break Sampling ADC Systems,* **EDN**, Oct. 1, 1992, pp. 97-105.

7. Walt Kester, *Layout, Grounding, and Filtering Complete Sampling ADC System,* **EDN**, Oct. 15, 1992, pp. 127-134.

8. Carl Moreland, *An 8-Bit 150MSPS Serial ADC*, **1995 ISSCC Digest of Technical Papers**, Vol. 38, p.272.

9. Roy Gosser and Frank Murden, *A 12-Bit 50MSPS Two-Stage A/D Converter*, **1995 ISSCC Digest of Technical Papers**, p. 278.

SECTION 5

UNDERSAMPLING APPLICATIONS

- Fundamentals of Undersampling

- Increasing ADC SFDR and ENOB using External SHAs

- Use of Dither Signals to Increase ADC Dynamic Range

- Effect of ADC Linearity and Resolution on SFDR and Noise in Digital Spectral Analysis Applications

- Future Trends in Undersampling ADCs

Undersampling Applications

SECTION 5

UNDERSAMPLING APPLICATIONS
Walt Kester

An exciting new application for wideband, low distortion ADCs is called *undersampling, harmonic sampling, bandpass sampling,* or *Super-Nyquist Sampling*. To understand these applications, it is necessary to review the basics of the sampling process.

The concept of discrete time and amplitude sampling of an analog signal is shown in Figure 5.1. The continuous analog data must be sampled at discrete intervals, t_s, which must be carefully chosen to insure an accurate representation of the original analog signal. It is clear that the more samples taken (faster sampling rates), the more accurate the digital representation, but if fewer samples are taken (lower sampling rates), a point is reached where critical information about the signal is actually lost. This leads us to the statement of Shannon's Information Theorem and Nyquist's Criteria given in Figure 5.2. Most textbooks state the Nyquist theorem along the following lines: *A signal must be sampled at a rate greater than twice its maximum frequency in order to ensure unambiguous data*. The general assumption is that the signal has frequency components from dc to some upper value, f_a. The Nyquist Criteria thus requires sampling at a rate $f_s > 2f_a$ in order to avoid overlapping aliased components. For signals which do not extend to dc, however, the minimum required sampling rate is a function of the *bandwidth* of the signal as well as its position in the frequency spectrum.

SAMPLING AN ANALOG SIGNAL

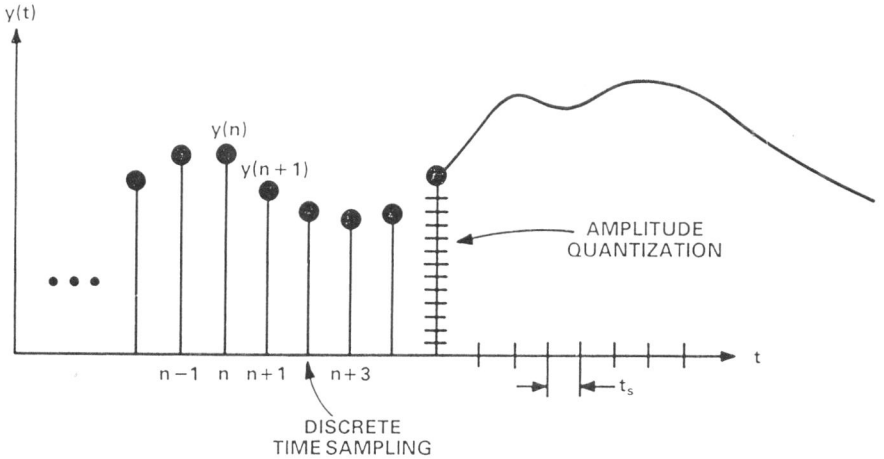

Figure 5.1

UNDERSAMPLING APPLICATIONS

SHANNON'S INFORMATION THEOREM AND NYQUIST'S CRITERIA

- **Shannon:**
  - An Analog Signal with a *Bandwidth* of f_a Must be Sampled at a Rate $f_s > 2f_a$ in Order to Avoid the Loss of Information.
  - The signal bandwidth may extend from DC to f_a (*Baseband Sampling*) or from f_1 to f_2, where $f_a = f_2 - f_1$ (*Undersampling, Bandpass Sampling, Harmonic Sampling, Super-Nyquist*)

- **Nyquist:**
  - If $f_s < 2f_a$, then a Phenomena Called *Aliasing* Will Occur.
  - *Aliasing* is used to advantage in undersampling applications.

Figure 5.2

In order to understand the implications of *aliasing* in both the time and frequency domain, first consider the four cases of a time domain representation of a sampled sinewave signal shown in Figure 5.3. In Case 1, it is clear that an adequate number of samples have been taken to preserve the information about the sinewave. In Case 2 of the figure, only four samples per cycle are taken; still an adequate number to preserve the information. Case 3 represents the ambiguous limiting condition where $f_s = 2f_a$. If the relationship between the sampling points and the sinewave were such that the sinewave was being sampled at precisely the zero crossings (rather than at the peaks, as shown in the illustration), then all information regarding the sinewave would be lost. Case 4 of Figure 5.3 represents the situation where $f_s < 2f_a$, and the information obtained from the samples indicates a sinewave having a frequency which is lower than $f_s/2$, i.e. the out-of-band signal is *aliased* into the Nyquist bandwidth between dc and $f_s/2$. As the sampling rate is further decreased, and the analog input frequency f_a approaches the sampling frequency f_s, the aliased signal approaches dc in the frequency spectrum.

TIME DOMAIN EFFECTS OF ALIASING

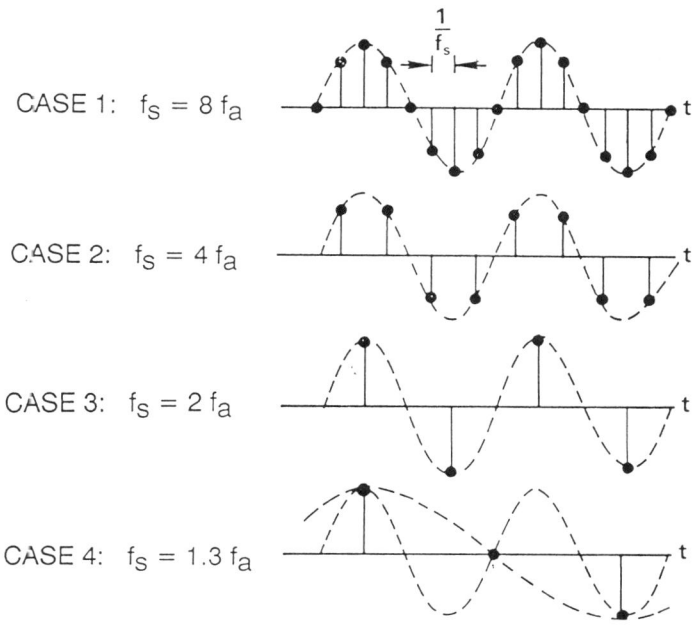

Figure 5.3

The corresponding frequency domain representation of the above scenario is shown in Figure 5.4. Note that sampling the analog signal f_a at a sampling rate f_s actually produces two alias frequency components, one at f_s+f_a, and the other at f_s-f_a. The upper alias, f_s+f_a, seldom presents a problem, since it lies outside the Nyquist bandwidth. It is the lower alias component, f_s-f_a, which causes problems when the input signal exceeds the Nyquist bandwidth, $f_s/2$.

From Figure 5.4, we make the extremely important observation that *regardless of where the analog signal being sampled happens to lie in the frequency spectrum (as long as it does not lie on multiples of $f_s/2$), the effects of sampling will cause either the actual signal or an aliased component to fall within the Nyquist bandwidth between dc and $f_s/2$.* Therefore, any signals which fall outside the bandwidth of interest, whether they be spurious tones or random noise, must be adequately filtered *before* sampling. If unfiltered, the sampling process will alias them back within the Nyquist bandwidth where they can corrupt the wanted signals.

Methods exist which use aliasing to our advantage in signal processing applications. Figure 5.5 shows four cases where a signal having a 1MHz bandwidth is located in different portions of the frequency spectrum. The sampling frequency must be chosen such that there is no overlapping of the aliased components. In general, the sampling frequency must be at least twice the signal bandwidth, and the sampled signal must not cross an integer multiple of $f_s/2$.

UNDERSAMPLING APPLICATIONS

FREQUENCY DOMAIN EFFECTS OF ALIASING

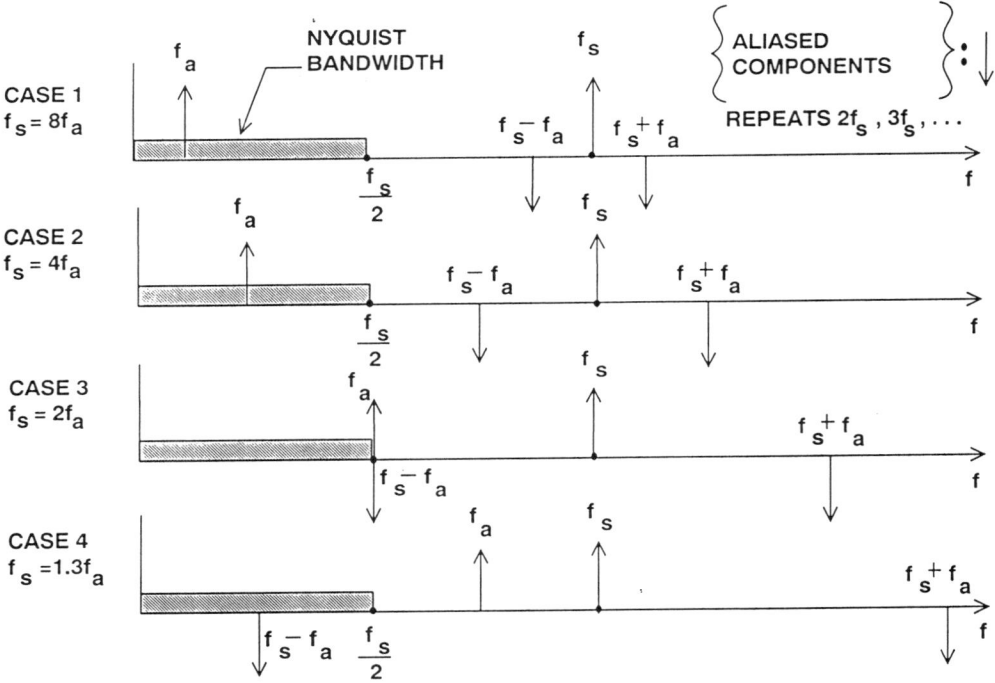

Figure 5.4

MINIMUM SAMPLING RATE REQUIRED FOR NON-OVERLAPPING ALIASING OF A 1MHz BANDWIDTH SIGNAL

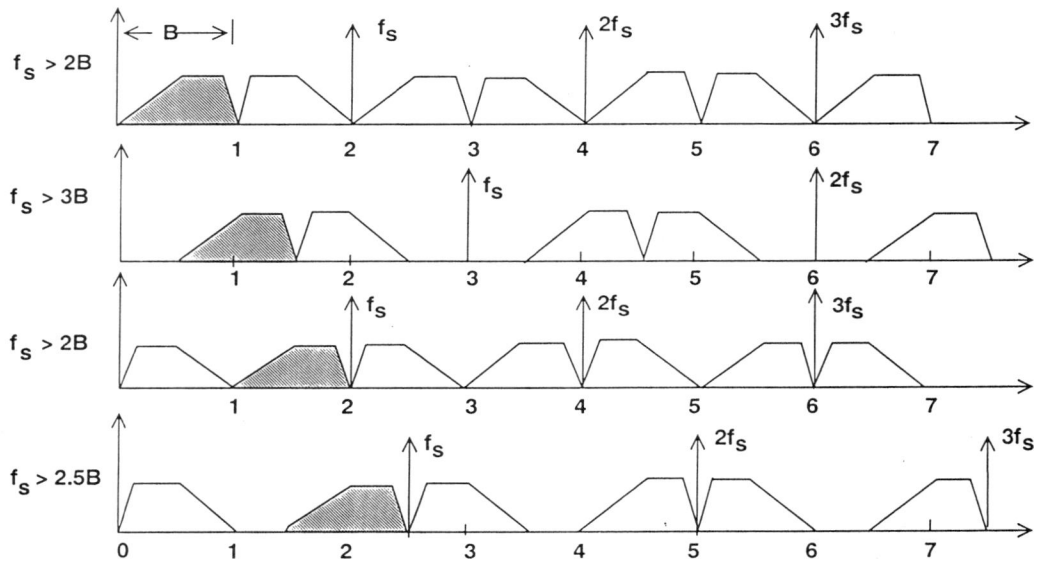

Figure 5.5

UNDERSAMPLING APPLICATIONS

In the first case, the signal occupies a band from dc to 1MHz, and therefore must be sampled at greater than 2MSPS. The second case shows a 1MHz signal which occupies the band from 0.5 to 1.5MHz. Notice that this signal must be sampled at a minimum of 3MSPS in order to avoid overlapping aliased components. In the third case, the signal occupies the band from 1 to 2MHz, and the minimum required sampling rate for no overlapping aliased components drops back to 2MSPS. The last case shows a signal which occupies the band from 1.5 to 2.5MHz. This signal must be sampled at a minimum of 2.5MSPS to avoid overlapping aliased components.

This analysis can be generalized as shown in Figure 5.6. The actual minimum required sampling rate is a function of the ratio of the highest frequency component, f_{MAX}, to the total signal bandwidth, B. Notice for large ratios of f_{MAX} to the bandwidth, B, the minimum required sampling frequency approaches 2B.

MINIMUM REQUIRED SAMPLING RATE AS A FUNCTION OF THE RATIO OF THE HIGHEST FREQUENCY COMPONENT TO THE TOTAL SIGNAL BANDWIDTH

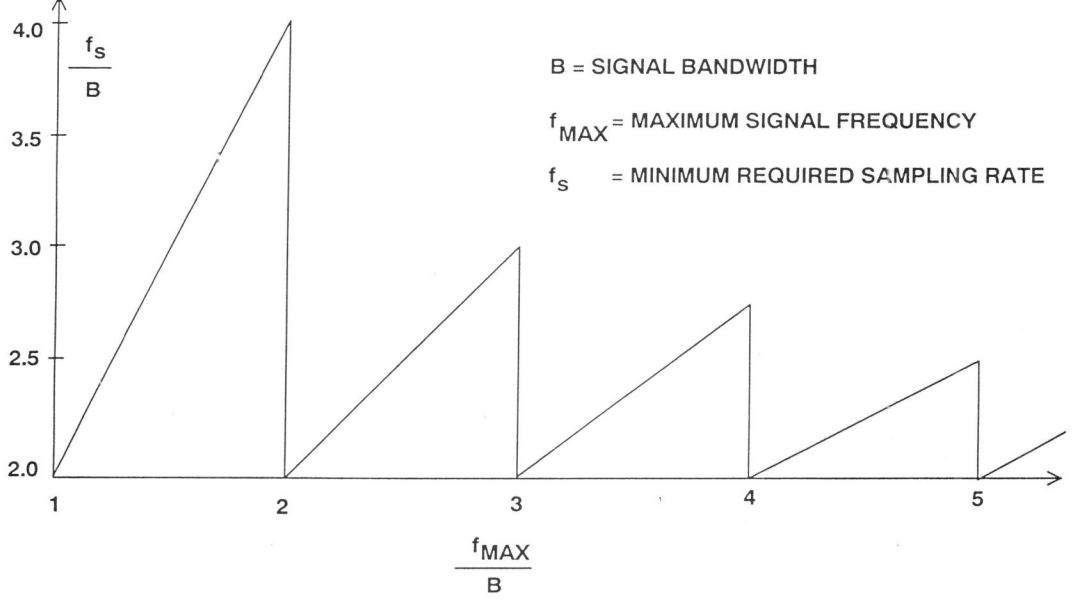

Figure 5.6

UNDERSAMPLING APPLICATIONS

Let us consider the case of a signal which occupies a *bandwidth* of 1MHz and lies between 6 and 7MHz as shown in Figure 5.7. Shannon's Information Theorem states that the signal (bandwidth = 1MHz) must be sampled at least at 2MSPS in order to retain all the information (avoid overlapping aliased components). Assuming that the ADC sampling rate, f_s, is 2MSPS, additional sampling frequencies are generated at all integer multiples of f_s: 4MHz, 6MHz, 8MHz, etc. The actual signal between 6 and 7MHz is aliased around each of these sampling frequency harmonics, f_s, $2f_s$, $3f_s$, $4f_s$, ..., hence the term *harmonic sampling*. Notice that any one of the aliased components is an accurate representation of the original signal (the frequency inversion which occurs for one-half of the aliased components can be removed in software). In particular, the component lying in the baseband region between dc and 1MHz is the one calculated using a Fast Fourier Transform, and is also an accurate representation of the original signal, assuming no ADC conversion errors. The FFT output tells us all the characteristics of the signal except for its original position in the frequency spectrum, which was apriori knowledge.

INTERMEDIATE FREQUENCY (IF) SIGNAL BETWEEN 6 AND 7 MHz IS ALIASED BETWEEN DC AND 1MHz BY SAMPLING AT 2MSPS

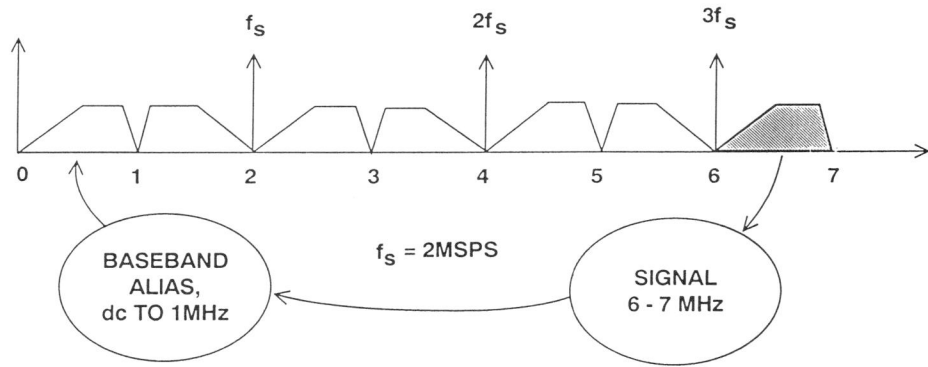

Figure 5.7

A popular application of undersampling is in digital receivers. A simplified block diagram of a traditional digital receiver using *baseband* sampling is shown in Figure 5.8. The mixer in the RF section of the receiver mixes the signal from the antenna with the RF frequency of the local oscillator. The desired information is contained in relatively small bandwidth of frequencies Δf. In actual receivers, Δf may be as high as a few megahertz. The local oscillator frequency is chosen such that the Δf band is centered about the IF frequency at the bandpass filter output. Popular IF frequencies are generally between 10 and 100MHz. The detector then translates the Δf frequency band down to baseband where it is filtered and processed by a baseband ADC. Actual receivers can have several stages of RF and IF processing, but the simple diagram serves to illustrate the concepts.

SIMPLIFIED DIGITAL RECEIVER USING BASEBAND SAMPLING

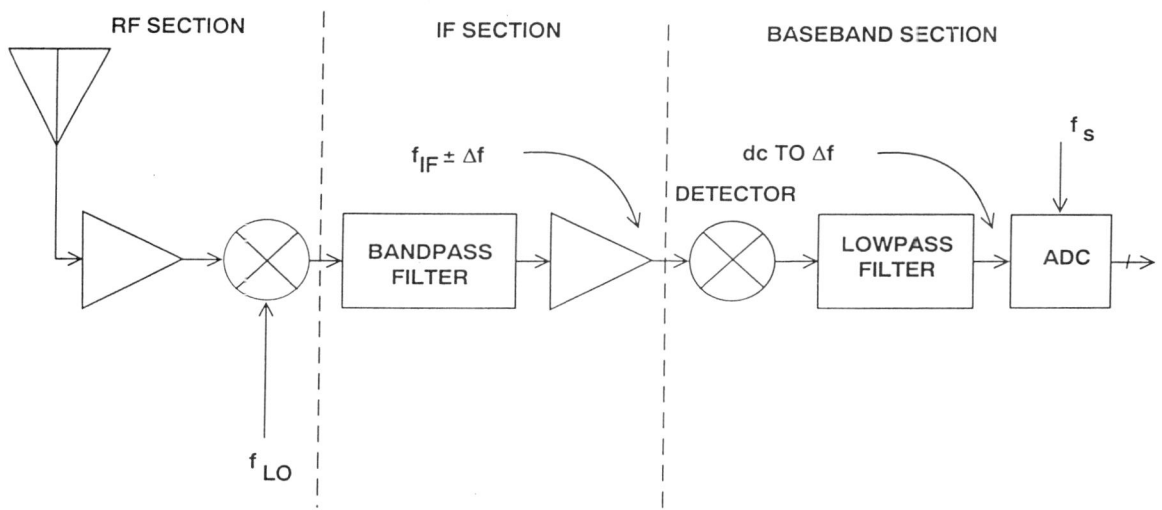

Figure 5.8

In a receiver which uses direct IF-to-digital techniques (often called *undersampling, harmonic, bandpass,* or *IF* sampling), the IF signal is applied directly to a wide bandwidth ADC as shown in Figure 5.9. The ADC sampling rate is chosen to be at least $2\Delta f$. The process of sampling the IF frequency at the proper rate causes one of the aliased components of Δf to appear in the dc to $f_s/2$ Nyquist bandwidth of the ADC output. DSP techniques can now be used to process the digital baseband signal. This approach eliminates the detector and its associated noise and distortion. There is also more flexibility in the DSP because the ADC sampling rate can be shifted to tune the exact position of the Δf signal within the baseband.

UNDERSAMPLING APPLICATIONS

The obvious problem with this approach is that the ADC must now be able to accurately digitize signals which are well outside the dc to $f_s/2$ Nyquist bandwidth which most ADCs were designed to handle. Special techniques are available, however, which can extend the dynamic range of ADCs to include IF frequencies.

SIMPLIFIED DIGITAL RECEIVER USING IF SAMPLING

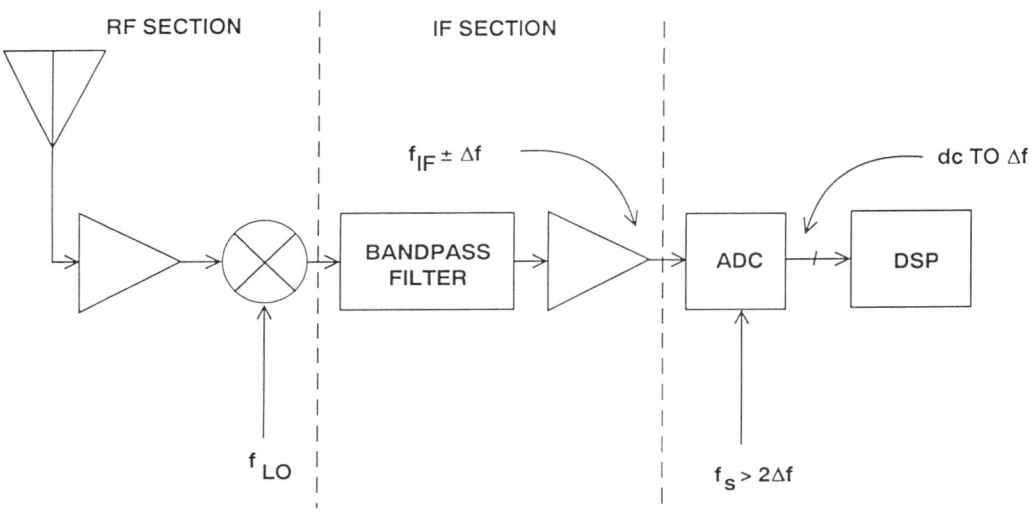

Figure 5.9

Let us consider a typical example, where the IF frequency is 72.5MHz, and the desired signal occupies a bandwidth of 4MHz (B=4MHz), centered on the IF frequency (see Figure 5.10). We know from the previous discussion that the minimum sampling rate must be greater than 8MHz, probably on the order of 10MHz in order to prevent dynamic range limitations due to aliasing. If we place the sampling frequency at the lower band-edge of 70MHz (72.5–2.5), we will definitely recover the aliased component of the signal in the dc to 5MHz baseband. There is, however, no need to sample at this high rate, so we may choose any sampling frequency 10MHz or greater which is an integer sub-multiple of 70MHz, i.e., 70÷2 = 35.000MHz, 70÷3 = 23.333MHz, 70÷4 = 17.500MHz, 70÷5 = 14.000MHz, 70÷6 = 11.667MHz, or 70÷7 = 10.000MHz. We will therefore choose the lowest possible sampling rate of 10.000MHz (70÷7).

UNDERSAMPLING APPLICATIONS

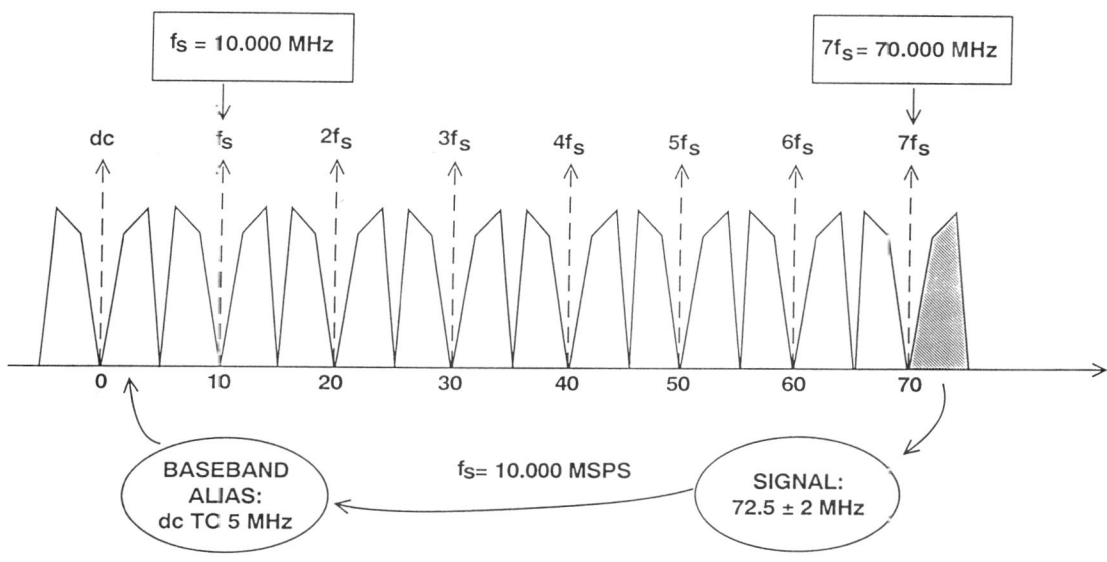

Figure 5.10

There is an advantage in choosing a sampling frequency which a sub-multiple of the lower band-edge in that there is no frequency inversion in the baseband alias as would be the case for a sampling frequency equal to a sub-multiple of the upper band-edge. (Frequency inversion can be easily dealt with in the DSP software should it occur, so the issue is not very important).

Undersampling applications such as the one just described generally require sampling ADCs which have low distortion at the high input IF input frequency. For instance, in the example just discussed, the ADC sampling rate requirement is only 10MSPS, but low distortion is required (preferably 60 to 80dB SFDR) at the IF frequency of 72.5MHz.

A large opportunity for bandpass sampling is in digital cellular radio base stations. For systems which have RF frequencies at 900MHz, 70MHz is a popular first-IF frequency. For systems using an RF frequency of 1.8GHz, first-IF frequencies between 200 and 240MHz are often used.

In broadband receiver applications, one ADC digitizes multiple channels in the receive path. Individual channel selection and filtering is done in the digital domain. Narrowband channel characteristics such as bandwidth, passband ripple, and adjacent channel rejection can be controlled with changes to digital parameters (i.e. filter coefficients). Such flexibility is not possible when narrowband analog filters are in the receive path.

UNDERSAMPLING APPLICATIONS

Figure 5.11 illustrates the kind of input spectrum an ADC must digitize in a multichannel design. The spectral lines represent narrowband signal inputs from a variety of signal sources at different received power levels. Signal "C" could represent a transmitter located relatively far away from the signal sources "A" and "B". However, the receiver must recover all of the signals with equal clarity. This requires that distortion from the front-end RF and IF signal processing components, including the ADC, not exceed the minimum acceptable level required to demodulate the weakest signal of interest. Clearly, third-order intermodulation distortion generated by "A" and "B" (2B - A, and 2A - B) will distort signals C and D if the nonlinearities in the front-end are severe. Strong out-of-band signals can also introduce distortion; signal "E" in Figure 5.11 shows a large signal that is partially attenuated by the antialiasing filter. In many systems, the power level of the individual transmitters is under control of the base station. This capability helps to reduce the total dynamic range required.

BROADBAND DIGITAL RECEIVER ADC INPUT

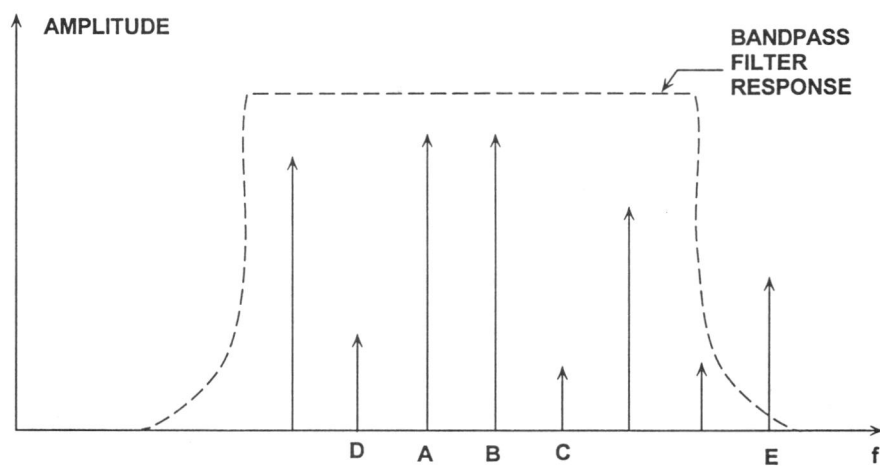

Figure 5.11

In broadband receiver applications (using an RF frequency of approximately 900MHz, and a first-IF frequency of 70MHz), SFDR requirements for the ADC are typically 70 to 80dBc. Signal bandwidths between 5MHz and 10MHz are common, requiring corresponding sampling rates of 10MSPS to 20MSPS.

Sampling ADCs are generally designed to process signals up to Nyquist ($f_s/2$) with a reasonable amount of dynamic performance. As we have seen, however, even though the input bandwidth of a sampling ADC is usually much greater than its maximum sampling rate, the SFDR and effective bit (ENOB) performance usually decreases dramatically for full scale input signals much above $f_s/2$. This implies that the selection criteria for ADCs used in undersampling applications is SFDR or ENOB at the IF frequency, rather than sampling rate.

The general procedure for selecting an ADC for an undersampling application is not straightforward. The signal bandwidth and its location within the frequency spectrum must be known. The bandwidth of the signal determines the minimum sampling rate required, and in order to ease the requirement on the antialiasing filter, a sampling rate of 2.5 times the signal bandwidth works well. After determining the approximate sampling frequency needed, select the ADC based on the required SFDR, S/(N+D), or ENOB at the IF frequency.

This is where the dilemma usually occurs. You will find that an ADC specified for a maximum sampling rate of 10MSPS, for instance, will not have adequate SFDR at the IF frequency (72.5MHz in the example above), even though its performance is excellent up to its Nyquist frequency of 5Mhz. In order to meet the SFDR, S/(N+D), or ENOB requirement, you will generally require an ADC having a much higher sampling rate than is actually needed.

Figure 5.12 shows the approximate SFDR versus input frequency for the AD9022/AD9023 (20MSPS), AD9026/AD9027 (31MSPS), and the AD9042 (40MSPS) series of low distortion ADCs. Notice that the AD9042 has superior SFDR performance.

SFDR COMPARISON BETWEEN 12-BIT SAMPLING ADCs

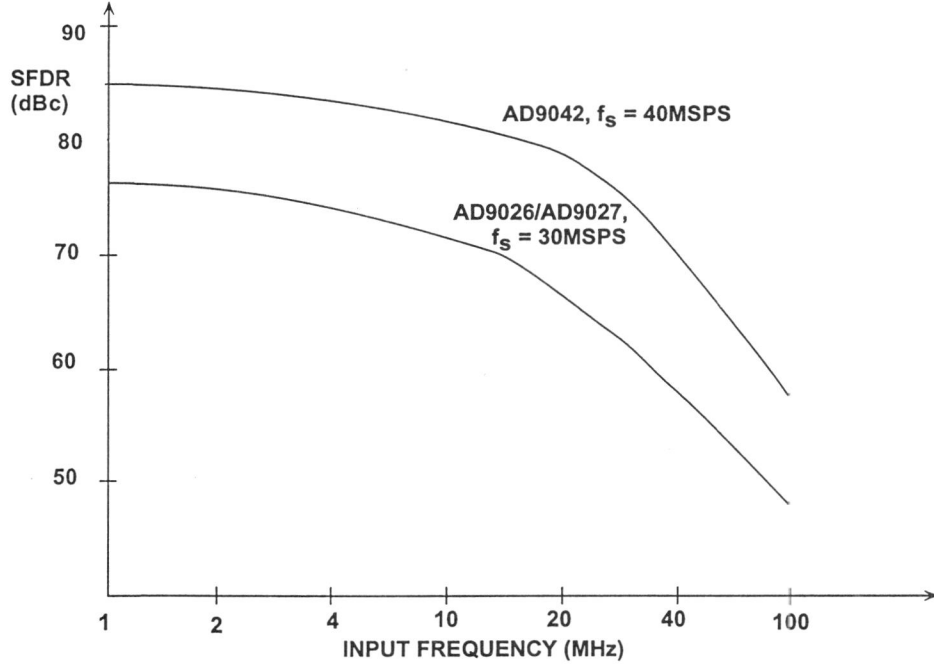

Figure 5.12

UNDERSAMPLING APPLICATIONS

The AD9042 is a state-of-the-art 12-bit, 40MSPS two stage subranging ADC consisting of a 6-bit coarse ADC and a 7-bit residue ADC with one bit of overlap to correct for any DNL, INL, gain or offset errors of the coarse ADC, and offset errors in the residue path. A block diagram is shown in Figure 5.13. A proprietary gray-code architecture is used to implement the two internal ADCs. The gain alignments of the coarse and residue, likewise the subtraction DAC, rely on the statistical matching of the process. As a result, 12-bit integral and differential linearity is obtained without laser trim. The internal DAC consists of 126 interdigitated current sources. Also on the DAC reference are an additional 20 interdigitated current sources to set the coarse gain, residue gain, and full scale gain. The interdigitization removes the requirement for laser trim. The AD9042 is fabricated on a high speed dielectrically isolated complementary bipolar process. The total power dissipation is only 575mW when operating on a single +5V supply.

BLOCK DIAGRAM OF AD9042 12-BIT, 40MSPS ADC

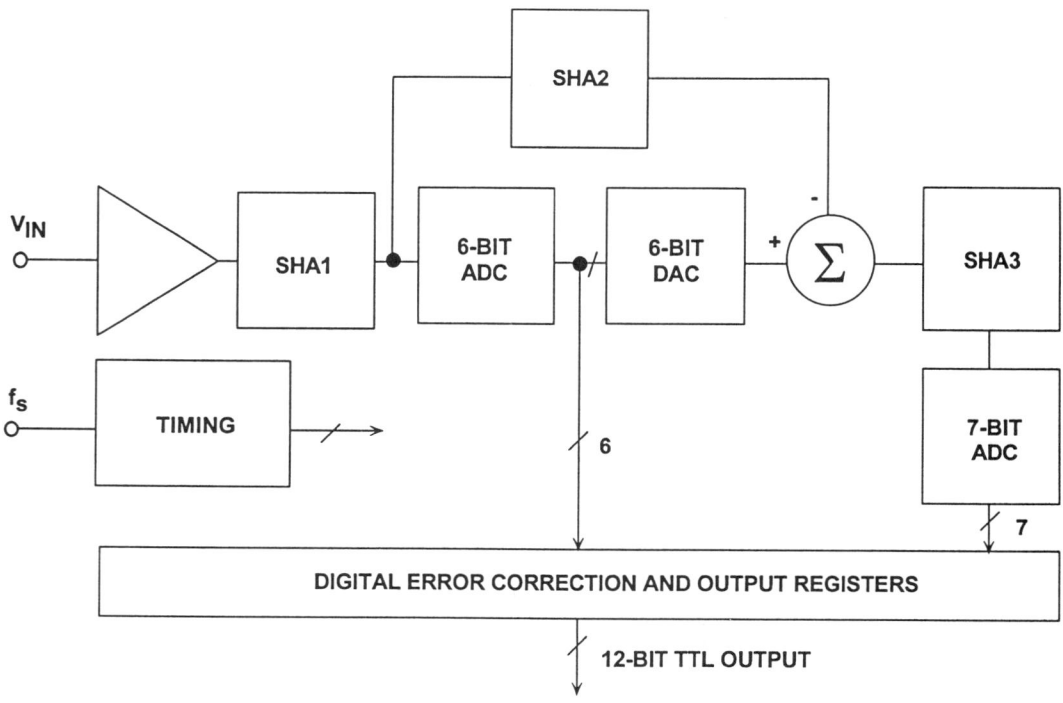

Figure 5.13

UNDERSAMPLING APPLICATIONS

AD9042 12-BIT, 40MSPS ADC KEY SPECIFICATIONS

- Input Range: 1V peak-to-peak, V_{cm} = +2.4V
- Input Impedance: 250Ω to V_{cm}
- Effective Input Noise: 0.33LSBs rms
- SFDR at 20MHz Input: 80dB
- S/(N+D) at 20MHz Input = 66dB
- Digital Outputs: TTL Compatible
- Power Supply: Single +5V
- Power Dissipation: 575mW
- Fabricated on High Speed Dielectrically Isolated Complementary Bipolar Process

Figure 5.14

The outstanding performance of the AD9042 is partly due to the use of differential techniques throughout the device. The low distortion input amplifier converts the single-ended input signal into a differential one. If maximum SFDR performance is desired, the signal source should be coupled directly into the input of the AD9042 without using a buffer amplifier. Figure 5.15 shows a method using capacitive coupling.

INPUT STRUCTURE OF AD9042 ADC IS DESIGNED TO BE DRIVEN DIRECTLY FROM 50Ω SOURCE FOR BEST SFDR

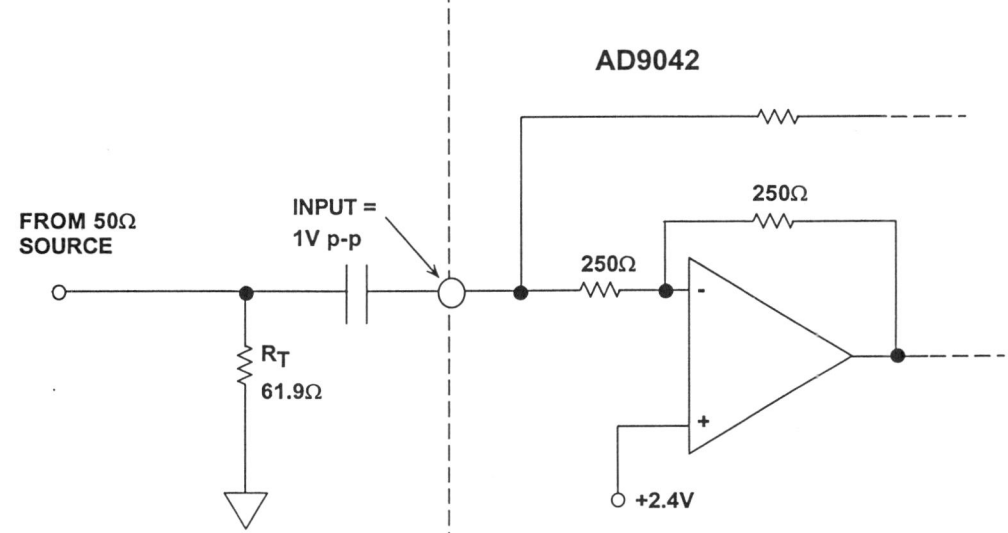

Figure 5.15

UNDERSAMPLING APPLICATIONS

INCREASING ADC SFDR AND ENOB USING EXTERNAL SHAs

An external SHA can increase the SFDR and ENOB of a sampling ADC for undersampling applications if properly selected and interfaced to the ADC. The SHA must have low hold-mode distortion at the frequency of interest. In addition, the acquisition time must be short enough to operate at the required sampling frequency. Figure 5.17 shows the effects of adding a SHA to an 8-bit flash converter.

The ADC is clocked at 20MSPS, and the input frequency to the ADC is 19.98MHz. The scope photo shows the "beat" frequency of 2kHz reconstructed with an 8-bit DAC. Notice that without the SHA, the ADC has non-linearities and missing codes. The addition of the SHA (properly selected and timed) greatly improves the linearity and reduces the distortion.

THE ADDITION OF AN EXTERNAL WIDEBAND LOW DISTORTION SHA EXTENDS THE LOW FREQUENCY PERFORMANCE OF THE ADC TO HIGHER FREQUENCIES

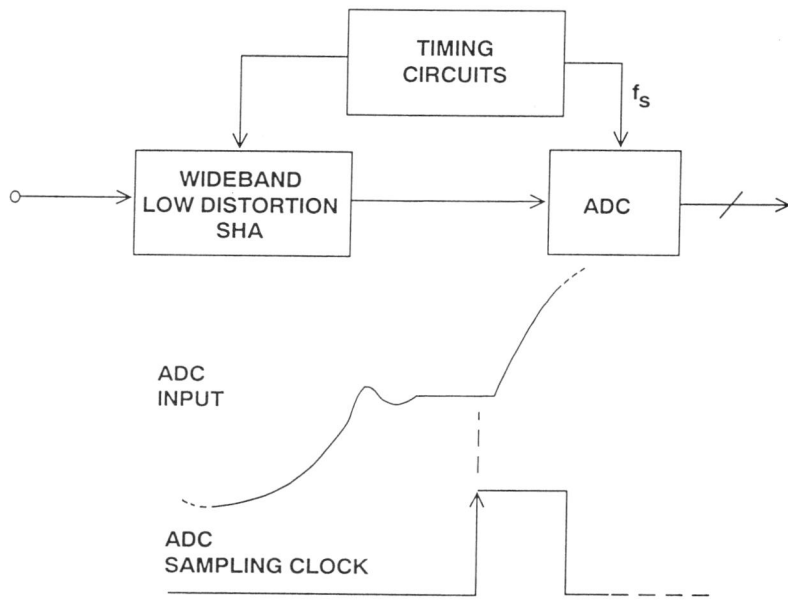

Figure 5.16

UNDERSAMPLING APPLICATIONS

EFFECTS OF EXTERNAL SHA ON FLASH ADC PERFORMANCE FOR f_{in} = 19.98MHz, f_s = 20.00MSPS

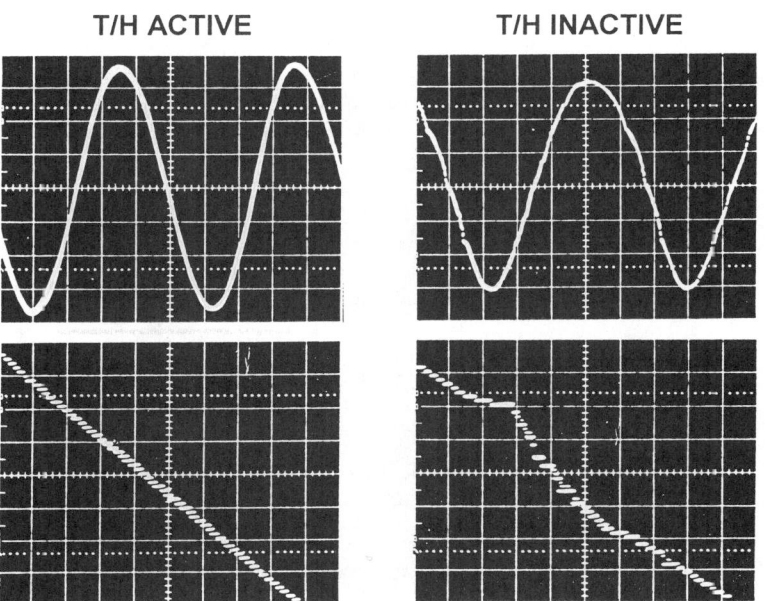

Figure 5.17

Most SHAs are specified for distortion when operating in the track mode. What is of real interest, however, is the signal distortion in the hold mode when the SHA is operating dynamically. The AD9100 (30MSPS) and AD9101(125MSPS) are ultra-fast SHAs and are specified in terms of hold-mode distortion. The measurement is done using a high performance low distortion ADC (such as the AD9014 14-bit, 10MSPS) to digitize the held value of the SHA output. An FFT is performed on the ADC output, and the distortion is measured digitally. For sampling rates greater than 10MSPS, the ADC is clocked at an integer sub-multiple of the SHA sampling frequency. This causes a frequency translation in the FFT output because of undersampling, but the distortion measurement still represents that of the SHA which is operating at the higher sampling rate. The AD9100 is optimized for low distortion up to 30MSPS, while the AD9101 will provide low distortion performance up to a sampling rate of 125MSPS. The low distortion performance of these SHAs is primarily due to the architecture which differs from the classical open-loop SHA architecture shown in Figure 5.18.

CLASSIC OPEN-LOOP SHA ARCHITECTURE

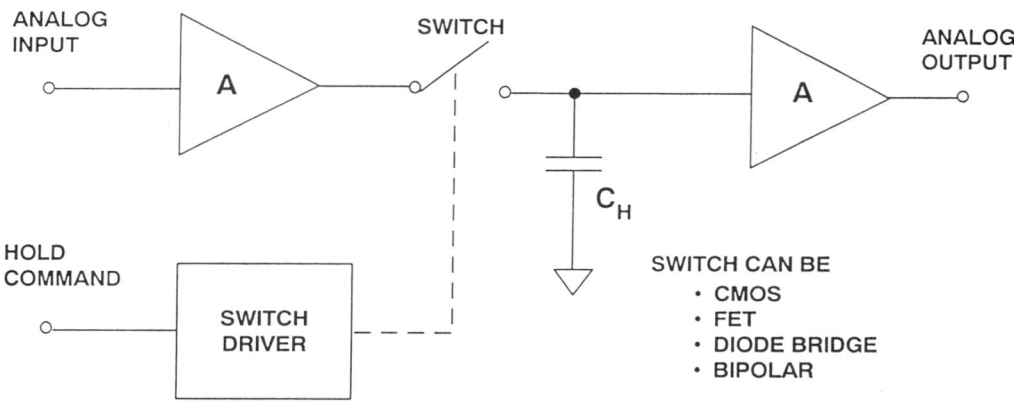

Figure 5.18

The sampling switch in the classic open-loop architecture is not within a feedback loop, and therefore distortion is subject to the non-linearity of the switch. The AD9100/AD9101 architecture shown in Figure 5.19 utilizes switches inside the feedback loop to achieve better than 12-bit AC and DC performance. The devices are fabricated on a high speed complementary bipolar process. In the track mode, S1 applies the buffered input signal to the hold capacitor, C_H, and S2 provides negative feedback to the input buffer. In the hold mode, both switches are disconnected from the hold capacitor, and negative feedback to the input buffer is supplied by S1. This architecture provides extremely low hold-mode distortion by maintaining high loop gains at high frequency. The output buffer can be configured to provide voltage gain (AD9101), which allows the switches to operate on lower common-mode voltage, thereby giving lower overall distortion.

UNDERSAMPLING APPLICATIONS

CLOSED-LOOP SHA ARCHITECTURE PROVIDES LOW DISTORTION AND HIGH SPEED
(AD9100, AD9101)

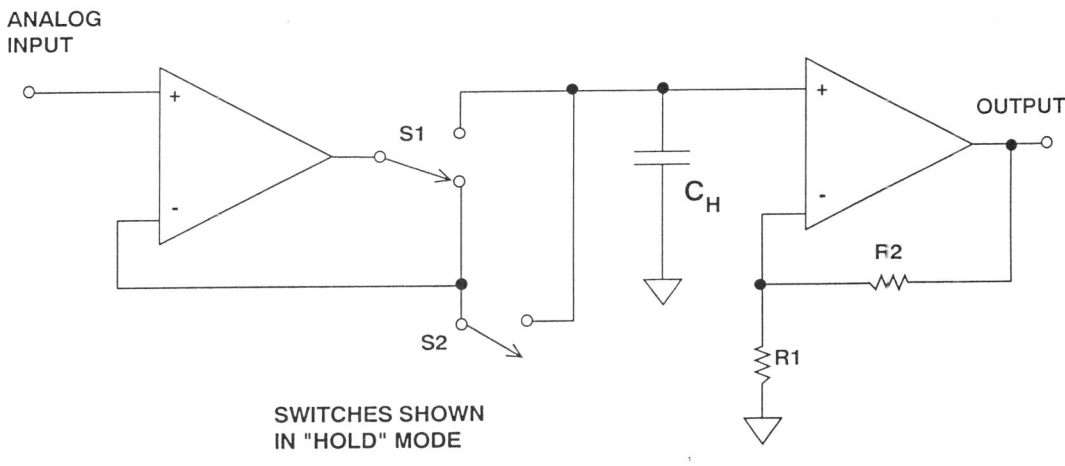

Figure 5.19

The hold-mode SFDR of the AD9100 is a function of the peak-to-peak input signal level and frequency as shown in Figure 5.20. Notice the data was taken as a sampling frequency of 10MSPS for three input amplitudes. The test configuration of Figure 5.21 was used to collect the data. For each input amplitude, the gain of the op amp between the AD9100 and the AD9014 ADC was adjusted such that the signal into the AD9014 was always full scale (2V peak-to-peak). Notice that optimum SFDR was obtained with a 200mV p-p input signal. Timing between the SHA and the ADC is critical. The SHA acquisition time should be long enough to achieve the desired accuracy, but short enough to allow sufficient hold-time for the ADC front-end to settle and yield a low-distortion conversion. For the test configuration shown, the optimum performance was achieved using an acquisition time of 20ns and a hold time of 80ns. The ADC is clocked close to the end of the SHA's hold time. Best performance in this type of application is always achieved by optimizing the timing in the actual circuit.

UNDERSAMPLING APPLICATIONS

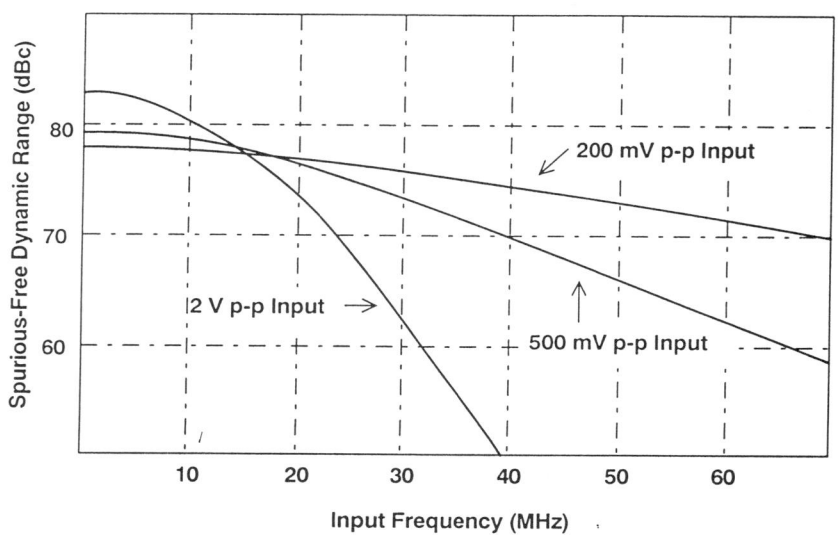

Figure 5.20

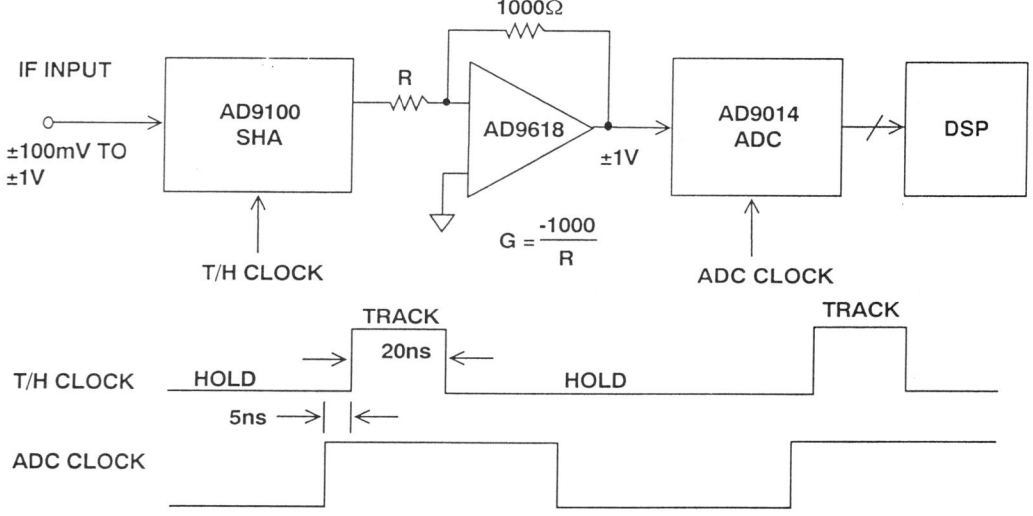

Figure 5.21

UNDERSAMPLING APPLICATIONS

Figure 5.22 shows the SFDR of the AD9100 superimposed on the SFDR of the AD9026/AD9027 and the AD9042 ADCs. These data indicate that the AD9100 will significantly improve the SFDR of the AD9026/AD9027 ADC at the higher input frequencies. The performance of the AD9042 indicates that SFDR improvements will only occur at input frequencies above 40MHz.

The performance of the AD9100 SHA driving the AD9026/AD9027 ADC at a sampling frequency of 10MSPS is shown in Figure 5.23. The input signal is a 200mV peak-to-peak 71.4MHz sinewave. The amplifier between the SHA and the ADC is adjusted for a gain of 10. The SFDR is 72dBc, and the SNR is 62dB.

Similar dynamic range improvements can be achieved with high speed flash converters at higher sampling rates with the AD9101 SHA. The architecture is similar to the AD9100, but the output buffer amplifier is optimized for a gain of 4 (see Figure 5.24). This configuration allows the front end sampler to operate at relatively low signal amplitudes, resulting in dramatic improvement in hold-mode distortion at high input frequencies and sampling rates up to 125MSPS. The AD9101 has an input bandwidth of 350MHz and an acquisition time of 7ns to 0.1% and 11ns to 0.01%.

A block diagram and a timing diagram is shown for the AD9101 driving the AD9002 8 bit flash converter at 125MSPS (Figure 5.25). The corresponding dynamic range with and without the AD9101 is shown in Figure 5.26.

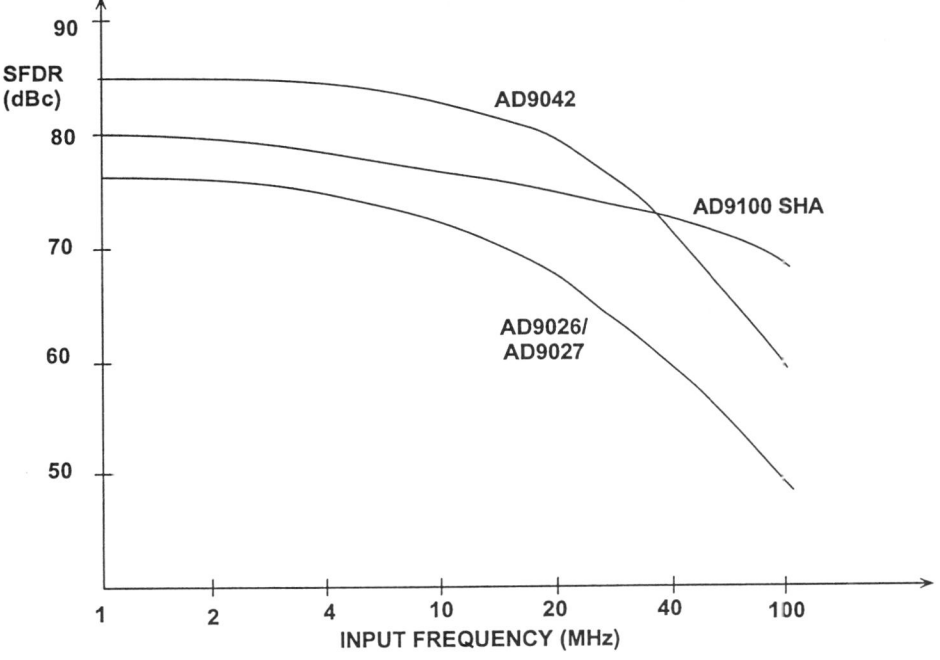

Figure 5.22

UNDERSAMPLING APPLICATIONS

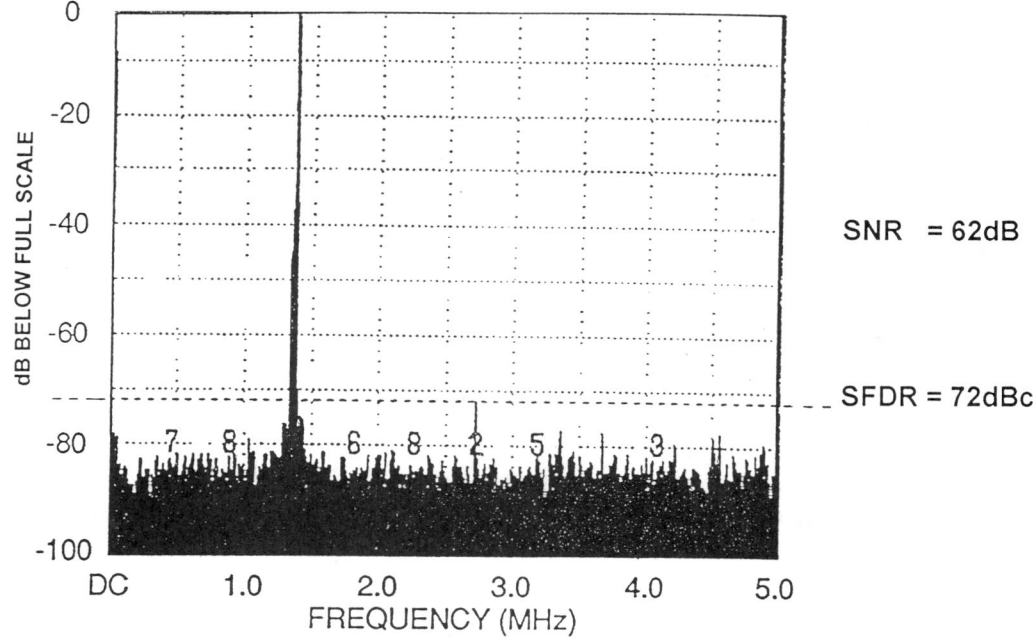

Figure 5.23

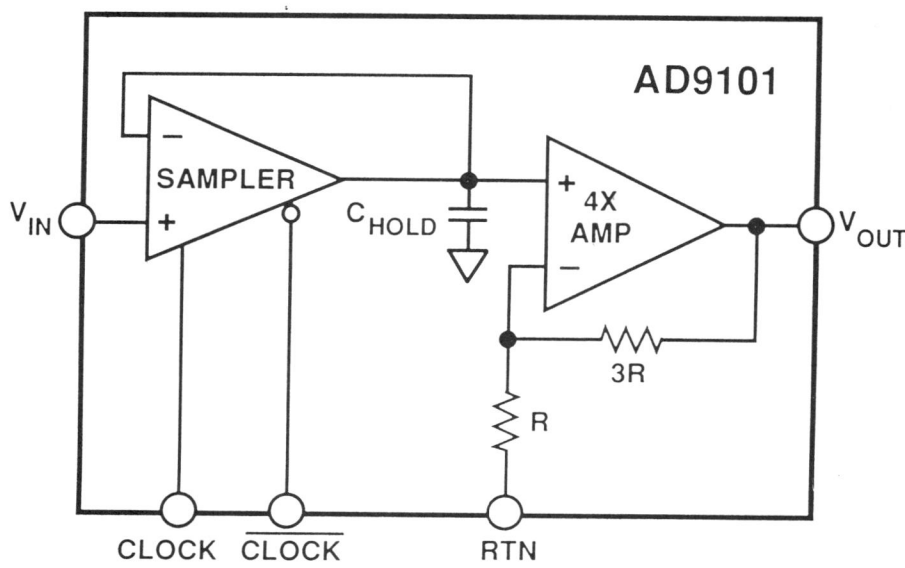

Figure 5.24

UNDERSAMPLING APPLICATIONS

AD9101 SHA DRIVING AD9002 8-BIT, 125MSPS FLASH CONVERTER FOR IMPROVED DYNAMIC RANGE

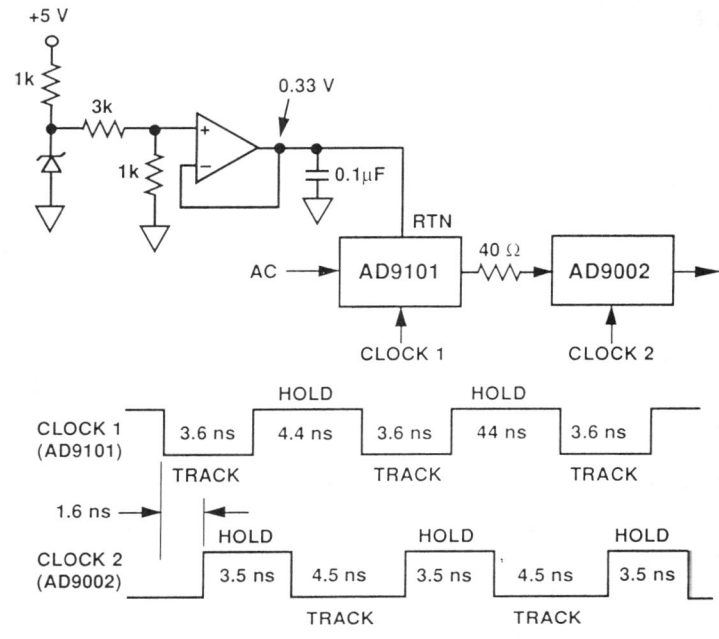

Figure 5.25

AD9002 DYNAMIC PERFORMANCE WITH AND WITHOUT AD9101 SHA

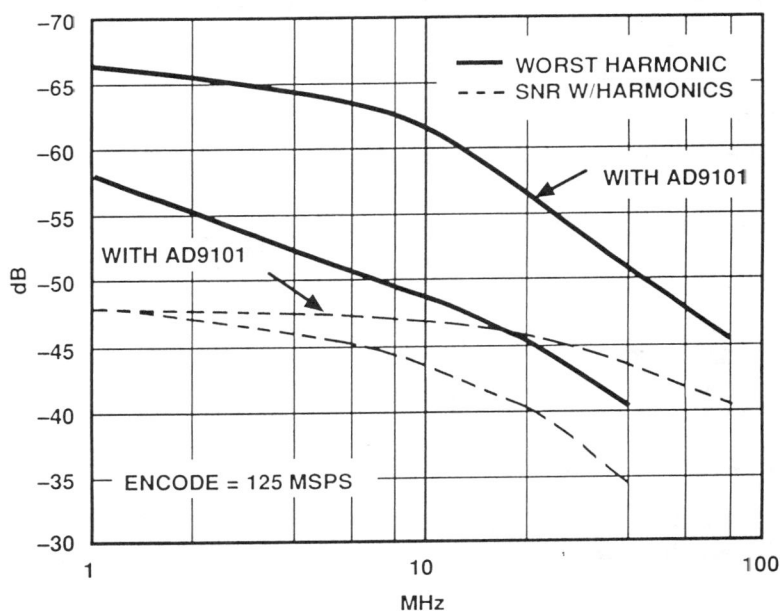

Figure 5.26

5-21

USE OF DITHER SIGNALS TO INCREASE ADC DYNAMIC RANGE

In the development of classical ADC quantization noise theory, the assumption is usually made that the quantization error signal is uncorrelated with the ADC input signal. If this is true, then the quantization noise appears as random noise spread uniformly over the Nyquist bandwidth, dc to $f_s/2$, and it has an rms value equal to $q/\sqrt{12}$. If, however, the input signal is locked to an non-prime integer sub-multiple of f_s, the quantization noise will no longer appear as uniformly distributed random noise, but instead will appear as harmonics of the fundamental input sinewave. This is especially true if the input is an exact even submultiple of f_s. Figure 5.27 illustrates the point using FFT simulation for an ideal 12 bit ADC. The FFT record length was chosen to be 4096. The spectrum on the left shows the FFT output when the input signal is an exact even submultiple (1/32) of the sampling frequency (the frequency was chosen so that there were exactly 128 cycles per record). The SFDR is approximately 78dBc. The spectrum on the right shows the output when the input signal is such that there are exactly 127 cycles per record. The SFDR is now about 92dBc which is an improvement of 14dB. Signal-correlated quantization noise is highly undesirable in spectral analysis applications, where it becomes difficult to differentiate between real signals and system-induced spurious components, especially when searching the spectrum for the presence of low-level signals in the presence of large signals.

EFFECTS OF SAMPLING A SIGNAL WHICH IS AN EXACT EVEN SUB-MULTIPLE OF THE ADC SAMPLING FREQUENCY (M = 4096, IDEAL 12-BIT ADC SIMULATION)

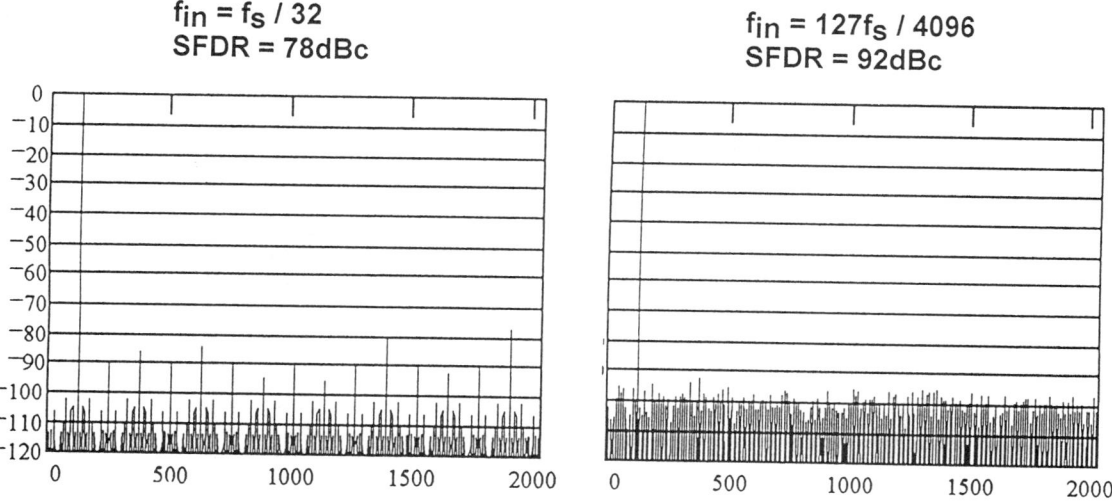

Figure 5.27

There are a number of ways to reduce this effect, but the easiest way is to add a small amount of broadband rms noise to the ADC input signal as shown in Figure 5.28. The rms value of this noise should be equal to about 1/2 LSB. The effect of this is to randomize the quantization noise and eliminate its possible signal-dependence. In most systems, there is usually enough random noise present on the input signal so that this happens automatically. This is especially likely when using high speed ADCs which have 12 or more bits of resolution and a relatively small input range of 2V p-p or less. The total noise at the ADC is composed of the noise of the input signal, the effective input noise of the ADC, and an additional component caused by the effects of the sampling clock jitter. In most cases, the rms value of the total ADC input noise is greater than 1/2 LSB.

THE ADDITION OF WIDEBAND GAUSSIAN NOISE TO THE ADC INPUT RANDOMIZES QUANTIZATION NOISE AND REMOVES INPUT SIGNAL DEPENDENCE

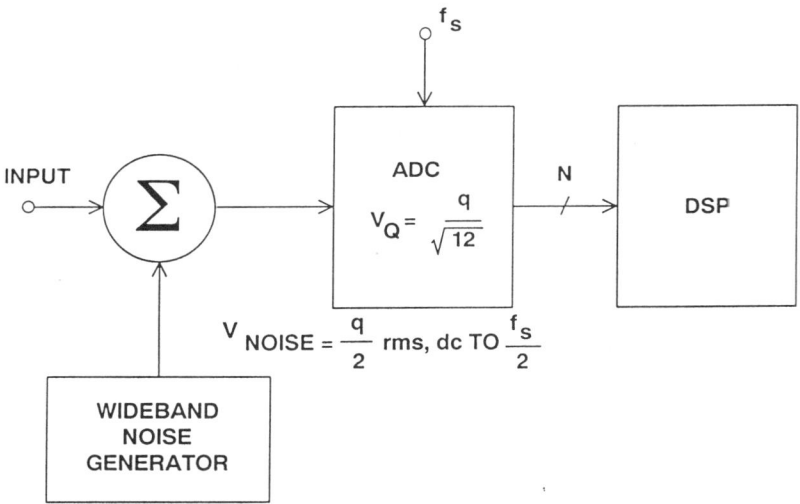

Figure 5.28

Effect of ADC Linearity and Resolution on SFDR and Noise in Digital Spectral Analysis Applications

In order to understand the relationships between ADC resolution, noise, and Spurious Free Dynamic Range (SFDR), it is first necessary to review the some of the issues relating to digital spectral analysis, specifically the FFT. The FFT takes a discrete number of time samples, M, and converts them into M/2 discrete spectral components. The spacing between the spectral lines is $\Delta f = f_s/M$. When a full scale sinewave signal is applied to an ADC having a resolution of N bits, the theoretical rms signal to rms noise ratio is $6.02N + 1.76\text{dB}$. If the quantization noise is uncorrelated with the signal, it appears as gaussian noise spread uniformly over the bandwidth dc to $f_s/2$. The FFT acts as a narrowband filter with a bandwidth of Δf, and the FFT noise floor is therefore $10\log_{10}(M/2)$ dB below the broadband quantization noise level $(6.02N + 1.76\text{dB})$. The FFT noise floor is pushed down by 3dB each time the FFT record length, M, is doubled (see Figure 5.29). This reduction in the noise floor is the same effect achieved by narrowing the bandwidth of an analog spectrum analyzer to a bandwidth of f_s/M.

For example, a 4096 point FFT has a noise floor which is 33dB below the theoretical broadband rms quantization noise floor of 74dB for a 12-bit ADC as shown in Figure 5.30 (where the average noise floor is about $74 + 33 = 107\text{dB}$ below full scale). Notice that there are random peaks and valleys around the average FFT noise floor. These peaks (due to quantization noise, FFT artifacts, and roundoff error) limit the ideal SFDR to about 92dBc.

RELATIONSHIP BETWEEN AVERAGE NOISE IN FFT BINS AND BROADBAND RMS QUANTIZATION NOISE LEVEL

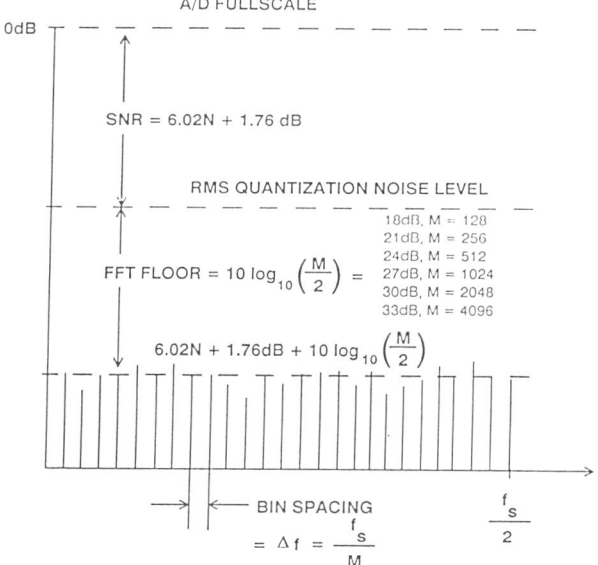

Figure 5.29

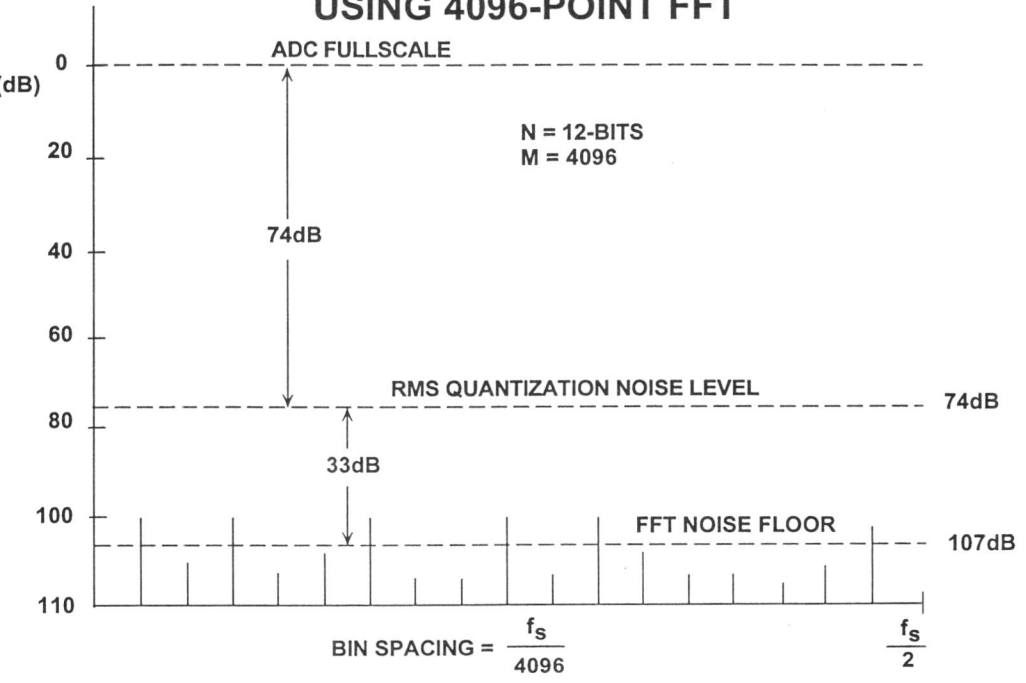

Figure 5.30

Approximately the same dynamic range could be achieved by reducing the resolution of the ADC to 11 bits and using a 16,384 point FFT. There is a tradeoff, however, because lower resolution ADCs tend to have quantization noise which is correlated to the input signal, thereby producing larger frequency spurs. Averaging the results of several FFTs will tend to smooth out the FFT noise floor, but does nothing to reduce the average noise floor.

Using more bits improves the SFDR only if the ADC AC linearity improves with the additional bits. For instance, there would be little advantage in using a 14-bit ADC which has only 12-bit linearity. The extra bits would only serve to slightly reduce the overall noise floor, but the improvement in SFDR would be only marginal.

Future Trends in Undersampling ADCs

Future ADCs specifically designed for undersampling applications will incorporate the previously discussed techniques in a single-chip designs. These ADCs will be characterized by their wide SFDR at input frequencies extending well above the Nyquist limit, $f_s/2$. The basic architecture of the digital IF receiver is shown in Figure 5.31. The addition of a low-distortion PGA under DSP control increases the dynamic range of the system. IF frequencies associated with 900MHz digital cellular base stations are typically around 70MHz with bandwidths between 5 and 10MHz. SFDR requirements are between 70 and 80dBc. On the other hand, 1.8GHz digital receivers typically have IF frequencies between 200 and 240MHz with bandwidths of 1MHz. SFDR requirements are typically 50dBc.

DIRECT IF TO DIGITAL RECEIVER USING PGA TO INCREASE SYSTEM DYNAMIC RANGE

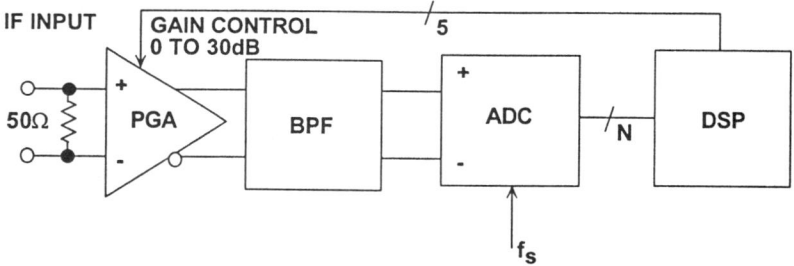

f_{IF}	BANDWIDTH	f_s	SFDR
40 - 70 MHz	5 - 10 MHz	10 - 20 MSPS	70 - 80dBc
200 - 240 MHz	1 MHz	2 - 3 MSPS	50 - 60 dBc

Figure 5.31

The bandpass sigma-delta architecture offers interesting possibilities in this area. Traditional sigma-delta ADCs contain integrators, which are lowpass filters. They have passbands which extend from DC, and the quantization noise is pushed up into the higher frequencies. At present, all commercially available sigma-delta ADCs are of this type (although some which are intended for use in audio or telecommunications contain bandpass filters to eliminate any DC response).

There is no particular reason why the filters of the sigma-delta modulator should be lowpass filters, except that traditionally ADCs have been thought of as being baseband devices, and that integrators are somewhat easier to construct than bandpass filters. If we replace the integrators in a sigma-delta ADC with bandpass filters as shown in Figure 5.32, the quantization noise is moved up and down in frequency to leave a virtually noise-free region in the passband (see Figure 5.33). If the digital filter is then programmed to have its passband in this noise-free region, we have a sigma-delta ADC with a bandpass, rather than a lowpass characteristic.

REPLACING INTEGRATORS WITH BANDPASS FILTERS GIVES A BANDPASS SIGMA-DELTA ADC

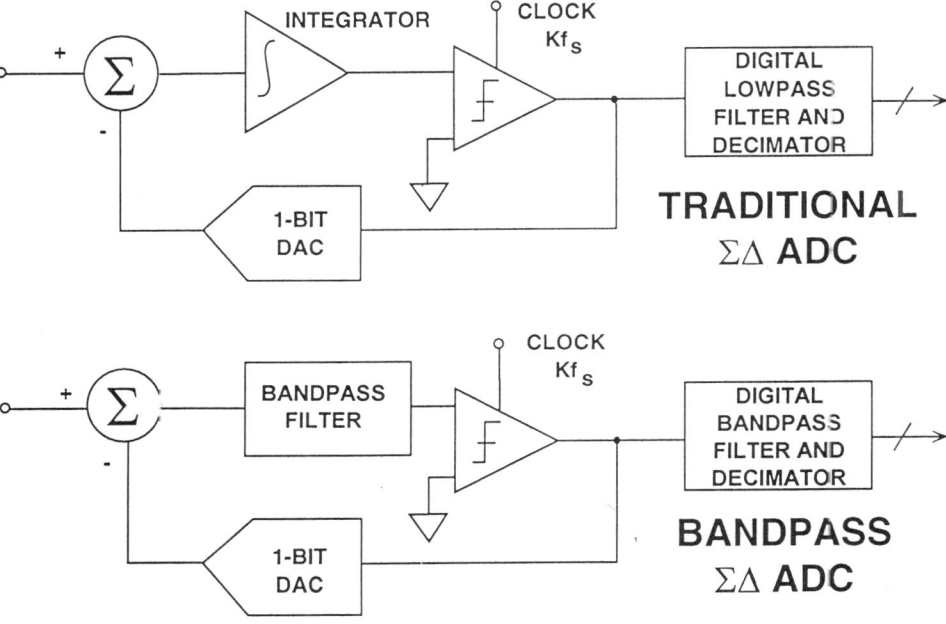

Figure 5.32

UNDERSAMPLING APPLICATIONS

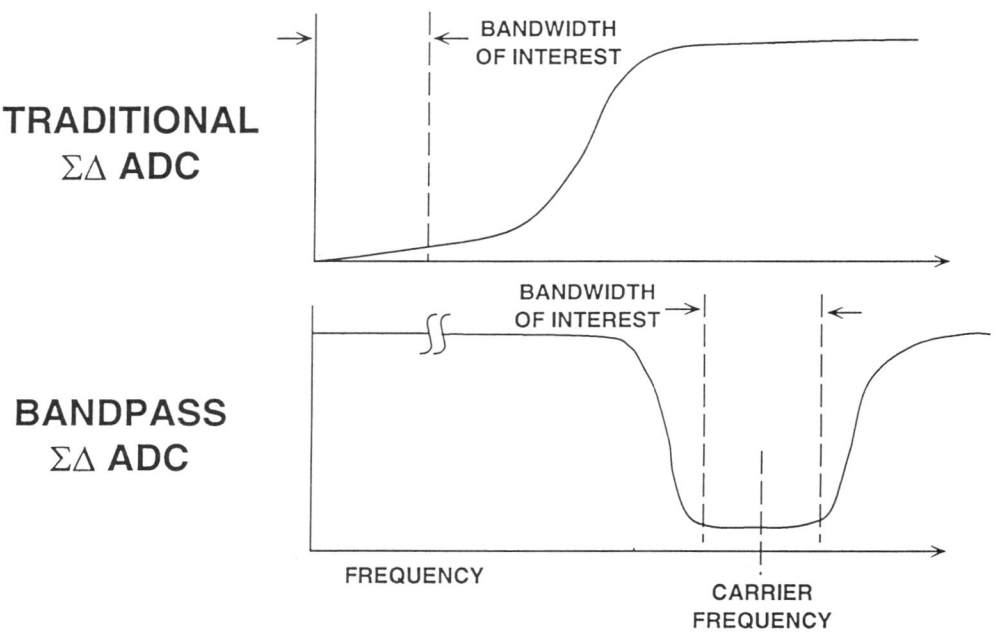

Figure 5.33

The theory is straightforward, but the development of a sigma-delta ADC is expensive, and there is no universal agreement on ideal characteristics for such a bandpass sigma-delta ADC, so developing such a converter from scratch to verify the theory would be unlikely to yield a commercial product. Researchers at Analog Devices and the University of Toronto (See References 16, 17, and 18) have therefore modified a commercial baseband (audio) sigma-delta ADC chip by rewiring its integrators as switched capacitor bandpass filters and reprogramming its digital filter and decimator. This has provided a fast, and comparatively inexpensive, proof of the concept, but at the expense of relative low Effective Bits (11-bits), the result of less than ideal bandpass filters. Nevertheless the results are extremely encouraging and open the way to the design of purpose-built bandpass sigma-delta ADC chips for specific ASIC applications, especially, but not exclusively, radio receivers.

The overall performance characteristics of the experimental ADC are shown in Figure 5.34. The device was designed to digitize the popular radio IF frequency of 455kHz.

SUMMARY OF RESULTS FOR EXPERIMENTAL BANDPASS SIGMA-DELTA ADC

- Center Frequency: 455kHz
- Bandwidth: 10kHz
- Sampling Rate: 1.852MSPS
- Oversampling Ratio: 91
- SNR in Specified Band: 65dB
- Supply: ±5V,   Power: 750mW
- Process: 3µm CMOS,   Active Area: 1.8 x 3.4mm

Figure 5.34

In the future it may be possible to have such bandpass sigma-delta ADCs with user-programmable digital filter coefficients, so that the passband of a receiver could be modified during operation in response to the characteristics of the signal (and the interference!) being received. Such a function is very attractive, but difficult to implement, since it would involve loading, and storing, several hundreds or even thousands of 16-22 bit filter coefficients, and would considerably increase the size, and cost, of the converter.

A feature which could be added comparatively easily to a sigma-delta ADC is a more complex digital filter with separate reference (I) and quadrature (Q) outputs. Such a feature would be most valuable in many types of radio receivers.

Technology exists today which should allow the bandpass sigma-delta architecture to achieve 16-bit resolution, SFDR of 70 to 80dBc, and an effective throughput rate of 10 to 20MSPS (input sampling rate = 100MSPS, corresponding to an oversampling ratio of 5 to 10). This would allow 40MHz IF with a 2MHz bandwidth to be digitized directly.

CHARACTERISTICS OF A BANDPASS SIGMA-DELTA ADC DESIGNED FOR IF-SAMPLING

- IF Frequency: 40 to 70MHz
- Signal Bandwidth: 2MHz
- Input Sampling Rate: 100MSPS
- Output Data Rate: 10 to 20MSPS
- Process: BiCMOS

Figure 5.35

EFFECTS OF SAMPLING CLOCK JITTER IN UNDERSAMPLING APPLICATIONS

The effects of sampling clock jitter on Signal-to-Noise Ratio (SNR) and Effective Bit (ENOB) performance discussed in Section 3 are even more dramatic in undersampling applications because of the higher input signal frequencies. Figure 5.36 shows the relationship between sampling clock jitter and SNR previously presented.

Consider the case where the IF frequency is 70MHz, and 12-bit dynamic range is required (70 to 80dB). From Figure 5.36, the rms sampling clock jitter required to maintain this SNR is approximately 1ps rms. This assumes an ideal ADC with no internal aperture jitter. ADC aperture jitter combines with the sampling clock jitter in an rms manner to further degrade the SNR.

The implications of this analysis are extremely important in undersampling applications. The ADC aperture jitter must be minimal, and the sampling clock generated from a low phase-noise quartz crystal oscillator. Furthermore, the oscillator should use discrete bipolar and FET devices in the circuits recommended by the crystal manufacturers. The popular oscillator design which uses a resistor, one or more logic gates, a quartz crystal, and a couple of capacitors should never be used! For very high frequency clocks, a surface acoustic wave (SAW) oscillator is preferable.

UNDERSAMPLING APPLICATIONS

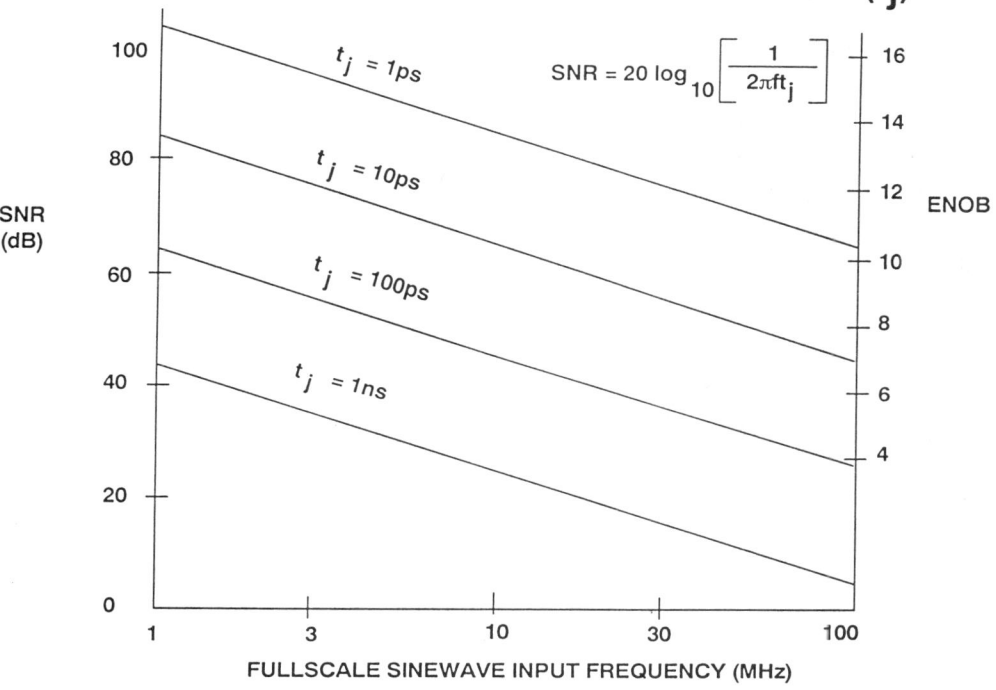

Figure 5.36

SAMPLING CLOCK OSCILLATORS

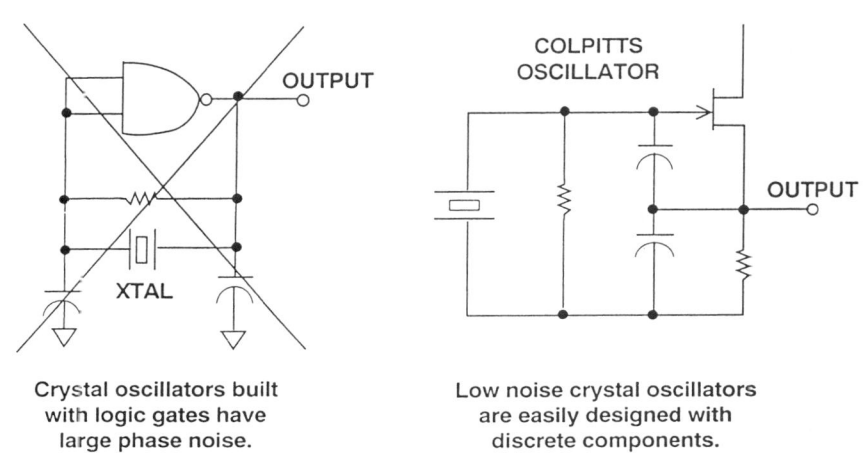

Crystal oscillators built with logic gates have large phase noise.

Low noise crystal oscillators are easily designed with discrete components.

Figure 5.37

The sampling clock should be isolated as much as possible from the noise present in the digital parts of the system. There should be few or no logic gates in the sampling clock path, as a single ECL gate has approximately 4ps rms timing jitter. The sampling clock generation circuitry should be on separate chips, perhaps with separately decoupled power supplies, from the remainder of the digital system, and the sampling clock signal lines should not be located where they can pick up digital noise from the rest of the system. All sampling clock circuitry should be grounded and decoupled to the analog ground plane, as would be the case for a critical analog component.

Of course, the sampling clock is itself a digital signal. It has as much potential for causing noise in the analog part of the system as any other digital signal.

We therefore see that a sampling clock is very inconvenient, as it must be isolated from both the analog and digital parts of the system.

Because the sampling clock jitter is wideband and therefore creates wideband random noise, digital filtering can be used to reduce its effects in a system. In the case of an FFT, however, doubling the FFT record length reduces the noise floor by only 3dB.

This discussion illustrates the basic fact that undersampling systems have fundamental limitations with respect to their ability to process wide dynamic range broadband signals, and system tradeoffs between broadband and narrowband approaches must ultimately be made in the design of such systems.

SAMPLING CLOCK NOISE

- Low phase-noise crystal (or SAW) oscillators mandatory in high frequency undersampling applications

- Ground and decouple sampling clock circuitry to the Analog Ground Plane!

- Route sampling clock away from digital and analog signals

- Digital filtering techniques can be used to reduce the effects of sampling clock phase noise

- Sampling clock noise can ultimately dictate the tradeoffs between broadband and narrowband digital receivers

Figure 5.38

REFERENCES

1. **System Applications Guide**, Analog Devices, 1993, Chapter 15.

2. Richard Groshong and Stephen Ruscak, *Undersampling Techniques Simplify Digital Radio*, **Electronic Design**, May 23, 1991, pp. 67-78.

3. Richard Groshong and Stephen Ruscak, *Exploit Digital Advantages In An SSB Receiver*, **Electronic Design**, June 13, 1991, pp. 89-96.

4. Richard G. Lyons, *How Fast Must You Sample?*, **Test and Measurement World**, November, 1988, pp. 47-57.

5. Richard C. Webb, *IF Signal Sampling Improves Receiver Detection Accuracy*, **Microwaves and RF**, March, 1989, pp. 99-103.

6. Jeff Kirsten and Tarlton Fleming, *Undersampling Reduces Data-Acquisition Costs For Select Applications*, **EDN**, June 21, 1990, pp 217-228.

7. Hans Steyskal and John F. Rose, *Digital Beamforming For Radar Systems*, **Microwave Journal**, January, 1989, pp. 121-136.

8. Tom Gratzek and Frank Murden, *Optimize ADCs For Enhanced Signal Processing*, **Microwaves and RF**, Vol. 30, No.3, March, 1991, pp. 129-136.

9. Howard Hilton, *10-MHz ADC With 110dB Linearity*, **High Speed ADC Conference**, Las Vegas, April 21-22, 1992.

10. Howard Hilton, *10MSample/Second ADC With Filter And Memory*, **High Speed ADC Conference**, Las Vegas, April 21-22, 1992.

11. Dan Asta, *Recent Dynamic Range Characterization of Analog-to-Digital Converters for Spectral Analysis Applications*, **Project Report AST-14**, MIT Lincoln Laboratory, Lexington, MA, July 9, 1991.

12. Fred H. Irons and T.A. Rebold, *Characterization of High-Frequency Analog-to-Digital Converters for Spectral Analysis Applications*, **Project Report AST-2**, MIT Lincoln Laboratory, November 28, 1986.

13. F. H. Irons, *Dynamic Characterization and Compensation of Analog-to-Digital Converters*, **IEEE International Symposium on Circuits and Systems**, May 1986, San Jose, CA. Catalog No. CH2255-8/86/0000-1273.

14. T.A. Rebold and F. H. Irons, *A Phase-Plane Approach to the Compensation of High-Speed Analog-to-Digital Converters*, **IEEE International Symposium on Circuits and Systems**, Philadelphia, PA, May, 1987.

15. Dan Asta and Fred H. Irons, *Dynamic Error Compensation of Analog-to-Digital Converters*, **The Lincoln Laboratory Journal**, Vol. 2, No. 2, 1989, pp. 161-182.

16. James M. Bryant, *Bandpass Sigma-Delta ADCs for Direct IF Conversion*, **DSP- The Enabling Technology for Communications**, RAI Congrescentrum, Amsterdam, Netherlands, 9-10 March 1993.

17. S.A. Jantzi, M. Snelgrove, P.F. Ferguson, Jr., *A 4th-Order Bandpass Sigma-Delta Modulator*, **Proceedings of the IEEE 1992 Custom Integrated Circuits Conference**, pp. 16.5.1-4.

18. S.A. Jantzi, R. Schreier, and M. Snelgrove, *A Bandpass Sigma-Delta Convertor for a Digital AM Receiver*, **Proceedings of the IEE International Conference on Analogue-to-Digital and Digital-to-Analogue Conversion**, Swansea, UK., September, 1991, pp. 75-80.

19. S.A. Jantzi, W. Martin Snelgrove, and Paul F. Ferguson, Jr., *A Fourth-Order Bandpass Sigma-Delta Modulator*, **IEEE Journal of Solid-State Circuits**, Vol. 38, No. 3, March 1993, pp. 282-291.

20. Barrie Gilbert, *A Low Noise Wideband Variable-Gain Amplifier Using an Interpolated Ladder Attenuator*, **IEEE ISSCC Technical Digest**, 1991, pages 280, 281, 330.

21. Barrie Gilbert, *A Monolithic Microsystem for Analog Synthesis of Trigonometric Functions and their Inverses*, **IEEE Journal of Solid State Circuits**, Vol. SC-17, No. 6, December 1982, pp. 1179-1191.

22. **1992 Amplifier Applications Guide**, Analog Devices, Norwood MA, 1992.

23. Carl Moreland, *An 8-Bit 150MSPS Serial ADC*, **1995 ISSCC Digest of Technical Papers**, Vol. 38, p.272.

24. Roy Gosser and Frank Murden, *A 12-Bit 50MSPS Two-Stage A/D Converter*, **1995 ISSCC Digest of Technical Papers**, p. 278.

SECTION 6

MULTICHANNEL APPLICATIONS

- Data Acquisition System Considerations
- Multiplexing
- Filtering Considerations for Data Acquisition Systems
- SHA and ADC Settling Time Requirements in Multiplexed Applications
- Complete Data Acquisition Systems on a Chip
- Multiplexing into Sigma-Delta ADCs
- Simultaneous Sampling Systems
- Data Distribution Systems using Multiple DACs

Multichannel Applications

SECTION 6

MULTICHANNEL APPLICATIONS
Walt Kester, Wes Freeman

DATA ACQUISITION SYSTEM CONFIGURATIONS

There are many applications for data acquisition systems in measurement and process control. All data acquisition applications involve digitizing analog signals for analysis using ADCs. In a measurement application, the ADC is followed by a digital processor which performs the required data analysis. In a process control application, the process controller generates feedback signals which typically must be converted back into analog form using a DAC.

Although a single ADC digitizing a single channel of analog data constitutes a data acquisition system, the term *data acquisition* generally refers to multi-channel systems. If there is feedback from the digital processor, DACs may be required to convert the digital responses into analog. This process is often referred to as data distribution.

Figure 6.1 shows a data acquisition/distribution process control system where each channel has its own dedicated ADC and DAC. An alternative configuration is shown in Figure 6.2, where analog multiplexers and demultiplexers are used with a single ADC and DAC. In most cases, especially where there are many channels, this configuration provides an economical alternative.

**DATA ACQUISITION SYSTEM
USING ADC / DAC PER CHANNEL**

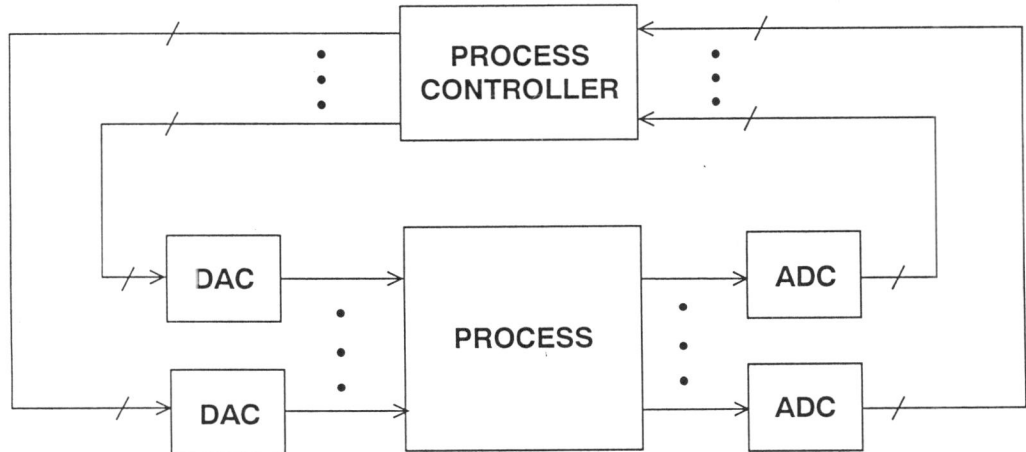

Figure 6.1

MULTICHANNEL APPLICATIONS

DATA ACQUISITION SYSTEM USING ANALOG MULTIPLEXING / DEMULTIPLEXING AND SINGLE ADC / DAC

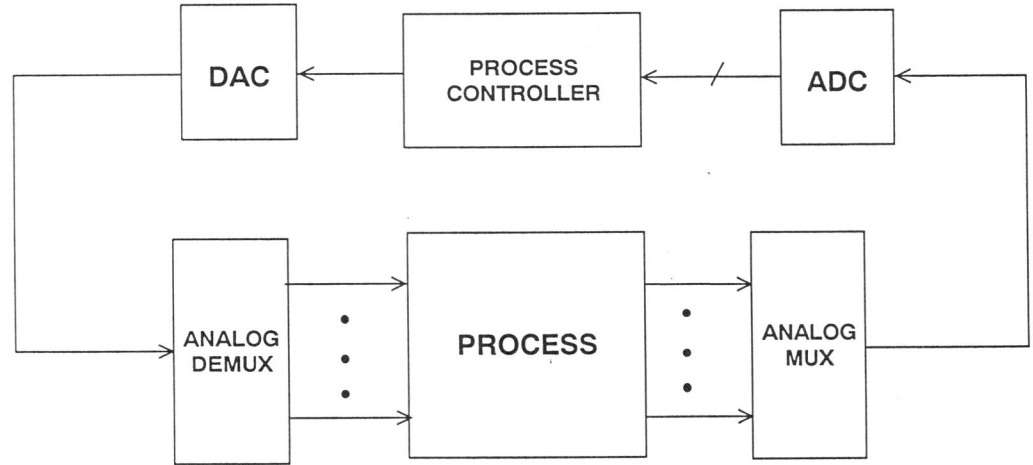

Figure 6.2

There are many tradeoffs involved in designing a data acquisition system. Issues such as filtering, amplification, multiplexing, demultiplexing, sampling frequency, and partitioning must be resolved.

MULTIPLEXING

Multiplexing is a fundamental part of a data acquisition system. Multiplexers and switches are examined in more detail in Reference 1, but a fundamental understanding is required to design a data acquisition system. A simplified diagram of an analog multiplexer is shown in Figure 6.3. The number of input channels typically ranges from 4 to 16, and the devices are generally fabricated on CMOS processes. Some multiplexers have internal channel-address decoding logic and registers, while with others, these functions must be performed externally. Unused multiplexer inputs *must* be grounded or severe loss of system accuracy may result. The key specifications are *switching time, on-resistance, on-resistance modulation,* and *off-channel isolation (crosstalk)*. Multiplexer switching time ranges from about 50ns to over 1µs, on-resistance from 25Ω to several hundred ohms, and off-channel isolation from 50 to 90dB. The use of trench isolation has eliminated latch-up in multiplexers while yielding improvements in speed at low supply voltages.

MULTICHANNEL APPLICATIONS

SIMPLIFIED DIAGRAM OF A TYPICAL ANALOG MULTIPLEXER

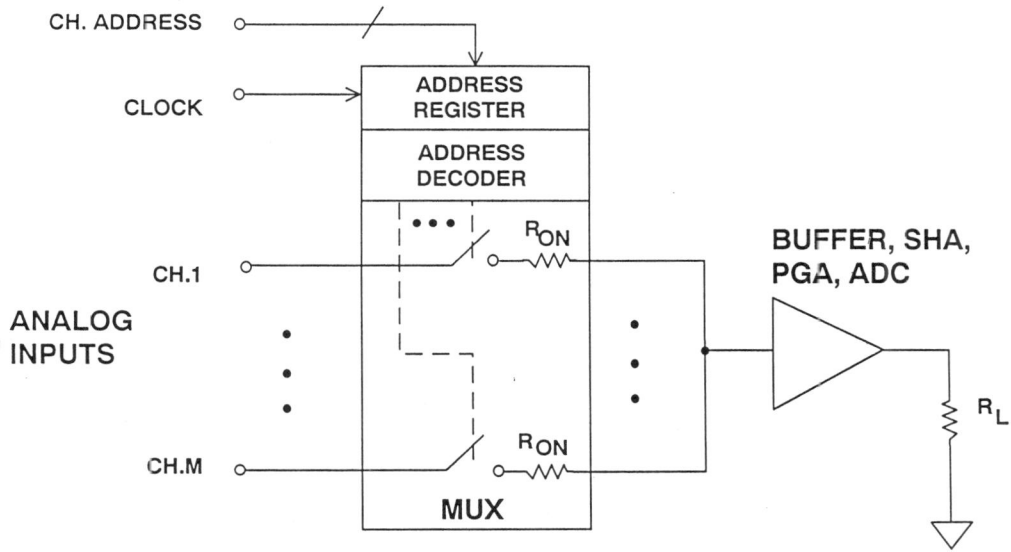

Figure 6.3

MULTIPLEXER KEY SPECIFICATIONS

- Switching Time: 50ns to >1μs
- On-Resistance: 25Ω to hundreds of Ω's
- On-Resistance Modulation (Ron change with signal level)
- Off-Channel Isolation: 50 to 90dB
- Overvoltage Protection

Figure 6.4

WHAT'S NEW IN MULTIPLEXERS?

- ■ Trench Isolation gives high speed, latch-up protection, and low-voltage operation

- ■ ADG511, ADG512, ADG513: +3.3V, +5V, ±5V specified
  Ron < 50Ω @ ±5V
  Switching Time: < 200ns @ ±5V

- ■ ADG411, ADG412, ADG413: ±15V, +12V specified
  Ron < 35Ω @ ±15V
  Switching Time: < 150ns @ ±15V

- ■ ADG508F, ADG509F, ADG528F: ±15V specified
  Ron < 300Ω
  Switching Time: < 250ns
  Fault-Protection on Inputs and Outputs

Figure 6.5

Multiplexer on-resistance is generally slightly dependent on the signal level (often called R_{on} modulation). This will cause signal distortion if the multiplexer must drive a load resistance, therefore the multiplexer output should therefore be isolated from the load with a suitable buffer amplifier. A separate buffer is not required if the multiplexer drives a high input impedance, such as a PGA, SHA or ADC - but beware! Some SHAs and ADCs draw high frequency pulse current at their sampling rate and cannot tolerate being driven by an unbuffered multiplexer. A detailed analysis of multiplexers can be found in Reference 1, Section 8, or Reference 2, Section 2.

An M-channel multiplexed data acquisition system is shown in Figure 6.6.

The multiplexer output drives a PGA whose gain can be adjusted on a per-channel basis depending on the channel signal level. This ensures that all channels utilize the full dynamic range of the ADC. The PGA gain is changed at the same time as the multiplexer is switched to a new channel. The ADC *Convert Command* is applied after the multiplexer and the PGA have settled to the required accuracy (1LSB). The maximum sampling frequency (when switching between channels) is limited by the multiplexer switching time t_{mux}, the PGA settling time t_{pga}, and the ADC conversion time t_{conv} as shown in the formula.

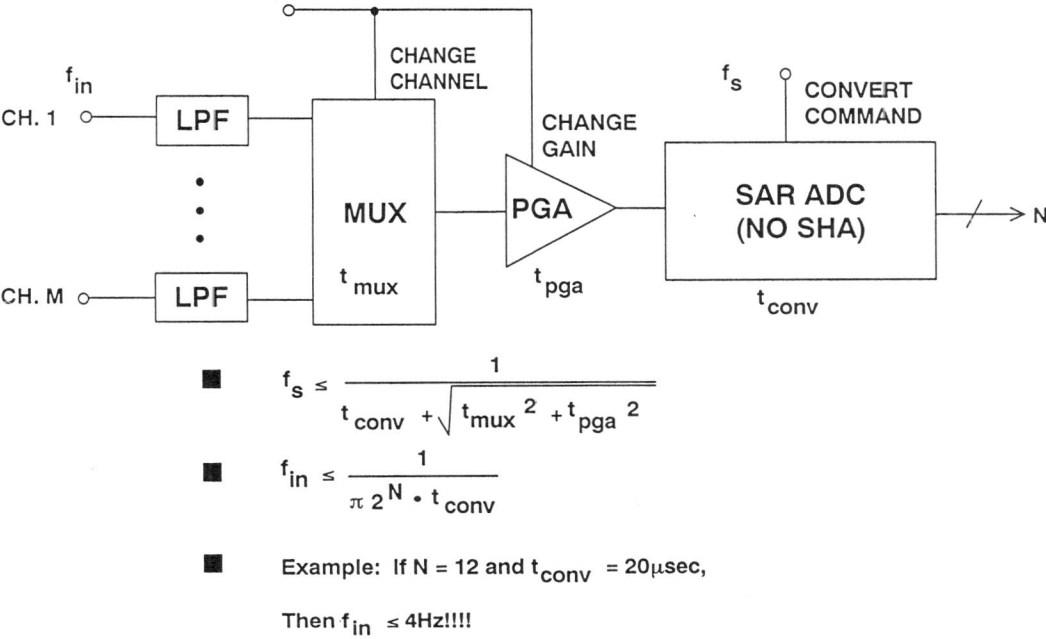

Figure 6.6

In a multiplexed system it is possible to have a positive fullscale signal on one channel and a negative fullscale signal on the other. When the multiplexer switches between these channels its output is a fullscale step voltage. All elements in the signal path must settle to the required accuracy (1LSB) before the conversion is made. The effect of inadequate settling is dc crosstalk between channels.

The SAR ADC chosen in this application has no internal SHA (similar to the industry-standard AD574-series), and therefore the input signal must be held constant (within 1LSB) during the conversion time in order to prevent encoding errors. This defines the maximum rate-of-change of the input signal:

$$\left.\frac{dv}{dt}\right|_{max} \leq \frac{1\,\text{LSB}}{t_{conv}}$$

The amplitude of a fullscale sinewave input signal is equal to $2^N/2$, or $2^{(N-1)}$, and its maximum rate-of-change is

$$\left.\frac{dv}{dt}\right|_{max} = 2\pi f_{max} \cdot 2^{N-1} = \pi f_{max} \cdot 2^N$$

Setting the two equations equal, and solving for f_{max},

$$f_{max} \leq \frac{1}{\pi \cdot 2^N t_{conv}}$$

For example, if the ADC conversion time is 20μsec (corresponding to a maximum sampling rate of slightly less than 50kSPS), and the resolution is 12-bits, then the maximum channel input signal frequency is limited to 4Hz. This may be adequate if the signals are DC, but the lack of a SHA function severely

MULTICHANNEL APPLICATIONS

limits the ability to process dynamic signals.

Adding a SHA function to the ADC as shown in Figure 6.7 allows processing of much faster signals with almost no increase in system complexity, since sampling ADCs such as the AD1674 have the SHA function on-chip.

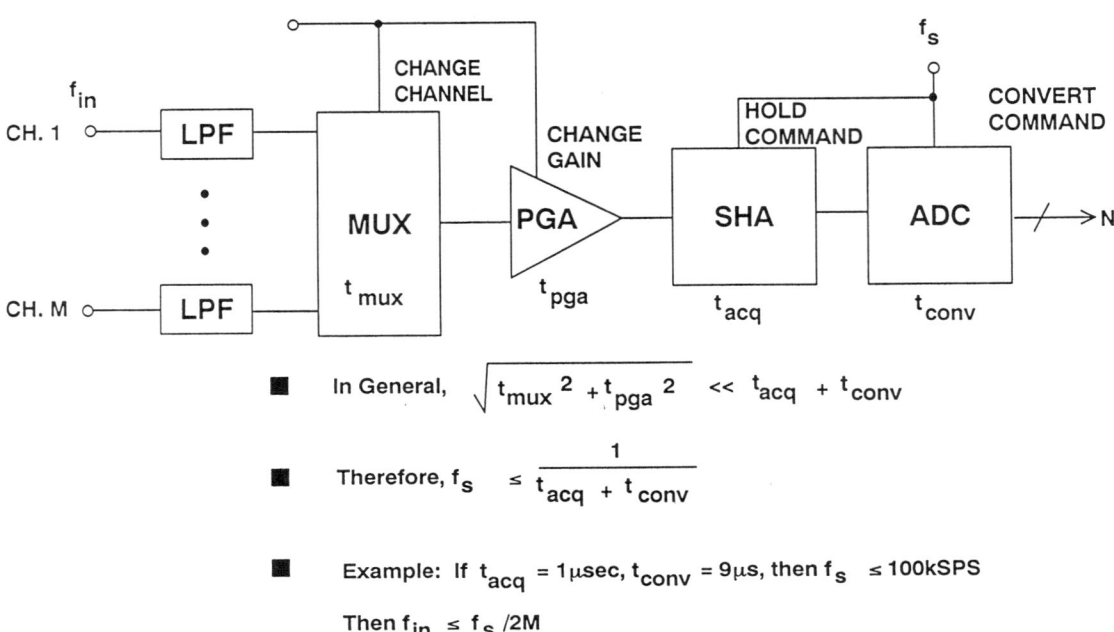

THE ADDITION OF A SHA FUNCTION TO THE ADC ALLOWS PROCESSING OF DYNAMIC INPUT SIGNALS

- In General, $\sqrt{t_{mux}^2 + t_{pga}^2} \ll t_{acq} + t_{conv}$
- Therefore, $f_s \leq \dfrac{1}{t_{acq} + t_{conv}}$
- Example: If $t_{acq} = 1\mu sec$, $t_{conv} = 9\mu s$, then $f_s \leq 100\text{kSPS}$

Then $f_{in} \leq f_s / 2M$

Figure 6.7

The timing is adjusted such that the multiplexer and the PGA are switched immediately following the acquisition time of the SHA. If the combined multiplexer and PGA settling time is less than the ADC conversion time (see Figure 6.8), then the maximum sampling frequency of the system is given by:

$$f_s \leq \frac{1}{t_{acq} + t_{conv}}$$

The AD1674 has a conversion time of 9μs, an acquisition time of 1μs to 12-bits, and a sampling rate of 100kSPS is possible, if all the channels are addressed. The per-channel sampling rate is obtained by dividing the ADC sampling rate by M.

Multichannel Applications

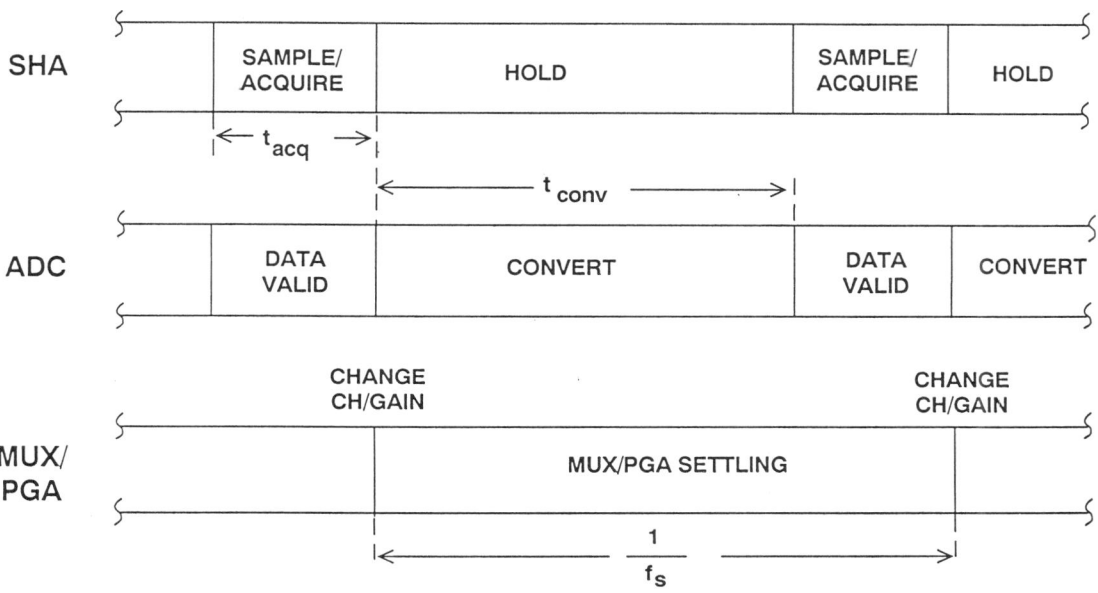

Figure 6.8

Filtering Considerations in Data Acquisition Systems

Filtering in data acquisition systems not only prevents aliasing of unwanted signals but also reduces noise by limiting bandwidth. In a multiplexed system, there are basically two places to put filters: in each channel, and at the multiplexer output.

The filter at the input of each channel is used to prevent aliasing of signals which fall outside the Nyquist bandwidth. The per-channel sampling rate (assuming each channel is sampled at the same rate) is f_s/M, and the corresponding Nyquist frequency is $f_s/2M$. The filter should provide sufficient attenuation at $f_s/2M$ to prevent dynamic range limitations due to aliasing.

A second filter can be placed in the signal path between the multiplexer output and the ADC, usually between the PGA and the SHA. The cutoff frequency of this filter must be carefully chosen because of its impact on settling time. In a multiplexed system such as shown in Figure 6.7, there can be a fullscale step voltage change at the multiplexer output when it is switched between channels. This occurs if the signal on one channel is positive fullscale, and the signal on the adjacent channel is negative fullscale. From the timing diagram shown in Figure 6.8, the signal from the filter has essentially the entire conversion period ($1/f_s$) to settle from the step voltage. The signal should settle to within 1LSB of the final value in order not to introduce a significant error. The settling time requirement therefore places a lower limit on the filter's cutoff frequency. The single-pole filter settling time required to maintain a given accuracy is shown in Figure 6.10. The settling time requirement is expressed in terms of the filter time constant and also the ratio of the filter cutoff frequency, f_{c2}, to the ADC sampling frequency, f_s.

As an example, assume that the ADC is a 12-bit one sampling at 100kSPS. From the table in Figure 6.10, 8.32 time constants are required for the filter to settle to 12-bit accuracy, and

$$\frac{f_{c2}}{f_s} \geq 1.32, \text{ or } f_{c2} \geq 132\text{kSPS}.$$

While this filter will help prevent wideband noise from entering the SHA, *it does not provide the same function as the antialiasing filters at the input of each channel.*

The above analysis assumes that the multiplexer/PGA combined settling time is significantly less than the filter settling time. If this is not the case, then the filter cutoff frequency must be larger, and in most cases it should be left out entirely in favor of per-channel filters.

FILTERING IN A DATA ACQUISITION SYSTEM

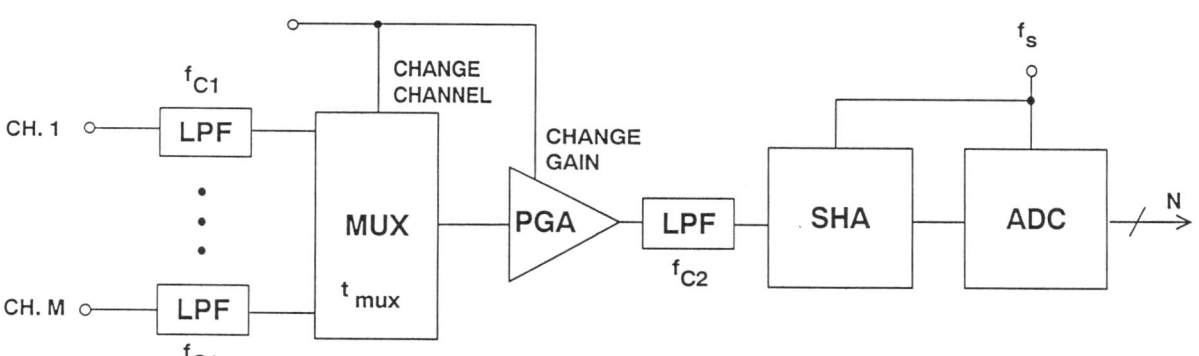

For Sequential Sampling, $f_{c1} < \dfrac{f_s}{2M}$

Figure 6.9

SINGLE-POLE FILTER SETTLING TIME TO REQUIRED ACCURACY

RESOLUTION, # OF BITS	LSB (%FS)	# OF TIME CONSTANTS	f_{c2}/f_s
6	1.563	4.16	0.67
8	0.391	5.55	0.89
10	0.0977	6.93	1.11
12	0.0244	8.32	1.32
14	0.0061	9.70	1.55
16	0.00153	11.09	1.77
18	0.00038	12.48	2.00
20	0.000095	13.86	2.22
22	0.000024	15.25	2.44

Figure 6.10

SHA AND ADC SETTLING TIME REQUIREMENTS IN MULTIPLEXED APPLICATIONS

We have discussed the importance of the fullscale settling time of the multiplexer/PGA/filter combination, but what is equally important is the ability of the ADC to acquire the final value of the step voltage input signal to the required accuracy. Failure of any link in the signal chain to settle will result in dc crosstalk between adjacent channels and loss of accuracy. If the data acquisition system uses a separate SHA and ADC, then the key specification to examine is the SHA *acquisition time*, which is usually specified as a the amount of time required to acquire a fullscale input signal to 0.1% accuracy (10-bits) or 0.01% accuracy (13-bits). In most cases, both 0.1% and 0.01% times are specified. If the SHA acquisition time is not specified for 0.01% accuracy or better, it should not be used in a 12-bit multiplexed application.

If the ADC is a sampling one (with internal SHA), the SHA acquisition time required to achieve a level of accuracy may still be specified, as in the case of the AD1674 (1µs to 12-bit accuracy). SHA acquisition time and accuracy are not directly specified for some sampling ADCs, so the *transient response* specification should be examined. The transient response of the ADC (settling time to within 1 LSB for a fullscale step input) must be less the $1/f_s$, where f_s is the ADC sampling rate. This often ignored specification may become the weakest link in the signal chain. In some cases neither the SHA acquisition time to specified accuracy nor the transient response specification may appear on the data sheet for the particular ADC, in which case it is probably not acceptable for multiplexed applications. Because of the difficulty in measuring and achieving better than 12-bit settling times using discrete components, the accuracy of most multiplexed data acquisition systems is limited to 12-bits. Designing multiplexed systems with greater accuracy is extremely difficult, and using a single ADC per channel should be strongly considered at higher resolutions.

SHA AND ADC CONSIDERATIONS IN MULTIPLEXED DATA ACQUISITION SYSTEMS

- Examine SHA Acquisition Time Specification to Required Accuracy :

    0.1% = 10-bits

    0.01% = 13-bits

- If Sampling ADC, SHA Acquisition Time may not be given, so examine Transient Response Specification

- Inadequate Settling Results in Loss of Accuracy and Causes DC Crosstalk Between Channels

- Multiplexing at greater than 12-bits Accuracy, or at Video Speeds is Extremely Difficult!

Figure 6.11

COMPLETE DATA ACQUISITION SYSTEMS ON A CHIP

VLSI mixed-signal processing allows the integration of large and complex data acquisition circuits on a single chip. Most signal conditioning circuits including multiplexers, PGAs, and SHAs, may now be manufactured on the same chip as the ADC. This high level of integration permits data acquisition systems to be specified and tested as a single complex function.

Such functionality relieves the designer of most of the burden of testing and calculating error budgets. The DC and AC characteristics of a complete data acquisition system are specified as a complete function, which removes the necessity of calculating performance from a collection of individual worst case device specifications. A complete monolithic system should achieve a higher performance at much lower cost than would be possible with a system built up from discrete functions. Furthermore, system calibration is easier and in fact many monolithic DASs are self calibrating.

With these high levels of integration, it is both easy and inexpensive to make many of the parameters of the device programmable. Parameters which can be programmed include gain, filter cutoff frequency, and even ADC resolution and conversion time, as well as the obvious digital/MUX functions of input channel selection, output data format, and unipolar/bipolar range.

The AD7890 is an example of a highly integrated monolithic data acquisition system. It has 8 multiplexed input channels, a SHA, an internal voltage reference, and a fast 12-bit ADC. Input scaling allows up to ±10V inputs when operating on a single +5V supply. Its block diagram is shown in Figure 6.12, and key specifications are summarized in Figure 6.13. Both AC and DC parameters are fully specified, simplifying the preparation of an error budget, and three types are available with three different standard input ranges:-

AD7890-10	±10 V
AD7890-5	0 to 5V
AD7890-2	0 to +2.5V

AD7890 8-CHANNEL, 12-BIT, 100kSPS COMPLETE DATA ACQUISITION SYSTEM

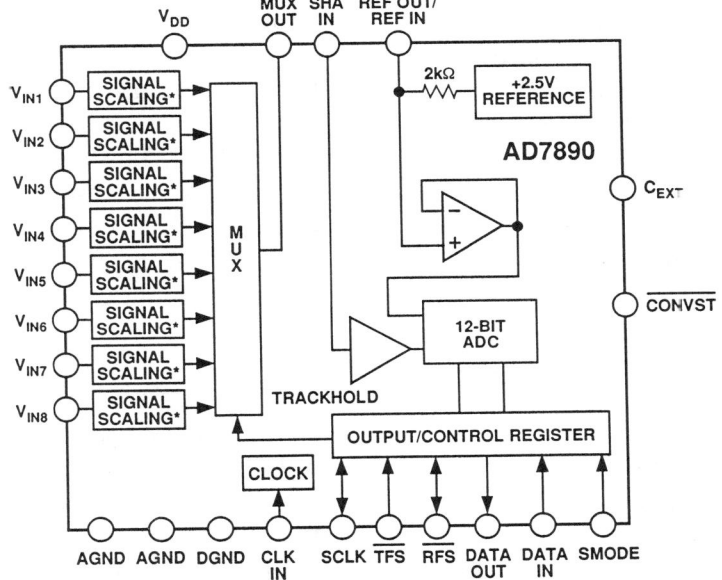

*NO SCALING ON AD7890-2

Figure 6.12

AD7890 SPECIFICATIONS

- ADC Conversion Time: 5.9μs
- SHA Acquisition Time: 2μs
- 117kSPS Throughput Rate (Includes 0.6μs Overhead)
- AC and DC Specifications
- Single +5V Operation
- Low Power Drain:
    - Operational: 30mW
    - Power Down Mode: 1mW
- Standard Input Ranges:
    - AD7890 - 10: ±10V
    - AD7890 - 5: 0 to +5V
    - AD7890 - 2: 0 to +2.5V

Figure 6.13

The input channel selection is via a serial input port. A total of 5 bits of data control the AD7890 via a serial port:- 3 address bits select the input channel, a CONV bit starts the A-D conversion, and 1 in the STBY register places the device in a power-down mode where its power consumption is under 1mW. All timing takes place on the chip and a single external capacitor controls the acquisition time of the internal track-and-hold. A-D conversion may also be initiated externally using the $\overline{\text{CONVST}}$ pin.

With the serial clock rate at its maximum of 10MHz, the achievable throughput rate for the AD7890 is 5.9μs (conversion time) plus 0.6μs (six serial clocks of internal overhead) plus 2μs (acquisition time). This results in a minimum throughput time of 8.5μs (equivalent to a throughput rate of 117kSPS). The AD7890 draws 30mW from a +5V supply.

The entire family of AD789X 12-bit data acquisition ADCs is shown in Figure 6.14. The AD7890 and AD7891 are complete 8-channel data acquisition systems, while the AD7892, AD7893, and AD7896 are designed for use on a single channel, or with an external multiplexer.

The AD785X 12-bit low power data acquisition ADCs have been designed and fully specified for either +3V or +5V operation. This family includes parallel and serial single and 8-channel versions. The devices have self or system calibration modes for offset, gain, and the internal SAR DAC.

MULTICHANNEL APPLICATIONS

AD789X SERIES OF 12-BIT ADCs FOR DATA ACQUISITION

MODEL	AD7890	AD7891	AD7892	AD7893	AD7896
T'Put (kSPS)	100	500	500/600	117	100
Pwr Supply	5V	5V	5V	5V	2.7 to 5V
Power	50mW	90mW	90mW	45mW	15mW @ 3.3V
Power Down	Yes	Yes	Yes	No	Yes
Interface	Serial	Parallel	Par / Ser	Serial	Serial
Channels	8	8	1	1	1
Pin Count	24	44	24	8	8

Figure 6.14

AD785X SERIES OF 3V / 5V 12-BIT ADCs

MODEL	AD7853 (AD7853L)	AD7854 (AD7854L)	AD7858 (AD7858L)	AD7859 (AD7859L)
T'Put (kSPS)	200 (100)	200 (100)	200 (100)	200 (100)
Pwr Supply	+3V, +5V	+3V, +5V	+3V, +5V	+3V, +5V
Power, +3V	15mW (5.5mW)	15mW (5.5mW)	15mW (5.5mW)	15mW (5.5mW)
Power Down	Yes	Yes	Yes	Yes
Interface	Serial	Parallel	Serial	Parallel
Channels	1	1	8	8
Pin Count	24	28	24	40 / 44

( ) VALUES FOR LOW POWER, L, VERSIONS

Figure 6.15

MULTICHANNEL APPLICATIONS

MULTIPLEXING INPUTS TO SIGMA-DELTA ADCS

As was discussed in Section 3, the digital filter is an integral part of a sigma-delta ADC. When the input to a sigma-delta ADC changes by a large step, the entire digital filter must fill with the new data before the output becomes valid, which is a slow process. This is why sigma-delta ADCs are sometimes said to be unsuitable for multi-channel multiplexed systems - they are not inherently so, but the time taken to change channels can be inconvenient.

As an example, the AD7710-family of ADCs contains an on-chip multiplexer (see Figure 6.16), and the digital filter (frequency response shown in Figure 6.17) requires three conversion cycles (300ms at a 10Hz throughput rate) to settle. It is thus possible to multiplex sigma-delta converters, provided adequate time is allowed for the internal digital filter to settle.

THE AD771X-SERIES PROVIDES A HIGH LEVEL OF INTEGRATION IN A 24-PIN PACKAGE

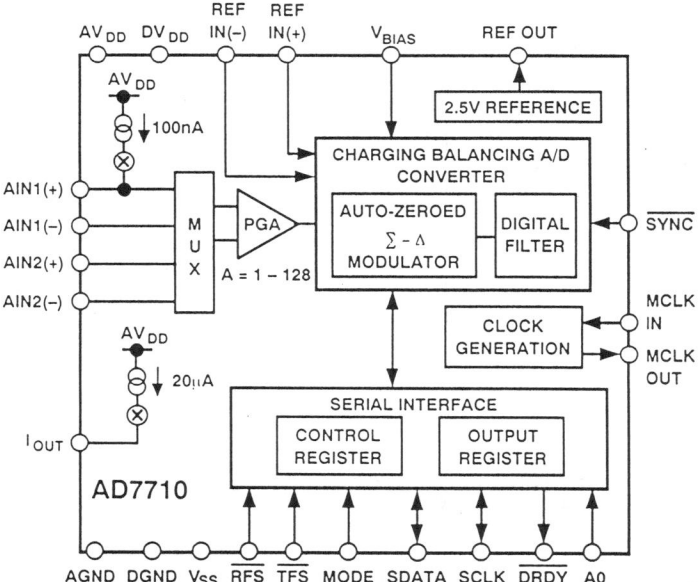

Figure 6.16

AD7710 DIGITAL FILTER FREQUENCY RESPONSE

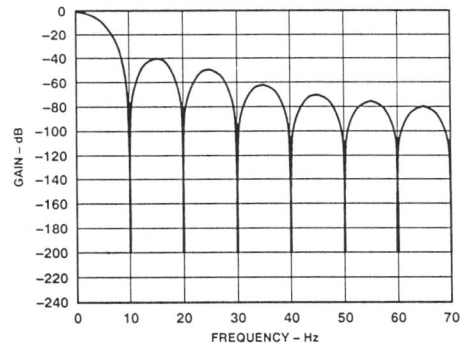

- Response Follows a $\text{sinc}^3 = \left(\dfrac{\sin x}{x}\right)^3$
- First Notch Frequency is Programmable and given by:
$$f_{notch} = \left(\dfrac{f_{clkin}}{512}\right)\left(\dfrac{1}{\text{Decimal Value of Digital Code}}\right)$$
- For $f_{clkin} = 10\text{MHz}$, $9.76\text{Hz} \le f_{notch} \le 1.028\text{kHz}$

Figure 6.17

THE RATE OF CONVERSION AND SETTLING TIME DEPENDS ON THE FILTER SETTING

	FILTER NOTCH FREQUENCY (Hz)								
	10	25	30	50	60	100	250	500	1k
CONVERSION TIME (ms)	100	40	33.3	20	16.7	10	4	2	1
MUX SWITCHING OR FULLSCALE WITH $\overline{\text{SYNC}}$, SETTLING TIME (ms)	300	120	100	60	50	30	12	6	3
ASYNCHRONOUS FULLSCALE SETTLING TIME (ms)	400	160	133.3	80	66.7	40	16	8	4

- Conversion Time = $\dfrac{1}{\text{Filter Notch Frequency}}$
- Digital Filter Requires Settling Time for Input Step Changes
- Use $\overline{\text{SYNC}}$ Input to Decrease Settling Time

Figure 6.18

MULTICHANNEL APPLICATIONS

In the case of the AD771X-series, four conversions must take place after a channel change before the output data is again valid (Figure 6.18). The $\overline{\text{SYNC}}$ input pin resets the digital filter and, if it is used, data is valid on the third output afterwards, saving one conversion cycle. When the internal multiplexer is switched, the $\overline{\text{SYNC}}$ is automatically operated.

If sigma-delta ADCs are used in multichannel applications, consider using one sigma-delta ADC per channel as shown in Figure 6.19. This eliminates the requirement for an analog multiplexer but requires that the outputs be synchronized in simultaneous sampling applications. Although the inputs are sampled at the same instant at a rate Kf_s, the decimated output frequency, f_s, is generally derived internally in each ADC by dividing the input sampling frequency by K (the oversampling rate). The output data must therefore be synchronized by the same clock at a frequency f_s.

SYNCHRONIZING MULTIPLE SIGMA-DELTA ADCs IN SIMULTANEOUS SAMPLING APPLICATIONS

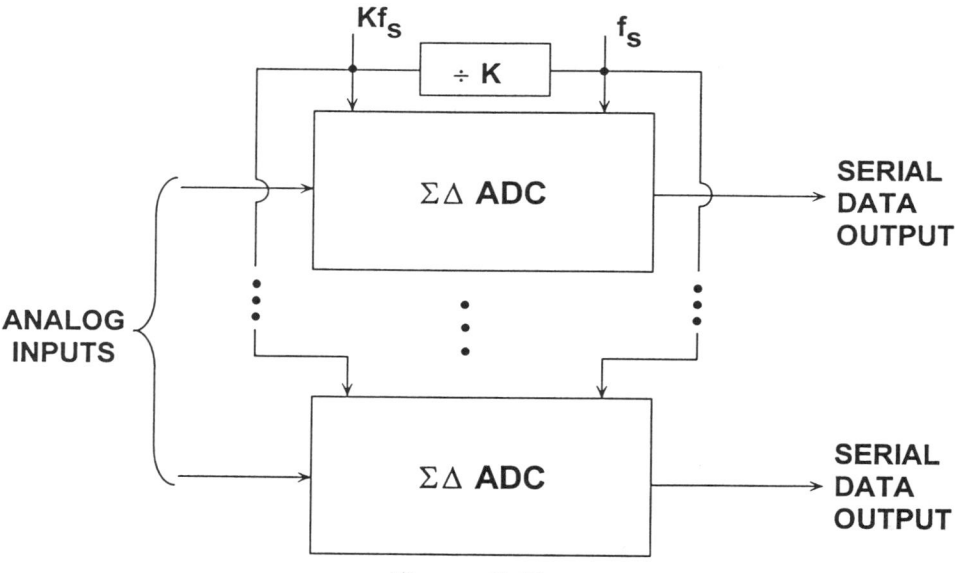

Figure 6.19

Products such as the AD7716 include multiple sigma-delta ADCs in a single IC, and provide the synchronization automatically. The AD7716 is a quad sigma-delta ADC with up to 22-bit resolution and an over-sampling rate of 570kSPS. A functional diagram of the AD7716 is shown in Figure 6.20 and some of its key features in Figure 6.21. The device does not have a "start conversion" control input, but samples continuously. The cutoff frequency of the digital filters (which may be changed during operation, but only at the cost of a loss of valid data for a short time while the filters clear) is programmed by data written to the DAS. The output register is updated at a rate which depends on the cutoff frequency chosen. The AD7716 contains an auto-zeroing system to minimize input offset drift.

AD7716 22-BIT QUAD SIGMA-DELTA ADC

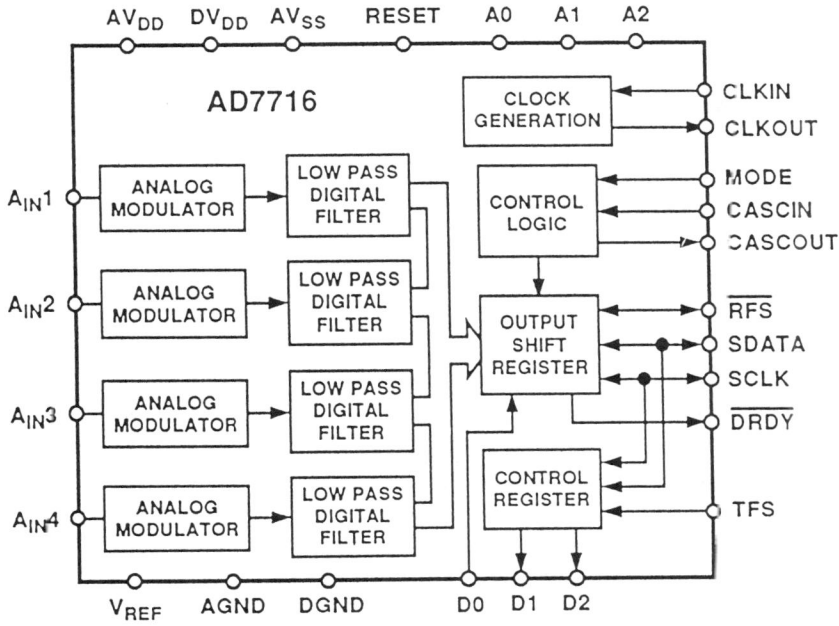

Figure 6.20

AD7716 QUAD SIGMA-DELTA ADC KEY FEATURES

- Up to 22-Bit Resolution, 4 Input Channels

- $\Sigma\Delta$ Architecture, 570kSPS Oversampling Rate

- On-Chip Lowpass Filter, Programmable from 36.5Hz to 584Hz

- Serial Input / Output Interface

- ±5V Power Supply Operation

- Low Power: 50mW

Figure 6.21

SIMULTANEOUS SAMPLING SYSTEMS

There are certain applications where it is desirable to sample a number of channels simultaneously such as in-phase and quadrature (I and Q) signal processing. A typical configuration is shown in Figure 6.22. Each channel requires its own filter and SHA. Each SHA is simultaneously placed in the *hold* mode by a common command signal. During the input SHAs' hold time the multiplexer is sequentially switched from channel to channel, and the single non-sampling ADC is used to digitize the signal on each channel. The maximum ADC sampling rate is the reciprocal of the sum of the multiplexer settling time, t_{mux}, and the ADC conversion time, t_{conv}.

$$f_{s2} \leq \frac{1}{t_{mux} + t_{conv}}$$

The maximum per-channel sampling frequency is determined by M, t_{mux}, t_{conv}, and the acquisition time of the simultaneous SHAs, t_{acq1}.

$$f_{s1} \leq \frac{1}{t_{acq1} + M(t_{mux} + t_{conv})}$$

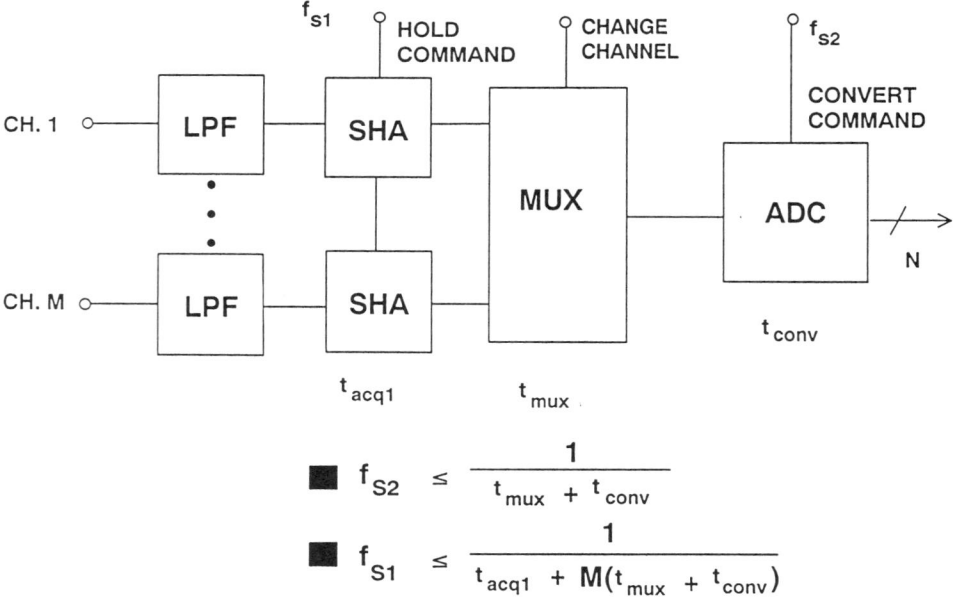

Figure 6.22

MULTICHANNEL APPLICATIONS

If a sampling ADC is used to perform the conversion (see Figure 6.23), the acquisition time of the second SHA, t_{acq2}, must be considered in determining the maximum ADC sampling rate, f_{s2}. The multiplexer should be switched to the next channel after the single SHA goes into the hold mode. If the multiplexer settling time is less than the ADC conversion time, then the maximum ADC sampling rate f_{s2} is the reciprocal of the sum of the SHA acquisition time and the ADC conversion time.

$$f_{s2} \leq \frac{1}{t_{acq2} + t_{conv}}$$

The maximum input sampling frequency is less than this value divided by M, where M is the number of channels. Additional timing overhead (t_{acq1}) is required for the simultaneous SHAs to acquire the signals.

$$f_{s1} < \frac{1}{t_{acq1} + M(t_{conv} + t_{acq2})}$$

SIMULTANEOUS SAMPLING DATA ACQUISITION SYSTEM USING SAMPLING ADC

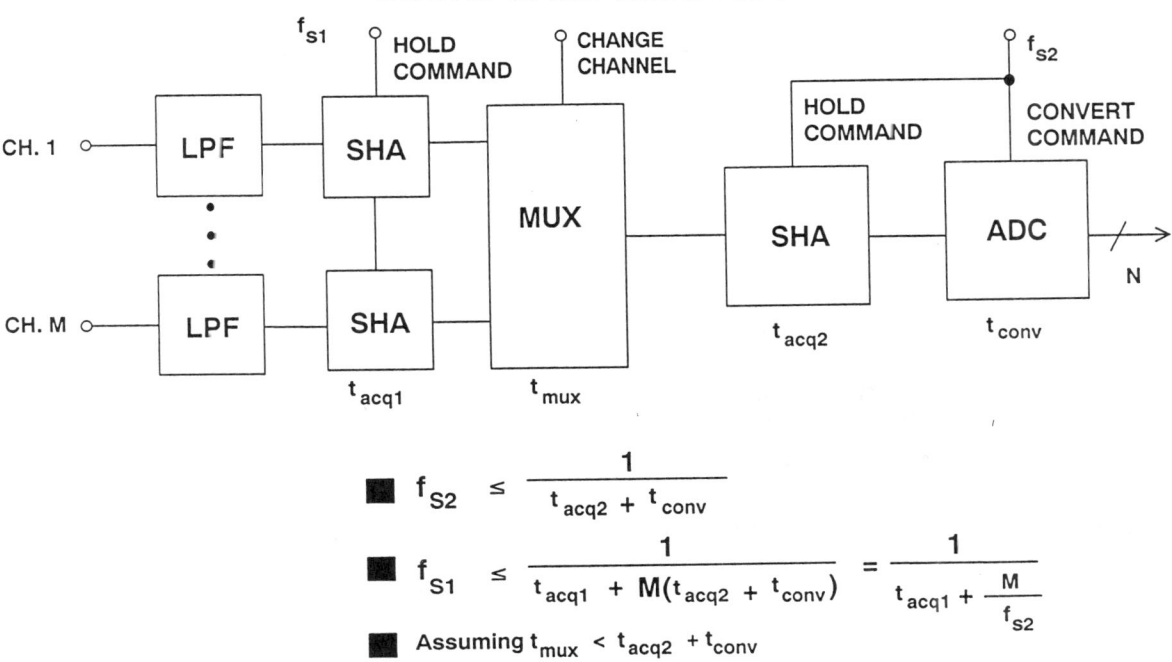

Figure 6.23

Using Multiple-DACs to Simplify Data Distribution Systems
Wes Freeman

In many industrial and process control applications, multiple programmable voltage sources are required. Traditionally, these applications have required a large number of components, but recent product developments have greatly reduced the parts count without compromising performance.

Multiple voltage outputs can either be derived by demultiplexing the output of a single DAC or by employing multiple DACs. These two approaches are shown in Figure 6.25. In the demultiplexed circuit, one DAC feeds the inputs of several sample-and-hold amplifiers (SHA). The equivalent digital value for the analog output is applied to the DAC, and the appropriate SHA is selected. After the DAC settling time and SHA acquisition time requirements have been met, the SHA can be deselected and the next channel updated. Once a SHA is deselected, the output voltage will begin to droop at a rate specified for the SHA. Thus, the SHA must be refreshed before the output voltage droop exceeds the required accuracy (typically 1/2 LSB).

DATA DISTRIBUTION SYSTEM USING MULTIPLE DACs

- Many systems require multiple, programmable voltages
    - Automatic Test Equipment (ATE)
    - Robotics
    - Industrial Automation
    - System Calibration
    - Ultrasound/Sonar Power Gain and Receiver Level

Figure 6.24

OPTIONS FOR ANALOG DATA DISTRIBUTION

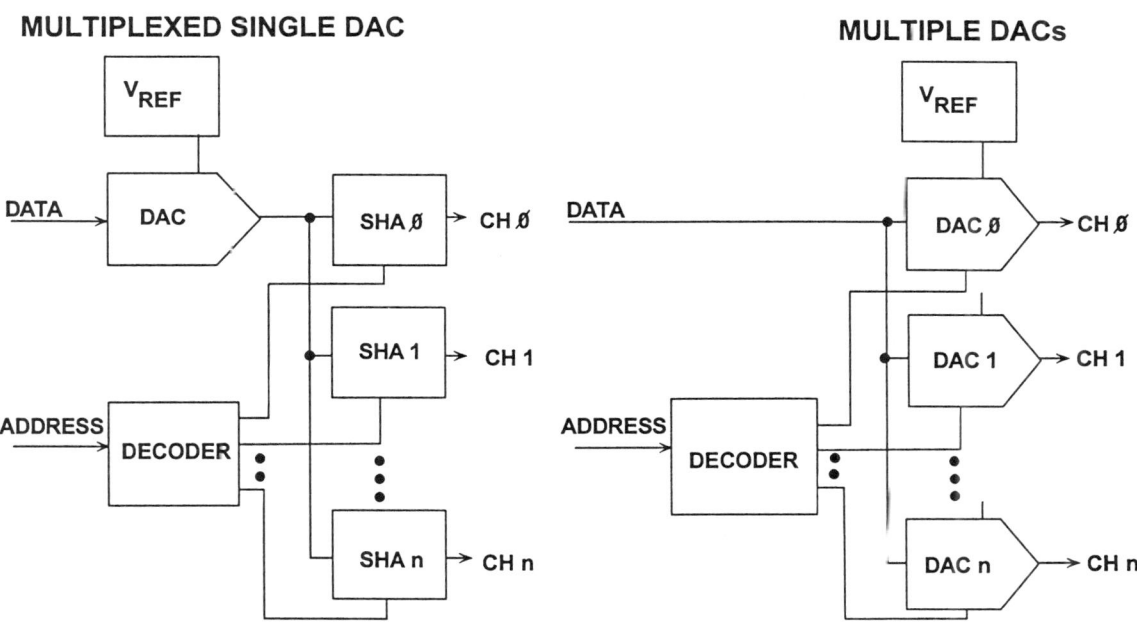

Figure 6.25

The multiple DAC application is straightforward. One DAC is provided for each channel, and an address decoder simply selects the appropriate DAC. No refresh is required.

The DAC plus SHA system evolved because, in the past, DACs were more expensive than SHAs. This situation was particularly true for DACs with resolution above 8 bits. In addition, multiple-SHAs with on-chip hold capacitors reduced the parts count, printed circuit board area, and cost of demultiplexed DAC systems. Finally, the demultiplexed DAC only requires one calibration step, since the same DAC provides the output voltage for each of the output channels. Of course, single-calibration is only valid if the SHA does not introduce unacceptable errors. For example, the SMP08 is an 8-channel SHA which exhibits a 10mV maximum offset voltage and is accurate to 1/2 LSB when demultiplexing an 8-bit, 5V full-scale DAC.

MULTICHANNEL APPLICATIONS

WHY DEMULTIPLEX A SINGLE DAC?

- Cost of DAC > Cost of SHA
- Multichannel SHAs (e.g. SMP-08/SMP-18) Reduce Parts Count
- Only One Calibration Required

Figure 6.26

A 16-CHANNEL 8-BIT DATA DISTRIBUTION SYSTEM USING A SINGLE MULTIPLEXED DAC

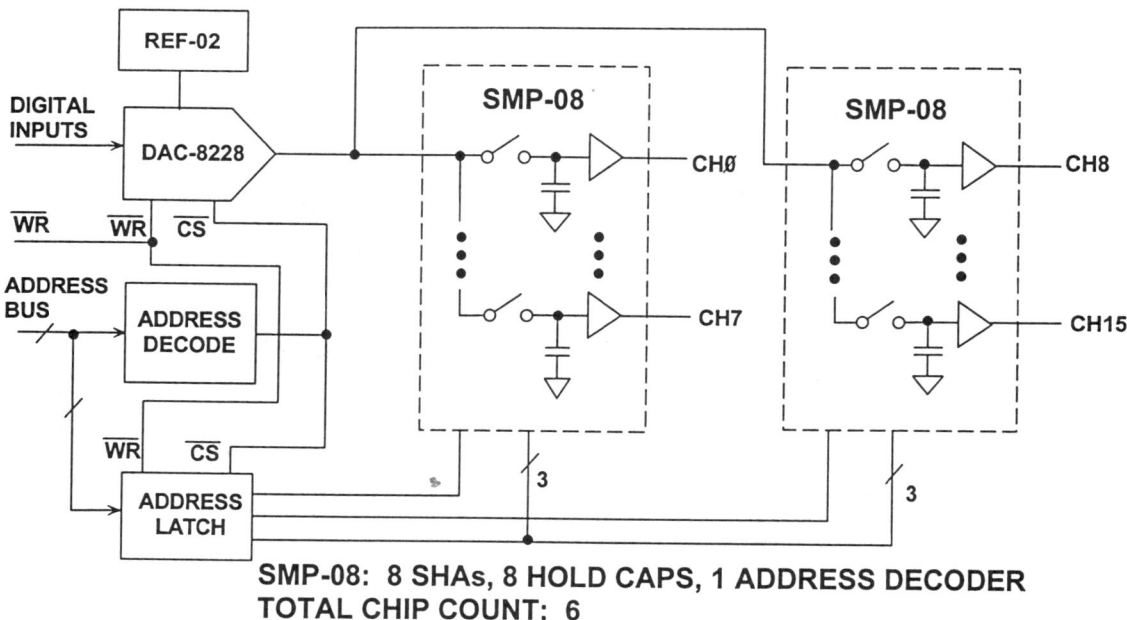

SMP-08: 8 SHAs, 8 HOLD CAPS, 1 ADDRESS DECODER
TOTAL CHIP COUNT: 6

Figure 6.27

Multichannel Applications

A representative 16-channel, 8-bit demultiplexed data distribution system is shown in Figure 6.27. The DAC8228 produces a voltage output which is applied to the input of 16 SHAs. The DAC digital value is written into the DAC8228 at the same time as the channel address is presented to the SHA. After the SHA's acquisition time requirement has been satisfied, the next channel can be refreshed. The SMP08's acquisition time to 0.1% is 7µs, so the maximum data transfer rate is the reciprocal of the acquisition time, or 140kHz.

Once a SHA channel is deselected, the input bias current of the amplifier begins to discharge the hold capacitor. The rate at which the capacitor voltage changes is specified on the SHA data sheet as the *droop rate*. With a maximum droop rate of 20mV/s, the SMP08 can maintain 8-bit accuracy (1/2 LSB at 5V full scale) provided refreshing occurs once each 500ms.

The SMP08 is a complete, single-chip 8-channel SHA, which incorporates analog switches, hold capacitors, amplifiers, and address decoders. Each SMP08 replaces 17 components (8 SHAs, 8 hold capacitors, and one decoder). Even with this level of integration, however, the complete data distribution circuit still requires six components.

Operation of the demultiplexed DAC is shown in Figure 6.28. With the DAC output at 5V, the system refreshes CH0 once each second. The upper trace is the address decode output which goes low to select CH0. The lower trace is the output of CH0's SHA. While the address input is low, the DAC's output is connected to the input of CH0's SHA. When the address input goes high, the SMP08 maintains less than 1/4 LSB error for one second. Thus, the droop rate of this particular SMP08 is about 2mV/sec. The photo also shows a 1mV hold step when switching from sample to hold mode.

At first glance, demultiplexed DAC systems may appear to be more cost-effective than multiple DACs. However, several factors combine to reduce the system cost when a multiple DAC is used (Figure 6.29). For example, advances in integrated circuit fabrication, trimming and testing have combined to improve yields and reduce costs. Process improvements produce transistors and resistors with better matching, which reduces trimming requirements. At the same time, improvements in laser trimming and testing permit more accurate matching of multiple devices on one die. While these advances also have an effect on the costs of demultiplexed systems, the economics of IC fabrication are such that the impact of improvements is relatively greater on multiple DAC devices. As a result of these improvements, multiple DACs are very cost competitive with demultiplexed systems on a per-channel basis.

MULTICHANNEL APPLICATIONS

REFRESHING THE MULTIPLEXED DAC

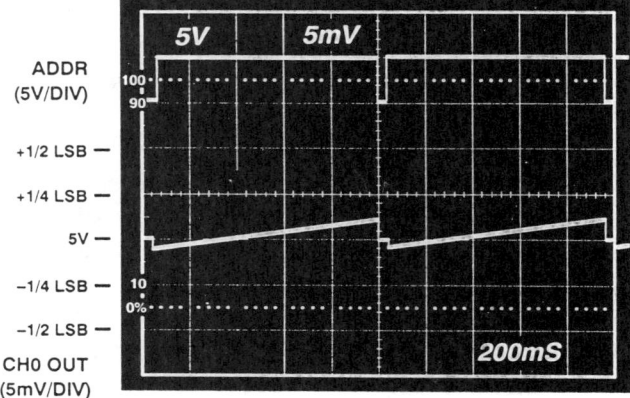

ADDR (5V/DIV)

CH0 OUT (5mV/DIV)

Upper Trace: $\overline{\text{CS}}$ of CH0 (Output of Address Decoder)
Lower Trace: Output of SHA for CH0

For 1/2 LSB Accuracy, Minimum Refresh Time is: $\dfrac{V_{FS}/2}{2^N \times \text{Droop Rate}}$

Where N is the DAC resolution in bits, V_{FS} is the DAC Full Scale Voltage

Figure 6.28

WHY USE A MULTIPLE-DAC SOLUTION?

- ■ Cost
  - ◆ Laser Trimming Aids Matching
  - ◆ Lower Parts Count
  - ◆ Reduced Design Time
- ■ Additional System Features
  - ◆ Less Microcontroller Overhead
  - ◆ Faster Update Rate
  - ◆ Synchronized Outputs
  - ◆ Power-On Reset
  - ◆ Read-Back Capability
  - ◆ Serial Interface

Figure 6.29

Another advantage of the multiple-DAC system is reduced parts count. While cost estimates may vary, the advantages of eliminating components in a design cannot be ignored. Among these advantages are reduced pin count, printed circuit board area, inventory, incoming inspection, and purchasing transactions.

Reducing design time also contributes to cost savings. From a design engineering point of view, design time is both a critical and a very visible factor in the "time-to-market" equation. Reducing parts count can have a major impact on design time by reducing device evaluation time, interface timing analyses, error budget analyses, printed circuit board layout, etc.

The real advantages of the single-chip multiple DAC, however, lie in the features that are difficult or impossible to add with a demultiplexed circuit. These advantages include:

1. *No refresh cycle is required.* As previously mentioned, the demultiplexed DAC must constantly be refreshed. The minimum time between refresh cycles is set by the required system accuracy (typically 1/2 LSB), SHA droop rate, and LSB voltage value. Specifically,

$$\text{Minimum Refresh Time} = \frac{V_{FS}/2}{2^N \times \text{Droop Rate}}$$

where V_{FS} = DAC full scale voltage and
N = DAC resolution in bits.

Constantly refreshing a demultiplexed DAC puts a software burden on the system. In particular, notice that the refresh time is halved for each additional bit of resolution. For example, if the DAC resolution is increased from 8-bits to 12-bits the DAC must be refreshed 16 times more often. Software burden vs. hardware savings tradeoffs must be evaluated carefully in high resolution systems.

2. *Faster data update rates are possible.* In addition to requiring constant refresh, the rate at which the multiplexed DAC can be updated is limited by the sequential architecture of the system. Thus, the acquisition time of the SHA is multiplied by the number of channels. For the circuit of Figure 6.27, the best-case time for updating all 16 DAC values, assuming a 7µs SHA acquisition time, will be:

$$t_s = 16 \cdot 7\mu s = 112\mu s$$

Multiple DACs, on the other hand, can usually be treated as memory or input/output locations, and updated at or near the maximum cycle time of the microcontroller. Consecutive DACs can be loaded with data while previously-loaded DACs are settling to their final values. Since the digital transfer rate is usually faster than the settling time of the DAC, the multiple-DAC system update rate is much faster than the demultiplexed DAC rate.

3. *Multiple DACs can offer synchronized outputs.* Servo, ATE, and other systems can benefit from multiple outputs which change simultaneously. Many multiple DACs, such as the quad 8-bit AD7225 and octal 12-bit AD7568, are double-buffered. Data can be loaded into storage latches sequentially, then transferred simultaneously to the DAC latches when a separate "load" pin is brought low. For higher throughput, new data can be loaded into the storage latches as soon as the "load" pin returns high. In this case, the new data is loaded during the settling time of the DACs, so the throughput rate is maximized.

4. *Multiple DACs usually have a power-on reset feature.* Many systems must assume a known state upon power-up. For example, a programmable power supply should not output random voltages when turned on. Several DACs, such as the quad 12-bit DAC8412/DAC8413 and octal 8-bit DAC8800, provide a reset feature which sets all of the DACs to a known state. In most cases, the DAC output is reset to zero. The DAC8412, however, will reset to mid-scale. This feature provides a zero-volt output at reset when the DAC is configured for a bipolar output.

5. *Some multiple DACs' registers feature data readback.* On some multiple DACs, the data bus is actually bi-directional, and the value written into the DAC register can also be read back by the controller. This feature provides the opportunity for hardware and software error checking. Since the DAC values do not have to be saved by the controller, additional data storage is also available for minimum-memory microcomputer applications.

6. *To save real estate and parts count, some multiple DACs use serial data interfaces.* Only two or three pins are required, at both the controller and the DAC, to transfer data. Creating a serial data interface for a demultiplexed-DAC system requires several additional packages to decode the serial address, even if a serial-input DAC is specified. Dramatic reductions in pin count can be obtained when a multiple DAC with serial input is specified. For example, the DAC8420 provides four, 12-bit voltage output DACs in one 16-pin package.

A serial data interface is inherently slower than a parallel interface. While this is usually a problem only in high speed systems, the slower serial data rate does exacerbate the demultiplexed-DAC's refresh requirements.

A two chip solution to the data distribution challenge is shown in Figure 6.30. The AD8600 is a multiple-channel DAC which combines 32 registers, 16 DACs, 16 voltage output buffers, control logic, and an address decoder in a single 44-pin package.

A 16-CHANNEL 8-BIT DATA DISTRIBUTION SYSTEM USING A 16-CHANNEL DAC

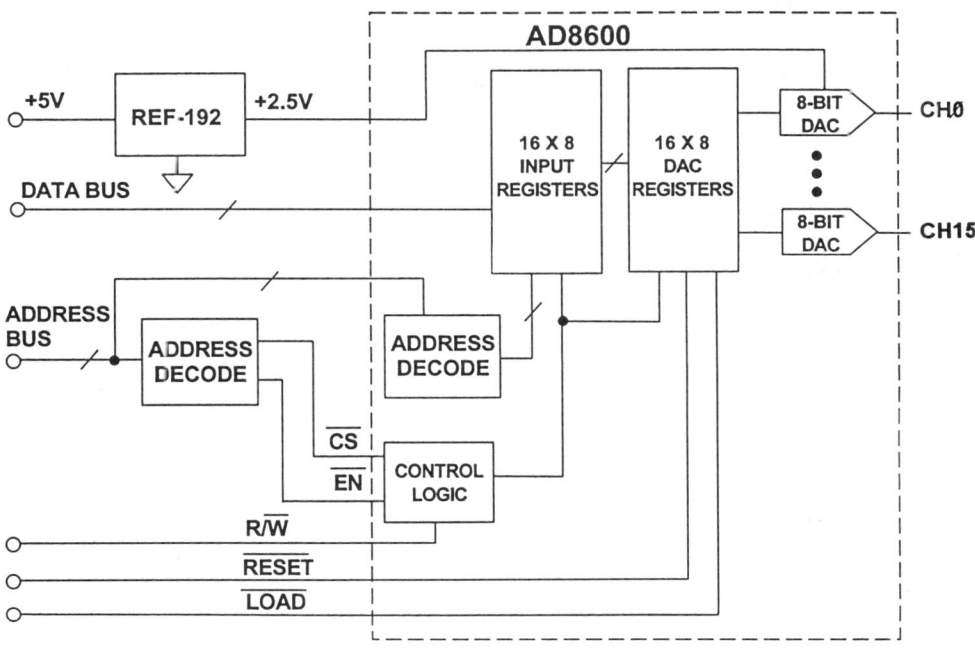

Figure 6.30

Interfacing to the AD8600 is easy. The DACs are simply treated as 16 memory or I/O locations. Updating the value of any DAC is merely a matter of writing the DAC value, via an 8-bit data bus, to the appropriate address. If desired, data can be written into the input registers without affecting the DAC values. When the $\overline{\text{LOAD}}$ input is pulled low, all DAC outputs will change simultaneously. Holding $\overline{\text{LOAD}}$ low causes the output of each DAC to change as soon as the $\overline{\text{WRITE}}$ command occurs.

This 16-channel, single chip DAC reduces parts count by 66% over the demultiplexed DAC solution. Inventory and assembly costs, printed circuit board area, and design time are all reduced. Refreshing the DAC values is not required, so software overhead is eliminated.

When evaluated solely on the cost of components, the demultiplexed single DAC approach is about 40% less expensive than the multiple DAC. On a system cost basis, however, the advantages of the multiple DAC make it the cost as well as the performance winner. A comparison of the two approaches is shown in Figure 6.31. The cost advantages of the DAC8300 solution include:

COMPARING REPRESENTATIVE 16-CHANNEL, 8-BIT DATA DISTRIBUTION SYSTEMS

FEATURE	SINGLE DEMULTIPLEXED DAC	MULTIPLE DAC
Design Time	Moderate	Low
Refresh Required	Yes	No
Refresh Ripple at Output	Dependent on Refresh Rate	None
Synchronized Outputs	No	Yes
Time Required to Change All 16 DACs	112 µs	3.3 µs
Power-On Reset	No	Yes
Read-Back Capability	No	Yes (Available on some)
Serial Data Interface	Additional Components	Yes, on Some DACs
Parts Count	6	2
PC Board Holes (Total Number of Pins)	74	52

Figure 6.31

1. *Reduced design time.* All of the DAC system specifications are defined in the multiple DAC data sheet. The demultiplexed DAC, on the other hand, requires analysis of the DC and AC parameters of both the DAC and SHA, as well as the relationship between these parameters. (For example, must the DAC settling time be added to the SHA acquisition time, or, since the DAC settling time is only 2µs compared to 7µs for the SHA, can the DAC settling time be ignored? Updating each channel in 7µs instead of 9µs reduces refresh overhead by about 20%, but this is only valid if system specifications are met. Clearly, prototype evaluation is required).

2. *Parts count is reduced by 66%.* As previously mentioned, costs for purchasing, inventory and assembly are reduced.

3. *The number of PC board holes (that is, the pin count) is reduced.* The number of holes that are required in the PC board is another measure of assembly cost. On this basis, the multiple DAC wins by about 30%

4. *No refresh is required.* This fact eliminates the software required to provide timing for the demultiplexed DAC, and may also free up the use of a timer on the microcontroller.

Although the cost savings listed above are significant, the multiple DAC's improvements to system performance are even more important in most applications. The system improvements include:

1. *Reduced output ripple.* The demultiplexed DAC output will always

have ripple, caused by the droop rate of the SHA.

2. *Faster update rate.* The delay time imposed by the SHA acquisition time is eliminated, so the AD8600 circuit can load new DAC values in as little as 80ns. To update 16 DACs, including an 80ns load pulse to change all DAC outputs simultaneously, and allowing 2µs for DAC settling, requires:

$$t_S = (16 + 1) \cdot 80\text{ns} + 2\mu\text{s} = 3.3\mu\text{s}.$$

The demultiplexed DAC, as was shown previously, is limited by the fact that the acquisition time of the SHA is multiplied by the sequential architecture of the system. The multiple DAC has reduced the refresh time from 112µs to only 3.3µs. In applications such as automatic test equipment, the effects of a 3300% improvement in test value update time are significant. The demultiplexed DAC circuit can be improved somewhat by specifying faster SHAs, but the increased droop rate of a faster SHA will demand that refresh be performed even more frequently. The increased refresh may offset the speed advantage of the faster SHA.

3. *Simultaneous DAC output changes.* With double-buffered data latches, each of the DAC values can be loaded sequentially, but the outputs will change simultaneously. If desired, of course, any of the DACs can be updated individually.

4. *A DAC reset input.* At system power up, or during power fault recovery, setting the reset input to a logic low will force all DAC registers into the zero state. This will asynchronously place zero-volts on all of the DAC outputs.

5. *Data readback.* The DAC has a bi-directional data bus, so that the value written into the DAC register can also be read back by the controller. This feature provides the opportunity for hardware and software error checking and can be especially useful during system debugging. Since the DAC values do not have to be saved in memory by the controller, an additional 16 bytes of data storage is also available for minimum-memory microcomputer applications.

Multiple DACs are offered in a wide variety of configurations, resolution, digital interface, and analog output. For example, dual, quad, and octal versions of both 8-bit and 12-bit devices are available. Serial data interfaces are also available, for reduced pin count. A 1994 selection guide (Reference 5) from Analog Devices contains 12 multiple DAC products which have a serial data interface.

Many multiple DACs are used in automatic calibration or system nulling applications. A typical device for these tasks is the AD8842, an octal 8-bit TrimDAC® (Figure 6.32). This device includes a serial data input for DAC data and address, eight DAC registers, and eight voltage-output DACs.

AD8842 OCTAL 8-BIT SERIAL INPUT TrimDAC®

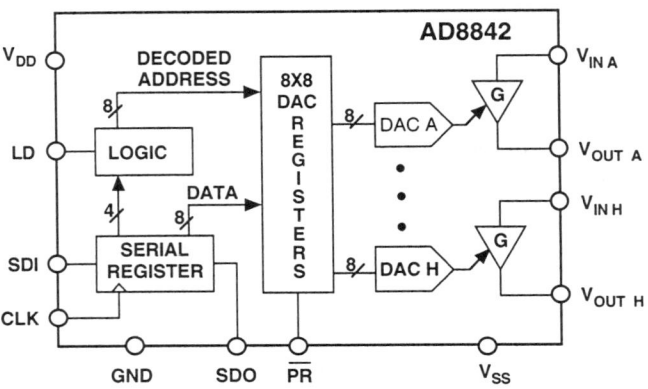

CURRENT-CONVEYOR
IMPLEMENTATION OF
MULTIPLYING DAC
CHANNEL

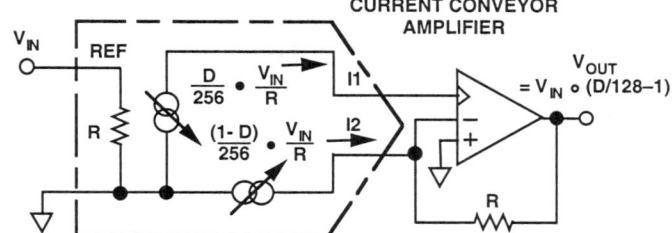

Figure 6.32

This device introduces a capability which is impossible with a demultiplexed DAC: *the ability to have different reference voltages for each DAC channel.* The output voltage of each channel of the AD8842 is simply:

$$V_{OUT} = V_{IN} \cdot \left(\frac{D}{128} - 1\right)$$

where D is a representation of the DAC input value (that is, 0 to 255 for an 8-bit DAC). The value of V_{IN} can be either positive or negative, DC or AC, allowing each DAC a 4-quadrant multiplying capability. Since V_{IN} can be an AC signal, the DACs can be also be used as a programmable gain control for signals up to 50kHz. V_{IN}, the analog input, exhibits a fixed input resistance of about 20kΩ, so driving it, either from a DC voltage reference or from an AC source, is easy.

The AD8842 is an example of the reduction in pin count which is possible when a serial data interface is used. It combines 8 voltage output DACs, under 3-wire serial interface and an asynchronous preset input, into a single 24-pin DIP or SOIC package. To update any DAC, the DAC output value and address are shifted into the serial input register. A logic LOW on the LD input then transfers the data to the appropriate DAC register. The last bit of the serial input shift register is also available on the serial data output, so that multiple AD8842s can be daisy-chained without additional logic. Although the digital data can be loaded at an 8MHz rate, each DAC output requires 5.4µs to settle to ± 1 LSB.

Other key features of the AD8842 are shown in Figure 6.33. The preset (PR) input can be used to force the outputs of all the DACs to 0V when power is first applied, or after a power fault. Placing a logic low on PR will force a value of 128_{10} (80_{16}) into all of the DAC registers. As can be seen from the output voltage equation shown in Figure 6.32, a DAC value of 128_{10} produces a 0V output.

AD8842 KEY FEATURES

- Eight Individual Channels
- 3-Wire Serial Data Input, 8MHz Loading Rate
- Asynchronous Reset Input
- 50kHz, 4-Quadrant Multiplying Bandwidth
- Replaces 8 Potentiometers
- Constant 20kΩ Input Resistance
- ±4V Output Swing
- Low Power: 95mW on ±5V Supplies
- 24-pin Narrow DIP or SO Package

Figure 6.33

The AD8842 TrimDAC® can also replace 8 potentiometers in voltage nulling applications. Unlike a trimming potentiometer, however, the TrimDAC® can generate both positive and negative voltages from a positive reference voltage. This is a significant advantage in many applications, because the nulling voltage can be derived from a very stable system reference voltage. Obtaining a bipolar nulling voltage with a potentiometer normally requires either providing two separate positive and negative reference voltages, or connecting one end of the pot to the negative power supply. Since power supplies are typically noisy and poorly regulated when compared to the system reference voltage, performance will be degraded if the power supply must be used.

The effect of a serial interface on pin count becomes even more evident as

MULTICHANNEL APPLICATIONS

the DAC resolution increases. For example, the AD8522 squeezes two 12-bit DACs, a 2.5V reference, a double-buffered serial input, and two reset control inputs into a single 14-pin package (Figure 6.34). A companion part, the AD8582, has similar features but requires 24 pins because it has a parallel interface.

The AD8522's 2.5V reference is available on the V_{REF} pin to provide a reference for other data acquisition portions of the system or for ratiometric applications. The reference output can also be used to create a virtual ground for applications where a quasi-bipolar output is desired. The V_{REF} pin cannot, however, be used as a reference input.

As with the AD8842 octal DAC, both AD8522 DAC registers can be asynchronously forced to a half-scale value (2048_{10} or 800_{16}). This action sets a 0V reading for quasi-bipolar applications. Both registers of the AD8522 can also be reset to a "000" value, which provides a 0V reset for unipolar applications. Other key features of the AD8522 are shown in Figure 6.35.

AD8522: WORLD'S SMALLEST (SO-14, PDIP-14) COMPLETE DUAL 12-BIT DAC

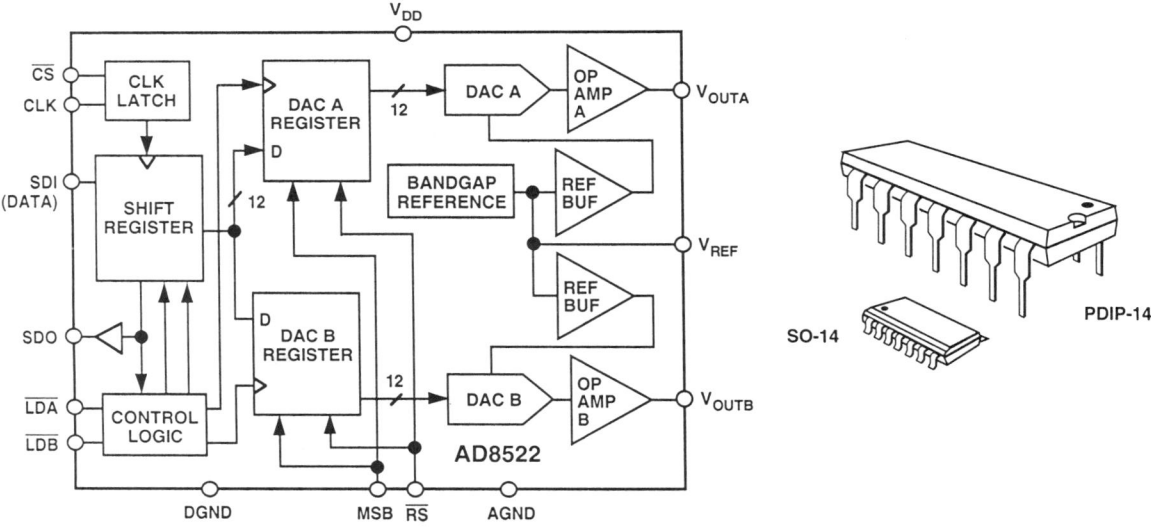

Figure 6.34

AD8522 KEY FEATURES

- Space-Saving SOIC-14 Package, only 1.5mm Height!
- Low Power: 10mW max
- No External Components, No Adjustments Necessary
- +5V Operation Guaranteed Down to 4.5V Minimum
- 4.095V Full Scale (1mV/LSB), Fully Trimmed
- Buffered Rail-to-Rail Voltage Outputs
- 2.5V VREF Output Pin - Useful for establishing virtual ground in bipolar applications
- Midscale or Zero-Scale Preset
- Double-Buffered 3-Wire Serial Data Input
- Software and Hardware A/B DAC Select

Figure 6.35

The AD8522 introduces two new concepts to the multi-channel data distribution discussion. The first new concept is the *complete* DAC (Figure 6.36). All of the systems mentioned previously have required a separate external reference, which had to be adjusted to set the DAC's full scale output voltage. The AD8522, on the other hand, has an on-chip bandgap reference which is laser-trimmed during production to provide a full scale output voltage of 4.095V (that is, 1mV/LSB). Since the voltage reference, DAC, and voltage output amplifiers are on a single chip, the entire DAC system specification is contained in the AD8522's data sheet specifications. This eliminates the necessity of evaluating several separate devices, as well as calculating the error contributions of each device, and further demonstrates the significant reduction in design time which results from reducing package count. In addition, system cost is reduced and reliability is improved because the calibration operation is eliminated.

EQUIVALENT SCHEMATIC OF AD8522 ANALOG SECTION

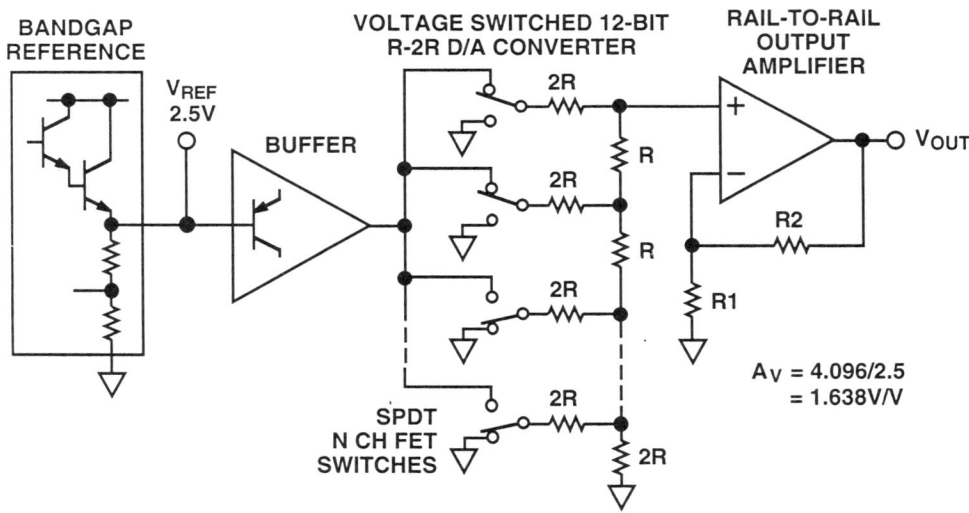

Figure 6.36

The other concept which has not been discussed previously is low-voltage, single supply operation. It is at this point that demultiplexed DAC systems rapidly become impractical. One effect of a low supply voltage is that the value of the least significant bit must be reduced. For single +5V operation, the practical limit on full scale voltage at 12-bits is typically 4.096V. This yields an LSB value of 1mV for a 12-bit DAC, which means that the total SHA error budget, including offset voltage, droop rate and hold step, must not exceed ±500µV in order to limit errors to 1/2 LSB.

Improving the droop rate and hold step errors of a multi-channel SHA is difficult because of the limited size of the on-chip capacitors. As the value of the hold capacitors increase, die size and cost rise rapidly. Adding individual SHAs with external hold capacitors is possible, but rapidly increases the component count. Again, the multi-channel DAC is a superior solution.

Another requirement for single-supply, low voltage operation is rail-to-rail operation. Single-supply bipolar amplifiers, using common-emitter output stages, are limited in some low voltage applications because the saturation voltage of the output transistors limits output voltage swing. The AD8522 employs P-channel and N-channel MOSFETs (Figure 6.37) to provide wide output voltage swing while operating from a single +5 V supply. The output of this type of stage (at a supply rail) looks like the on-resistance of the P or N-channel FET connected to the rail. Obviously, this on-resistance begins to limit the output swing as the output load current is increased.

MULTICHANNEL APPLICATIONS

AD8522 RAIL-TO-RAIL OUTPUT PERFORMANCE YIELDS 5mA WITH 60mA SHORT CIRCUIT CURRENT

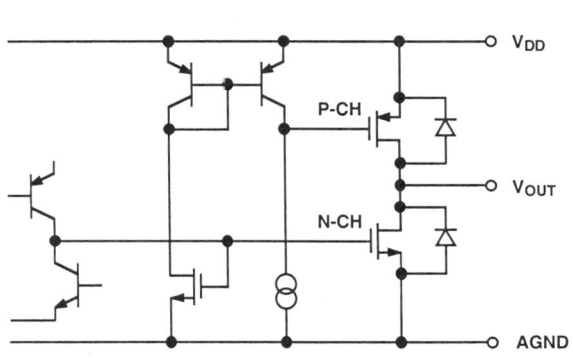
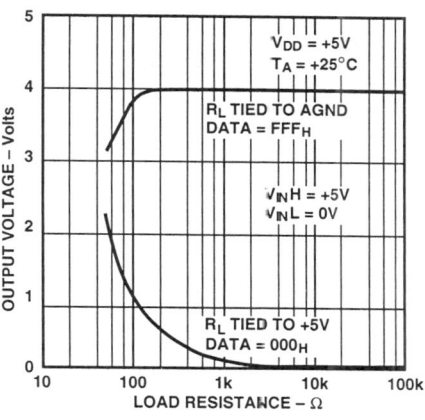

Figure 6.37

The effect of a serial interface on pin count is demonstrated by comparing the AD8522 with the similar AD8582.

The latter part has a 12-bit parallel interface, and requires 24 pins versus 14 pins for the serial data version.

AD8582 COMPLETE DUAL 12-BIT SINGLE +5V SUPPLY DAC

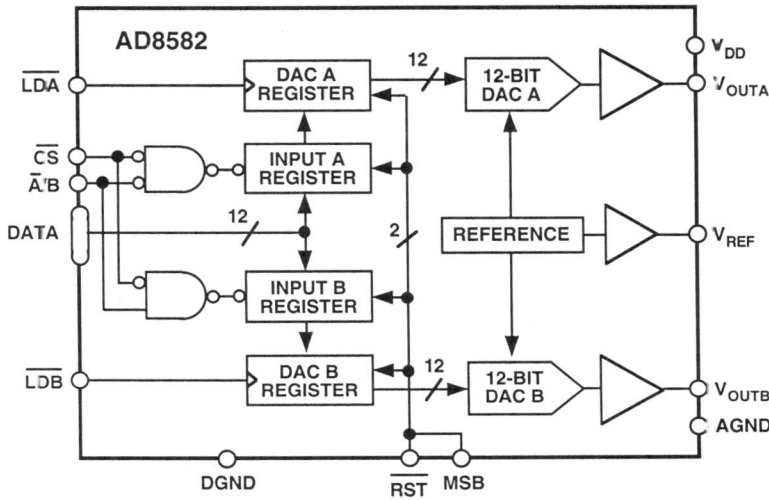

Figure 6.38

MULTICHANNEL APPLICATIONS

One significant advantage of the parallel interface is, of course, higher speed. The parallel device can update both 12-bit DACs in 100ns. The serial data version, on the other hand, requires 32 clock cycles to enter the data for two DACs. With a maximum clock frequency of 14MHz, this results in a data update period of 2.2µs. The parallel-data version also includes complete dual-rank data latches, so that both DACs can be updated simultaneously (Figure 6.39).

AD8582 KEY FEATURES

- Complete Dual 12-bit DAC
- No External Components
- Single +5V Operation
- 1mV/bit with 4.095V Full Scale
- True Voltage Output, ±5mA Drive
- Parallel Input Register with Fast 30ns Chip Select
- Double-Buffered for Simultaneous A and B Output Update
- Reset Pin Forces Output to Zero Volts or half Scale, Depending on MSB Pin
- Low Power: 5mW

Figure 6.39

The logical extension of a serial data interface and low voltage CMOS technology is expressed in Figure 6.40. While not a multi-channel device, the AD8300 does pack a complete 12-bit voltage output DAC into a single 8-pin package. No external components are required, except for supply bypass capacitors. Since it is capable of operating from a single 3V supply, this device is ideal for battery-powered applications.

MULTICHANNEL APPLICATIONS

A 12-BIT, +3V DAC SYSTEM BASED ON AD8300

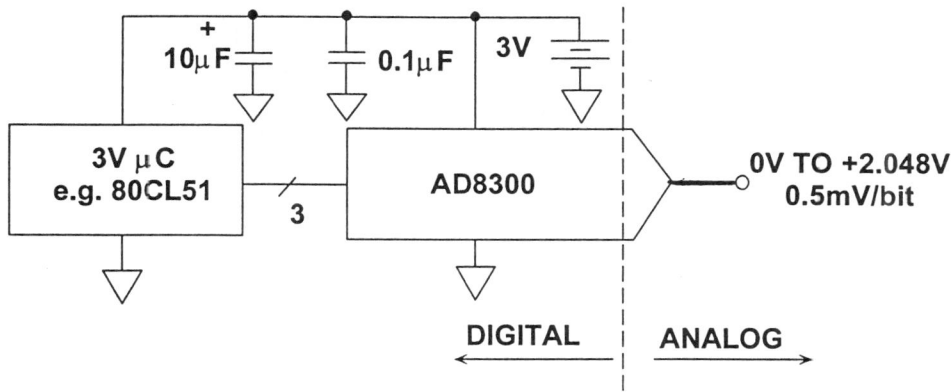

- No Adjustments or Trimming
- Single-Chip Error Budget Analysis

Figure 6.40

The AD8300 has only one analog pin, which is the DAC's voltage output. With only one analog pin, the AD8300 is an ideal device for designers whose experience is mainly digital. All analog circuitry, except for the analog voltage output pin, is transparent to the user. Double-buffered data latches prevent the DAC output from changing while new data is being shifted into the AD8300.

AD8300 BLOCK DIAGRAM

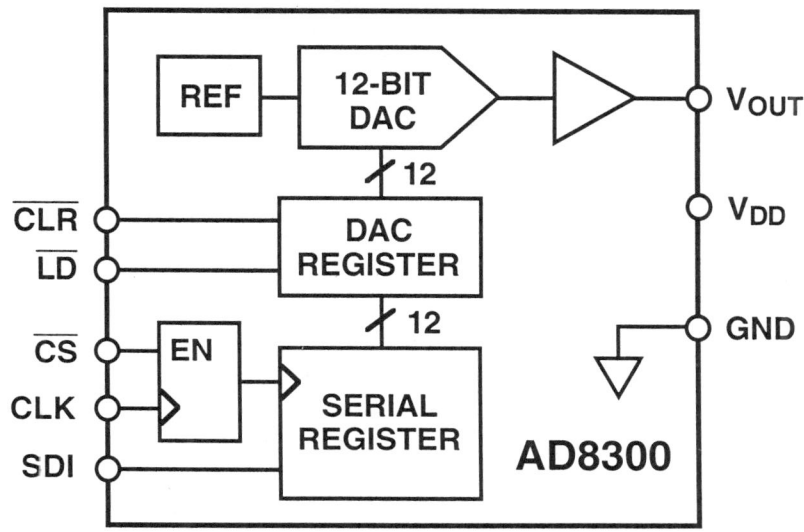

Figure 6.41

The AD8300 is laser-trimmed during production, so no calibration or other adjustments are required. The DAC value is simply shifted into the serial data input, and the analog output responds with a pre-trimmed value of:

$$V_{OUT} = D \cdot 0.5 mV,$$

where D is a decimal number which represents the DAC input value (that is, 0 to 4095). The full-scale output voltage is 2.0475V, which is appropriate for the minimum supply voltage of 2.7V. The DAC output can be asynchronously set to 0V with the $\overline{CLR}$ input, if a power-on reset is required.

KEY FEATURES OF THE AD8300 DAC

- Complete Voltage Output 12-bit DAC
- Single +3V Operation
- No External Components
- 2.0475V Full Scale Output (0.5mV/bit)
- 6 µs Output Settling Time
- Serial Data Input
- Asynchronous Clear Input
- Low Power: 3.6mW
- PCMCIA-Compatible SO-8 Package (1.5mm package height)

Figure 6.42

The data distribution systems discussed so far have illustrated the tradeoffs between demultiplexing a single DAC and using a single chip, multiple-channel DAC. Both of these concepts assume that the system being designed requires several analog voltages in close physical proximity to each other. The design considerations change, however, for systems where different circuit elements are not physically close. Analog signal traces should be kept as short as possible to reduce error sources such as leakage and noise pickup. Digital signals have more noise immunity than analog signals, so the rule of thumb is to locate the DAC as close to its associated analog circuitry as possible.

A complete DAC with serial input can utilize its small size to place the analog voltages near the circuits they control with minimal impact on overall PC board area. Two AD8300s, for example, have a total of only 16 pins. This is only two pins more than the previously-discussed dual-12-bit DAC, yet the two single DACs can be located directly adjacent to subsequent circuits (Figure 6.43). This eliminates long PC board traces (for the analog signal) and reduces noise pickup. Multiple AD8300s can be accessed with one serial data bus, and the outputs of the DACs can change either synchronously or asynchronously.

REMOTE, MULTI-CHANNEL DATA DISTRIBUTION

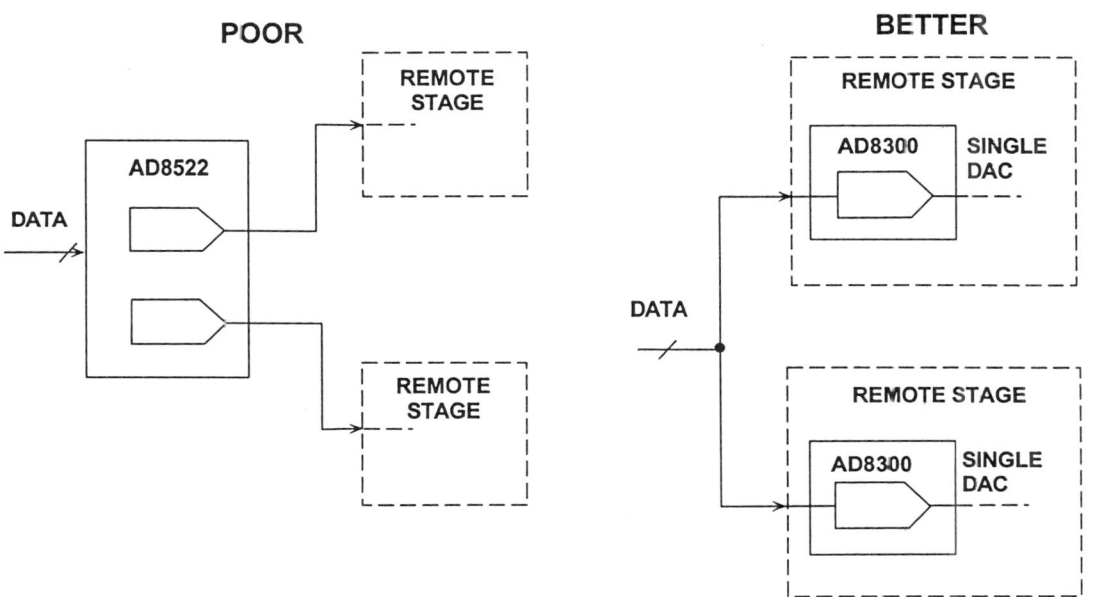

Figure 6.43

REFERENCES

1. **Linear Design Seminar**, Analog Devices, 1995, Chapter 7.

2. **System Applications Guide**, Analog Devices, 1993, Chapter 2.

3. Dan Sheingold, **Analog-Digital Conversion Handbook, Third Edition**, Prentice-Hall, 1986.

4. Adolfo A. Garcia, *Applications of the SMP-04 and the SMP-08/SMP-18, Quad and Octal Sample-and-Hold Amplifiers*, **Application Note AN-204**, Analog Devices, 1991.

5. *A Selection Guide for Serial DACs*, Document Number G1982, Analog Devices, 1994.

SECTION 7

OVERVOLTAGE EFFECTS ON ANALOG INTEGRATED CIRCUITS

- Amplifier Input Stage Overvoltage

- Amplifier Output Voltage Phase Reversal

- Understanding and Protecting Integrated Circuits from Electrostatic Discharge (ESD)

Overvoltage Effects on Analog Integrated Circuits

SECTION 7

OVERVOLTAGE EFFECTS ON ANALOG INTEGRATED CIRCUITS
Adolfo Garcia, Wes Freeman

One of the most commonly asked applications questions is: "What happens if external voltages are applied to an analog integrated circuit with the supplies turned off?" This question describes situations that can take on many different forms: from lightning strikes on cables which propagate very high transient voltages into signal conditioning circuits, to walking across a carpet and then touching a printed circuit board full of sensitive precision circuits. Regardless of the situation, the general issue is the effect of overvoltage stress (and, in some cases, abuse) on analog integrated circuits. The discussion which follows will be limited in general to operational amplifiers, because it is these devices that most often interface to the outside world. The principles developed here can and should be applied to all analog integrated circuits which are required to condition or digitize analog waveforms. These devices include (but are not limited to) instrumentation amplifiers, analog comparators, sample-and-hold amplifiers, analog switches and multiplexers, and analog-to-digital converters.

AMPLIFIER INPUT STAGE OVERVOLTAGE

In real world signal conditioning, sensors are often used in hostile environments where faults can and do occur. When these faults take place, signal conditioning circuitry can be exposed to large voltages which exceed the power supplies. The likelihood for damage is quite high, even though the components' power supplies may be turned on. Published specifications for operational amplifier absolute maximum ratings state that applied input signal levels should never exceed the power supplies by more than 0.3V or, in some devices, 0.7V. Exceeding these levels exposes amplifier input stages to potentially destructive fault currents which flow through internal metal traces and parasitic p-n junctions to the supplies. Without some type of current limiting, unprotected input differential pairs (BJTs or FETs) can be destroyed in a matter of microseconds. There are, however, some devices with built-in circuitry that can provide protection beyond the supply voltages, but in general, absolute maximum ratings must still be observed.

OVERVOLTAGE EFFECTS ON ANALOG INTEGRATED CIRCUITS

INPUT STAGE OVERVOLTAGE

- INPUT SHOULD NOT EXCEED ABSOLUTE MAXIMUM RATINGS
  (Usually Specified With Respect to Supply Voltages)

- A Common Specification Requires the Input Signal < $|V_S|$ ± 0.3V

- Input Voltage Should be Held Near Zero in the Absence of Supplies

- Input Stage Conduction Current Needs to be Limited
  (Rule of Thumb: ≤ 5mA)

- Avoid Reverse Bias Junction Breakdown in Input Stage
  Base-Emitter Junctions

- Differential and Common-Mode Ratings may Differ

- No Two Amplifiers are exactly the Same

- Some Op Amps Contain Input Protection (Voltage Clamps,
  Current Limits, or Both), but Absolute Maximum Ratings Must
  Still be Observed

Figure 7.1

Although more recent vintage operational amplifiers designed for single-supply or rail-to-rail operation are now including information with regard to input stage overvoltage effects, there are very many amplifiers available today without such information provided by the manufacturer. In those cases, the circuit designer using these components must ascertain the input stage current-voltage characteristic of the device in question before steps can be taken to protect it. All amplifiers will conduct current to the positive/negative supply, provided the applied input voltage exceeds some internal threshold. This threshold is device dependent, and can range from 0.7V to 30V, depending on the internal construction of the input stage. Regardless of the threshold level, externally generated fault currents should be limited to no more than ±5mA.

Many factors contribute to the current-voltage characteristic of an amplifier's input stage: internal differential clamping diodes, current-limiting series resistances, substrate potential connections, and differential input stage topologies (BJTs or FETs). Input protection diodes used as differential input clamps are typically constructed from the base-emitter junctions of NPN transistors. These diodes usually form a parasitic p-n-junction to the negative supply when the applied input voltage exceeds the negative supply. Current-limiting series resistances used in the input stages of operational amplifiers can be fabricated from three types of material: n- or p-type diffusions, polysilicon, or thin-films (SiCr, for example). Polysilicon and thin-film resistors are fabricated over thin layers of oxide which provide an insulating barrier to the substrate; as such, they do not exhibit any parasitic p-n junctions to either supply. Diffused resistors, on the other hand, exhibit p-n junctions to the supplies because they are constructed from either p- or n-type diffusion regions. The substrate potential of the amplifier is the most critical component, for it will determine the sensitivity of an amplifier's input current-voltage characteristic to supply voltage.

The configuration of the amplifier's input stage also plays a large role in the current-voltage characteristic of the amplifier. Input differential pairs of operational amplifiers are constructed from either bipolar transistors (NPN or PNP) or field-effect transistors (junction or MOS, N- or P-channel). While the bipolar input differential pairs do not have any direct path to either supply, FET differential pairs do. For example, an n-channel JFET forms a parasitic p-n junction between its backgate and the p-substrate that energizes when $V_{IN} + 0.7V < V_{NEG}$. As mentioned previously, many manufacturers of analog integrated circuits do not provide any details with regard to the behavior of the device's input structure. Either simplified schematics are not provided or, if they are shown, the behavior of the input stage under an overvoltage condition is omitted. Therefore, other measures must be taken in order to identify the conduction paths.

A standard transistor curve tracer can be configured to determine the current-voltage characteristic of any amplifier regardless of input circuit topology. As shown in Figure 7.2, both amplifier supply pins are connected to ground, and the collector output drive is connected to one of the amplifier's inputs. The curve tracer applies a DC ramp voltage and measures the current flowing through the input stage. In the event that a transistor curve tracer is not available, a DC voltage source and a multimeter can be substituted for the curve tracer. A 10kΩ resistor should be used between the DC voltage source and the amplifier input for additional protection. Ammeter readings from the multimeter at each applied DC voltage will yield the same result as that produced by the curve tracer. Although either input can be tested (both inputs should), it is recommended that the unused input is left open; otherwise, additional junctions could come into play and would complicate matters further. Evaluations of current feedback amplifier input stages are more difficult because of the lack of symmetry between the inputs. As a result, both inputs should be characterized for their individual current-voltage characteristics.

OVER-VOLTAGE CURVE TRACER TEST SETUP

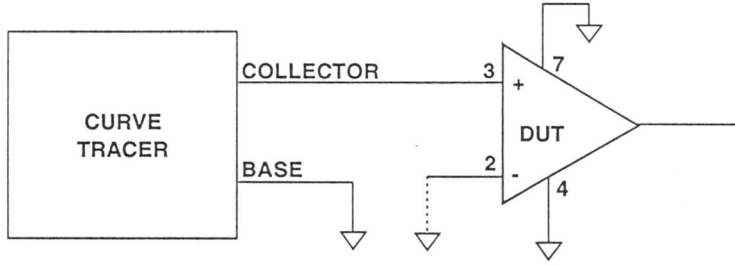

- Test Both Inputs of Op Amp -- Results are Identical for Voltage Feedback Types, but Not Current Feedback
- Force a Voltage Using the Collector Output
- Display Collector Current Versus Voltage
- Results May Differ Depending on Whether the Other Input is Open or Grounded

Figure 7.2

Once the input current-voltage characteristic has been determined for the device in question, the next step is to determine the minimum level of resistance required to limit fault currents to ± 5mA. Equation 7.1 illustrates the computation for R_s when the input overvoltage level is known:

$$R_s = \frac{V_{IN(MAX)} - V_{SUPPLY}}{5 \text{ mA}} \quad \text{Eq. 7.1}$$

The worst case condition for overvoltage would be when the power supplies are initially turned off or disconnected. In this case, V_{SUPPLY} is equal to zero. For example, if the input overvoltage could reach 100V under some type of fault condition, then the external resistor should be no smaller than 20kΩ. Most operational amplifier applications only require protection at one input; however, there are a few configurations (difference amplifiers, for example) where both inputs can be subjected to overvoltage and both must be protected. The need for protection on both inputs is much more common with instrumentation amplifiers.

OVERVOLTAGE EFFECTS

- Junctions may be <u>Forward Biased</u> if the Current is Limited

- In General a Safe Current Limit is 5mA

- <u>Reverse Bias</u> Junction Breakdown is Damaging Regardless of the Current Level

- When in Doubt, Protect with External Diodes and Series Resistances

- Curve Tracers Can be Used to Check the Overvoltage Characteristics of a Device

- Simplified Equivalent Circuits in Data Sheets do not tell the Entire Story!!!

Figure 7.3

AMPLIFIER OUTPUT VOLTAGE PHASE REVERSAL

Some operational amplifiers exhibit output voltage phase reversal when one or both of their inputs exceeds their input common-mode voltage range. Phase reversal is usually associated with JFET (n- or p-channel) input amplifiers, but some bipolar devices (especially single-supply amplifiers operating as unity-gain followers) may also be susceptible. In the vast majority of applications, output voltage phase reversal does not harm the amplifier nor the circuit in which the amplifier is used. Although a number of operational amplifiers suffer from phase reversal, it is rarely a problem in system design. However, in servo loop applications, this effect can be quite hazardous. Fortunately, this is only a temporary condition. Once the amplifier's inputs return to within its normal operating common-mode range, output voltage phase reversal ceases. It may still be necessary to consult the amplifier manufacturer, since phase reversal information rarely appears on device data sheets. Summarized as follows is a list of recent vintage Analog Devices amplifier products that are now including output voltage phase reversal characterization/commentary:

Single-Supply/ Rail-to-Rail	Dual Supply
OP295/OP495	OP282/OP482
OP113/OP213/OP413	OP285
OP183/OP283	OP467
OP292/OP492	OP176
OP191/OP291/OP491	BUF04
OP279	
AD820/AD822/AD824	
OP193/OP293/OP493	

OVERVOLTAGE EFFECTS ON ANALOG INTEGRATED CIRCUITS

In BiFET operational amplifiers, phase reversal may be prevented by adding an appropriate resistance in series with the amplifier's input to limit the current. Bipolar input devices can be protected by using a Schottky diode to clamp the input to within a few hundred millivolts of the negative rail. For a complete description of the output voltage phase reversal effect, please consult Reference 1.

Rail-to-rail operational amplifiers present a special class of problems to the integrated circuit designer, because these types of devices should not exhibit any abnormal behavior throughout the entire input common-mode range. In fact, it is desirable that devices used in these applications also not exhibit any abnormal behavior if the applied input voltages exceed the power supply range. One of the more recent vintage rail-to-rail input/output operational amplifiers, the OPX91 family (the OP191, the OP291, and the OP491), includes additional components that prevent overvoltage and damage to the device. As shown in Figure 7.5, the input stage of the OPX91 devices use six diodes and two resistors to clamp the input terminals to each other and to the supplies. D1 and D2 are base-emitter NPN diodes which are used to protect the bases of Q1-Q2 and Q3-Q4 against avalanche breakdown when the applied differential input voltage to the device exceeds 0.7V. Diodes D3-D6 are diodes formed from substrate PNP transistors that clamp the applied input voltages on the OPX91 to the supply rails.

BEWARE OF AMPLIFIER OUTPUT PHASE REVERSAL

- Sometimes Occurs in FET and Bipolar Input (Especially Single-Supply) Op Amps when Input Exceeds Common Mode Range

- Does Not Harm Amplifier, but may be Disastrous in Servo Systems!

- Not Usually Specified on Data Sheet, so Amplifier Must be Checked

- Easily Prevented:

    BiFETs: Add Appropriate Input Series Resistance (Determined Empirically, Unless Provided in Data Sheet)

    Bipolars: Use Schottky Diode Clamps to the Supply Rails.

Figure 7.4

A CLOSER LOOK AT THE OP-X91 INPUT STAGE REVEALS ADDITIONAL DEVICES

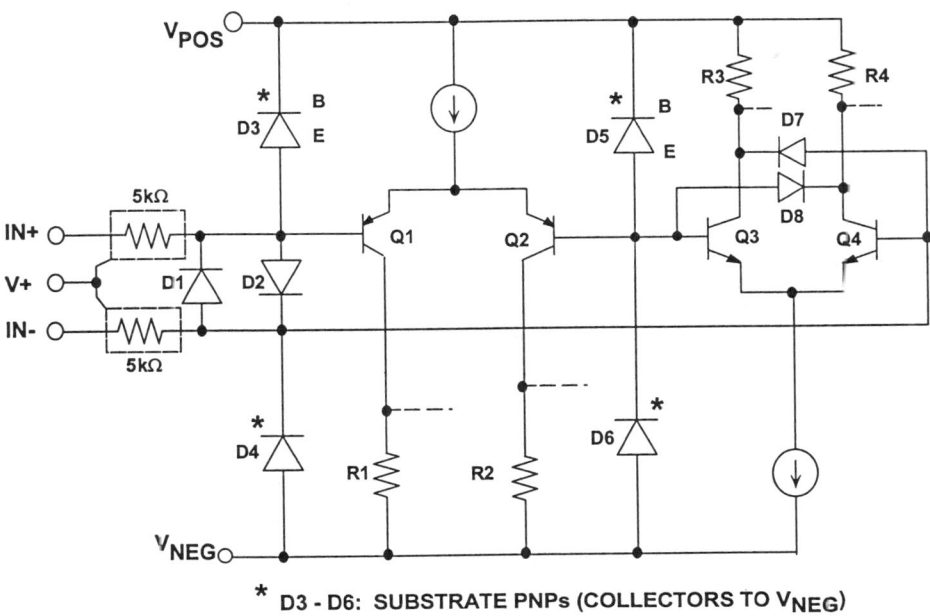

* D3 - D6: SUBSTRATE PNPs (COLLECTORS TO V_{NEG})

Figure 7.5

An interesting benefit from using substrate PNPs as clamp diodes is that their collectors are connected to the negative supply; thus, when the applied input voltage exceeds either supply rail, the diodes energize, and the fault currents are diverted directly to the supply and not through or into the device's input stage. There are also 5kΩ resistors in series with each of the inputs to the OPX91 to limit the fault current through D1 and D2 when the differential input voltage exceeds 0.7V. Note that these 5kΩ resistors are p-type diffusions placed inside an n-well, which is then connected to the positive supply. When the applied input voltage exceeds the positive supply, some of the fault current generated is also diverted to V_{POS} and away from the input stage. As a result of these measures, the input overvoltage characteristic of the OPX91 is well behaved as shown in Figure 7.6. Note that the combination of the 5kΩ resistors and clamp diodes safely limits the input current to less than 2mA, even when the inputs of the device exceed the supply rails by 10V.

INTERNAL 5kΩ RESISTORS PLUS INPUT CLAMP DIODES COMBINE TO PROTECT OP-X91 DEVICES AGAINST OVERVOLTAGE

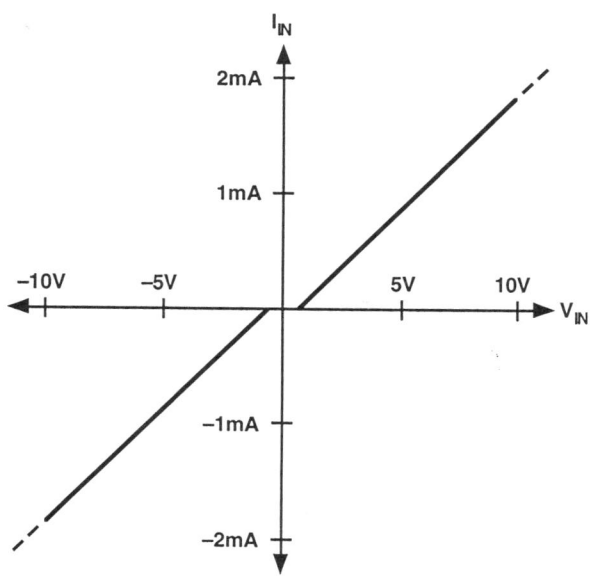

Figure 7.6

As an added safety feature, an additional pair of diodes is used in the input stage across Q3 and Q4 to prevent subsequent stages internal to the OPX91 from collapsing (that is, forced into cutoff). If these stages were forced into cutoff, then the amplifier would undergo output voltage phase reversal when the inputs exceeded the positive input common mode voltage. An illustration of the diodes' effectiveness is shown in Figure 7.7. Here, the OPX91 family can safely handle a 20Vp-p input signal on ±5V supplies without exhibiting any sign of output voltage phase reversal or other anomalous behavior. With these amplifiers, no external clamping diodes are required.

For those amplifiers where external protection is clearly required against both overvoltage abuse and output phase reversal, a common technique is to use a series resistance, R_S, to limit fault current, and Schottky diodes to clamp the input signal to the supplies, as shown in Figure 7.8.

ADDITION OF TWO CLAMP DIODES PROTECTS OP-X91 DEVICES AGAINST OUTPUT PHASE REVERSAL

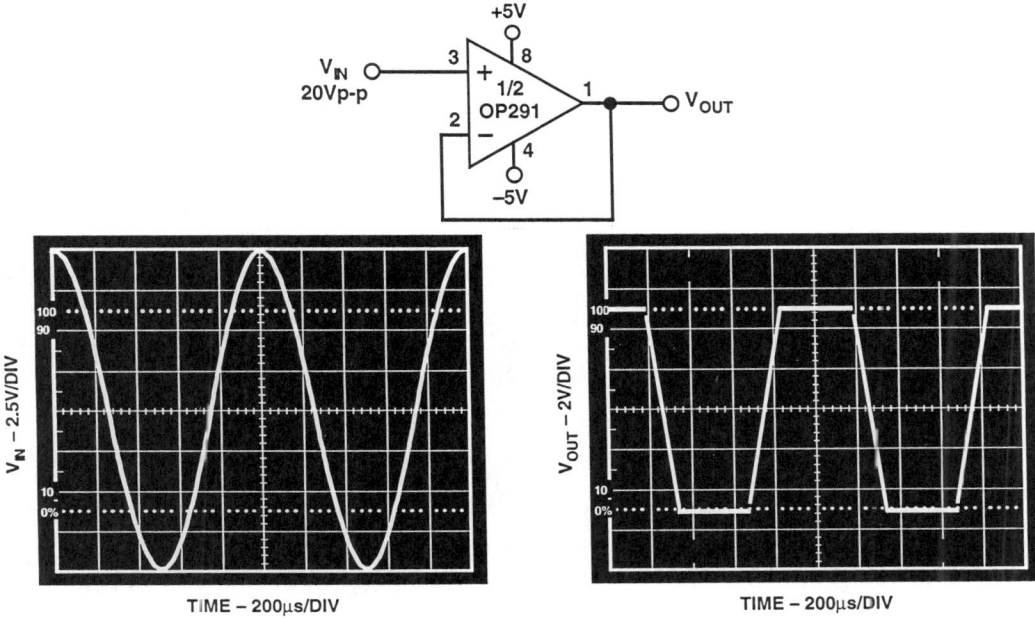

Figure 7.7

GENERALIZED EXTERNAL PROTECTION SCHEME AGAINST INPUT OVERVOLTAGE ABUSE AND OUTPUT VOLTAGE PHASE REVERSAL IN SINGLE-SUPPLY OP AMPS

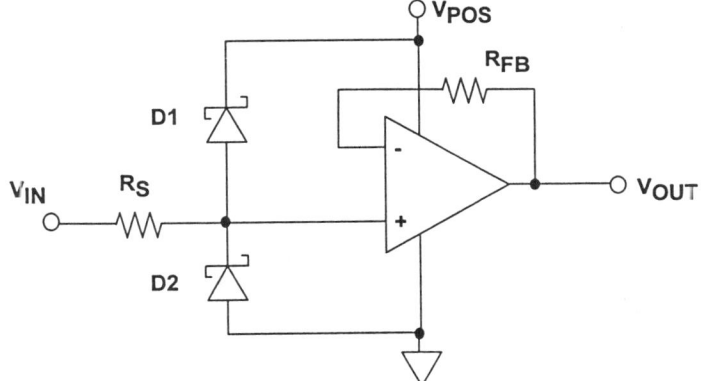

- Value for R_S provided by manufacturer or determined empirically
- R_{FB} may be required for high bias current devices
- D1 and D2 can be Schottky diodes (Check their capacitance and leakage current first)

Figure 7.8

The external input series resistance, R_S, will be provided by the manufacturer of the amplifier, or determined empirically by the user with the method previously shown in Figure 7.2 and Eq. 7.1. More often than not, the value of this resistor will provide enough protection against output voltage phase reversal, as well as limiting the fault current through the Schottky diodes.

It is evident that whenever resistance is added in series with an amplifier's input, its offset and noise performance will be affected. The effects of this series resistance on circuit noise can be calculated using the following equation.

$$E_{n,total} = \sqrt{(e_{n,op\ amp})^2 + (e_{n,R_S})^2 + (R_S \cdot i_{n,op\ amp})^2}$$

The thermal noise of the resistor, the voltage noise due to amplifier noise current flowing through the resistor, and the input noise voltage of the amplifier are added together (in root-sum-square manner, since the noise voltages are uncorrelated) to determine the total input noise and may be compared with the input voltage noise in the absence of the protection resistor.

A protection resistor in series with an amplifier input will also produce a voltage drop due to the amplifier bias current flowing through it. This drop appears as an increase in the circuit offset voltage (and, if the bias current changes with temperature, offset drift). In amplifiers where bias currents are approximately equal, a resistor in series with each input will tend to balance the effect and reduce the error. The effects of this additional series resistance on the circuit's overall offset voltage can be calculated:

$$V_{os(total)} = V_{os} + I_b R_S$$

For the case where $R_{FB} = R_S$ or where the source impedance levels are balanced, then the total circuit offset voltage can be expressed as:

$$V_{os(total)} = V_{os} + I_{os} R_S$$

To limit the additional noise of R_{FB}, it can be shunted with a capacitor.

When using external clamp diodes to protect operational amplifier inputs, the effects of diode junction capacitance and leakage current should be evaluated in the application. Diode junction capacitance and R_S will add an additional pole in the signal path, and diode leakage currents will double for every 10°C rise in ambient temperature. Therefore, low leakage diodes should be used such that, at the highest ambient temperature for the application, the total diode leakage current is less than one-tenth of the input bias current for the device at that temperature. Another issue with regard to the use of Schottky diodes is the change in their forward voltage drop as a function of temperature. These diodes do not, in fact, limit the signal to ±0.3V at all ambient temperatures, but if the Schottky diodes are at the same temperature as the op amp, they will limit the voltage to a safe level, even if they do not limit it at all times to within the data sheet rating. This is true if over-voltage is only possible at turn-on, when the diodes and the op amp will always be at the same temperature. If the op amp is warm when it is repowered, however, steps must be taken to ensure that diodes and op amp are at the same temperature.

Understanding and Protecting Integrated Circuits from Electrostatic Discharge (ESD)
Wes Freeman

Integrated circuits can be damaged by the high voltages and high peak currents that can be generated by electrostatic discharge. Precision analog circuits, which often feature very low bias currents, are more susceptible to damage than common digital circuits, because the traditional input-protection structures which protect against ESD damage also increase input leakage.

The keys to eliminating ESD damage are: (1) awareness of the sources of ESD voltages, and (2) understanding the simple handling steps that will discharge potential voltages safely.

The basic definitions relating to ESD are given in Figure 7.9. Notice that the *ESD Failure Threshold* level relates to any of the IC data sheet limits, and not simply to a *catastrophic failure* of the device. Also, the limits apply to each pin of the IC, not just to the input and output pins.

ESD DEFINITIONS

- ESD (Electrostatic Discharge):
  - A single fast, high current transfer of electrostatic charge.
    - Direct contact between two objects at different potentials.
    - A high electrostatic field between two objects when they are in close proximity.

- ESD Failure Threshold:
  - The highest voltage level at which all pins on a device can be subjected to ESD zaps without failing any 25°C data sheet limits.

Figure 7.9

OVERVOLTAGE EFFECTS ON ANALOG INTEGRATED CIRCUITS

The generation of static electricity caused by rubbing two substances together is called the *triboelectric effect*. Static charge can be generated either by dissimilar materials (for example, rubber-soled shoes moving across a rug) or by separating similar materials (for example, pulling transparent tape off of a roll).

A wide variety of common human activities can create high electrostatic charge. Some examples are given in Figure 7.10. The values shown will occur with a fairly high relative humidity. Low humidity, such as can occur indoors during cold weather, can generate voltages 10 times (or more) greater than the values shown.

In an effort to standardize the testing and classification of integrated circuits for ESD robustness, ESD models have been developed (Figure 7.11). These models attempt to simulate the source of ESD voltage. The assumptions underlying the three commonly-used models are different, so results are not directly comparable.

The most-often encountered ESD model is the Human Body Model (HBM). This model simulates the approximate resistance and capacitance of a human body with a simple RC network. The capacitor is charged through a high voltage power supply (HVPS) and then discharged (using a high voltage switch) through a series resistor. The RC values for different individuals will, of course, vary. However, the HBM has been standardized by MIL-STD-883 Method 3015 Electrostatic Discharge Sensitivity Classification, which specifies R-C combinations of 1.5kΩ and 100pF. (R, C, and L values for all three ESD models are shown in Figure 7.12.)

EXAMPLES OF ELECTROSTATIC CHARGE GENERATION

- Person walks across a typical carpet.
    - 1000 - 1500V generated
- Person walks across a typical vinyl floor.
    - 150 - 250V generated
- Person handles instructions protected by clear plastic covers.
    - 400 - 600V generated
- Person handles polyethylene bags.
    - 1000 - 1200V generated
- Person pours polyurethane foam into a box.
    - 1200 - 1500V generated
- An IC slides down a grounded handler chute.
    - 50 - 500V generated
- An IC slides down an open conductive shipping tube.
    - 25 - 250V generated

Note: Above values can occur in a high ($\approx$60%) RH environment. For low RH ($\approx$30%), generated voltages can be >10 times those listed above!

Figure 7.10

MODELING ELECTROSTATIC POTENTIAL

- Three Models:

    1. Human Body Model (HBM)

    2. Machine Model (MM)

    3. Charged Device Model (CDM)

- Model Correlation:

    ◆ Low - Assumptions are Different

Figure 7.11

ESD MODELS APPLICABLE TO ICs

- Human Body Model (HBM)

    Simulates the discharge event that occurs when a person charged to either a positive or negative potential touches an IC at a different potential.

    RLC: $R = 1.5k\Omega$, $L \approx 0nH$, $C = 100pF$

- Machine Model (MM)

    Non-real-world Japanese model based on worst-case HBM.

    RLC: $R \approx 0\Omega$, $L \approx 500nH$, $C = 200pF$

- Charged Device Model (CDM)

    Simulates the discharge that occurs when a pin on an IC, charged to either a positive or negative potential, contacts a conductive surface at a different (usually ground) potential.

    RLC: $R = 1\Omega$, $L \approx 0nH$, $C = 1 - 20pF$

Figure 7.12

The Machine Model (MM) is a worst-case Human Body Model. Rather than using an *average* value for resistance and capacitance of the human body, the MM assumes a worst-case value of 200pF and 0Ω. The 0Ω output resistance of the MM is also intended to simulate the discharge from a charged conductive object (for example, a charged DUT socket on an automatic test system) to an IC pin, which is how the Machine Model earned its name. However, the MM does not simulate many known real-world ESD events. Rather, it models the ESD event resulting from a ideal voltage source (in other words, with no resistance in the discharge path). EIAJ Specification ED-4701 Test Method C-111 Condition A and ESD Association Specification S5.2 provide guidelines for MM testing.

The Charged Device Model (CDM) originated at AT&T. This model differs from the HBM and the MM, in that the source of the ESD energy is the IC itself. The CDM assumes that the integrated circuit die, bond wires, and lead frame are charged to some potential (usually positive with respect to ground). One or more of the IC pins then contacts ground, and the stored charge rapidly discharges through the leadframe and bond wires. Typical examples of triboelectric charging followed by a CDM discharge include:

1. An IC slides down a handler chute and then a corner pin contacts a grounded stop bar.

2. An IC slides down an open conductive shipping tube and then a corner pin contacts a conductive surface.

The basic concept of the CDM is different than the HBM and MM in two ways. First, the CDM simulates a charged IC discharging to ground, while the HBM and MM both simulate a charged source discharging into the IC. Thus, current flows out of the IC during CDM testing, and into the IC during HBM and MM testing. The second difference is that the capacitor in the CDM is the capacitance of the package, while the HBM and MM use a fixed external capacitor.

Unlike the HBM and MM, CDM ESD thresholds may vary for the same die in different packages. This occurs because the device under test (DUT) capacitance is a function of the package. For example, the capacitance of an 8-pin package is different than the capacitance of a 14-pin package. CDM capacitance values can vary from about 1 to 20pF. The device capacitance is discharged through a 1Ω resistor.

Schematic representations of the three models are shown in Figure 7.13. Notice that C1 in the HBM and MM are external capacitors, while C_{PKG} in the CDM is the internal capacitance of the DUT.

The HBM discharge waveform is a predicable unipolar RC pulse, while the MM discharge shows ringing because of the parasitic inductance in the discharge path (typically 500nH.). Ideally, the CDM waveform is also a single unipolar pulse, but the parasitic inductance in series with the 1Ω resistor slows the rise time and introduces some ringing.

The significant features of each ESD model are summarized in Figure 7.14. The peak currents shown for each model are based on a test voltage of 400V. Peak current is lowest for the HBM because of the relatively high discharge resistance. The CDM discharge has low energy because device capacitance is only in the range of 1pF to 20pF, but peak current is high. The MM has the highest energy discharge, because it has the highest capacitance value (Power = $0.5 CV^2$).

SCHEMATIC REPRESENTATION OF ESD MODELS AND TYPICAL DISCHARGE WAVEFORMS

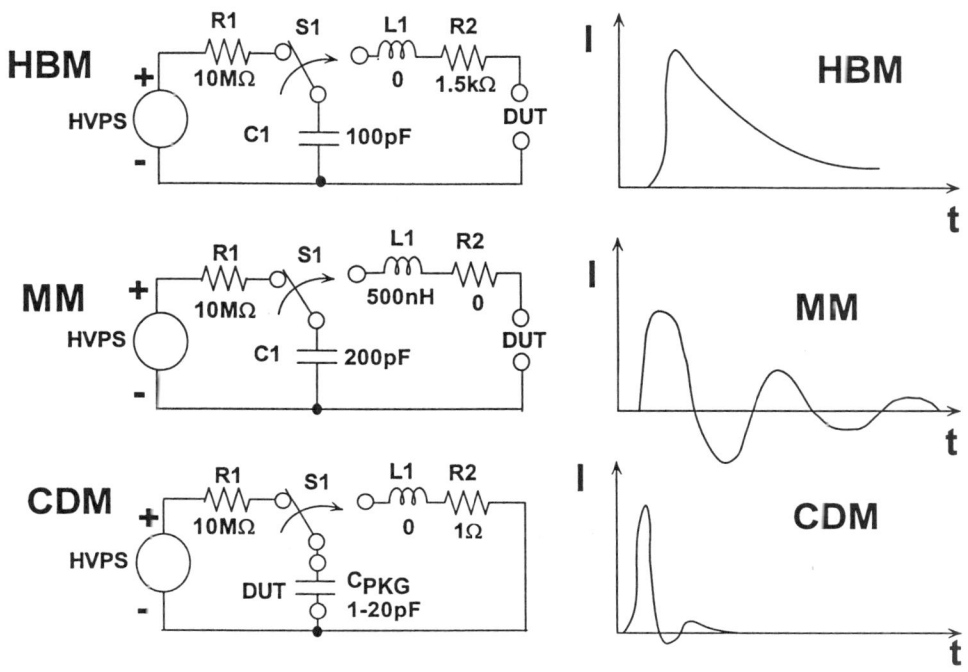

Figure 7.13

COMPARISON OF HBM, MM, AND CDM ESD MODELS

MODEL:	HBM	MM	SOCKETED CDM
Simulate:	Human Body	Machine	Charged Device
Origin:	US Military, Late 1960s	Japan, 1976	AT&T, 1974
Real World?	Yes	Generally	Yes
RC:	1.5kΩ, 100pF	0Ω, 200pF	1Ω, 1 - 20pF
Rise Time	<10ns (6-9ns typ)	14ns*	400ps**
Ipeak at 400V	0.27A	5.8A*	2.1A**
Energy:	Moderate	High	Low
Package Dependent:	No	No	Yes
Standard:	MIL-STD-883 Method 3015	ESD Association Std. S5.2; EIAJ Std. ED-4701, Method C-111	ESD Association Draft Std. DS5.3

* These values per ESD Association Std. S5.2.
  EIAJ Std. ED-4701, Method C-111 includes no waveform specifications.
** These values are for the direct charging (socketed) method.

Figure 7.14

Figure 7.15 compares 400V discharge waveforms of the CDM, MM, and HBM, with the same current and time scales.

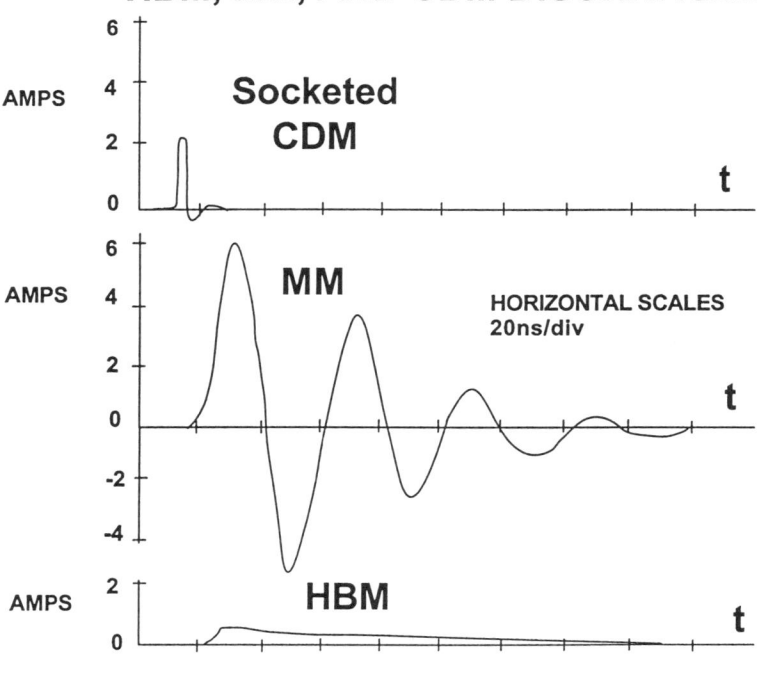

Figure 7.15

The CDM waveform corresponds to the shortest known real-world ESD event. The waveform has a rise time of <1ns, with the total duration of the CDM event only about 2ns. The CDM waveform is essentially unipolar, although some ringing occurs at the end of the pulse that results in small negative-going peaks. The very short duration of the overall CDM event results in an overall discharge of relatively low energy, but peak current is high.

The MM waveform consists of both positive- and negative-going sinusoidal peaks, with a resonance frequency of 10MHz to 15MHz. The initial MM peak has a typical rise time of 14ns, and the total pulse duration is about 150ns. The multiple high current, moderate duration peaks of the MM result in an overall discharge energy that is by far the highest of the three models for a given test voltage.

The risetime for the unipolar HBM waveform is typically 6-9ns, and the waveform decays exponentially towards 0V with a fall time of approximately 150ns. (Method 3015 requires a rise time of <10ns and a delay time of 150ns ± 20ns, with decay time defined as the time for the waveform to drop from 100% to 36.8% of peak current). The peak current for the HBM is 400V/1500Ω, or 0.267A, which is much lower than is produced by 400V CDM and MM events. However, the relatively long duration of the total HBM event still results in an overall discharge of moderately high energy.

As previously noted, the MM waveform is bipolar while HBM and CDM waveforms are primarily unipolar. However, HBM and CDM testing is done with both positive and negative polarity pulses. Thus all three models stress the IC in both directions.

MIL-STD-883 Method 3015 classifies ICs for ESD failure threshold. The classification limits, shown in Figure 7.16, are derived using the HBM shown in Figure 7.13. Method 3015 also mandates a marking method to denote the ESD classification. All military grade Class 1 and 2 devices have their packages marked with one or two "Δ" symbols, respectively, while class 3 devices (with a failure threshold >4kV) do not have any ESD marking. Commercial and industrial grade IC packages may not be marked with any ESD classification symbol.

CLASSIFYING AND MARKING ICs FOR ESD PER MIL-883C, METHOD 3015

HBM ESD CLASS	FAILURE THRESHOLD	MARKING
1	<2kV	Δ
2	2kV - <4kV	Δ Δ
3	>4kV	None

Note: Commercial and Industrial Grade ICs are not marked for ESD

Figure 7.16

Notice that the Class 1 limit includes all devices which do not pass a 2kV threshold. However, a Class 1 rating does not imply that all devices within that class will pass 1,999V. In any event, the emphasis must be placed on eliminating ESD exposure, not on attempting to decide how much ESD exposure is 'safe.'

A detailed discussion of IC failure mechanisms is beyond the scope of this seminar, but some typical ESD effects are shown in Figure 7.17.

UNDERSTANDING ESD DAMAGE

- ESD Failure Mechanisms:

    ◆ Dielectric or junction damage
    ◆ Surface charge accumulation
    ◆ Conductor fusing.

- ESD Damage Can Cause:

    ◆ Increased leakage
    ◆ Reduced performance
    ◆ Functional failures of ICs.

- ESD Damage is often Cumulative:

    ◆ For example, each ESD "zap" may increase junction damage until, finally, the device fails.

Figure 7.17

For the design engineer or technician, the most common manifestation of ESD damage is a catastrophic failure of the IC. However, exposure to ESD can also cause increased leakage or degrade other parameters. If a device appears to not meet a data sheet specification during evaluation, the possibility of ESD damage should be considered.

Special care should be taken when breadboarding and evaluating ICs. The effects of ESD damage can be cumulative, so repeated mishandling of a device can eventually cause a failure. Inserting and removing ICs from a test socket, storing devices during evaluation, and adding or removing external components on the breadboard should all be done while observing proper ESD precautions. Again, if a device fails during a prototype system development, repeated ESD stress may be the cause.

The key word to remember with respect to ESD is *prevention*. There is no way to un-do ESD damage, or to compensate for its effects.

Since ESD damage can not be undone, the only cure is prevention. Luckily, prevention is a simple two-step process. The first step is recognizing ESD-sensitive products, and the second step is understanding how to handle these products.

All static sensitive devices are shipped in protective packaging. ICs are usually contained in either conductive foam or in antistatic tubes. Either way, the container is then sealed in a static-dissipative plastic bag. The sealed bag is marked with a distinctive sticker, such as is shown in Figure 7.20, which outlines the appropriate handling procedures.

THE MOST IMPORTANT THING TO REMEMBER ABOUT ESD DAMAGE

- **ESD DAMAGE *CANNOT* BE "CURED" !!!**

- Circuits cannot be *tweaked*, *nulled*, *adjusted*, etc., to compensate for ESD damage.

ESD DAMAGE MUST BE *PREVENTED*!

Figure 7.18

PREVENTING ESD DAMAGE TO ICs

Two key elements in protecting circuits from ESD damage are:

- **Recognizing ESD-sensitive products**

- **Always handling ESD-sensitive products at a *grounded* workstation.**

Figure 7.19

RECOGNIZING ESD-SENSITIVE DEVICES

All static sensitive devices are sealed in protective packaging and marked with special handling instructions

CAUTION
SENSITIVE ELECTRONIC DEVICES

DO NOT SHIP OR STORE NEAR STRONG
ELECTROSTATIC, ELECTROMAGNETIC,
MAGNETIC OR RADIOACTIVE FIELDS

CAUTION
SENSITIVE ELECTRONIC DEVICES

DO NOT OPEN EXCEPT AT
APPROVED FIELD FORCE
PROTECTIVE WORK STATION

Figure 7.20

Once ESD-sensitive devices are identified, protection is easy. Obviously, keeping ICs in their original protective packaging as long as possible is the first step. The second step is to discharge potential ESD sources before damage to the IC can occur. The HBM capacitance is only 100pF, so discharging a potentially dangerous voltage can be done quickly and safely through a high impedance. Even with a source resistance of 10MΩ, the 100pF will be discharged in less than 100milliseconds.

The key component required for safe ESD handling is a workbench with a static-dissipative surface, as shown in Figure 7.21. This surface is connected to ground through a 1MΩ resistor, which dissipates static charge while protecting the user from electrical shock hazards caused by ground faults. If existing bench tops are nonconductive, a static-dissipative mat should be added, along with a discharge resistor.

OVERVOLTAGE EFFECTS ON ANALOG INTEGRATED CIRCUITS

WORKSTATION FOR HANDLING ESD-SENSITIVE DEVICES

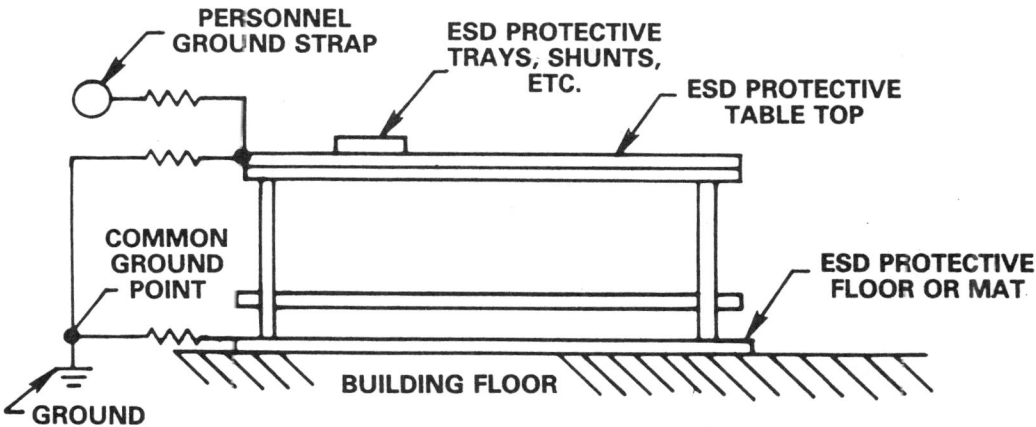

Note: Conductive table top sheet resistance ≈ $10^6\ \Omega/\square$

Figure 7.21

Notice that the surface of the workbench has a moderately high sheet resistance. It is neither necessary nor desirable to use a low-resistance surface (such as a sheet of copper-clad PC board) for the work surface. Remember, the CDM assumes that a high peak current will flow if a charged IC is discharged through a low impedance. This is precisely what happens when a charged IC contacts a grounded copper clad board. When the same charged IC is placed on the surface shown in Figure 7.21, however, the peak current is not high enough to damage the device.

A conductive wrist strap is also recommended while handling ESD-sensitive devices. The wrist strap ensures that normal tasks, such as peeling tape off of packages, will not cause damage to ICs. Again, a 1MΩ resistor, from the wrist strap to ground, is required for safety.

When building prototype breadboards or assembling PC boards which contain ESD-sensitive devices, all passive components should be inserted and soldered before the ICs. This procedure minimizes the ESD exposure of the sensitive devices. The soldering iron must, of course, have a grounded tip.

Protecting ICs from ESD requires the participation of both the IC manufacturer and the customer. IC manufacturers have a vested interest in providing the highest possible level of ESD protection for their products. IC circuit designers, process engineers, packaging specialists and others are constantly looking for new and improved circuit designs, processes, and packaging methods to withstand or shunt ESD energy (Figure 7.22)

ANALOG DEVICES' COMMITMENT

■ Analog Devices is committed to helping our customers *prevent* ESD damage by:

◆ Building products with the highest level of ESD protection commensurate with performance requirements

◆ Protecting products from ESD during shipment

◆ Helping customers to avoid ESD exposure during manufacture

Figure 7.22

A complete ESD protection plan, however, requires more than building ESD protection into ICs. Users of ICs must also provide their employees with the necessary knowledge of and training in ESD handling procedures (Figure 7.23).

The **ESD Protection Manual** is available from Analog Devices' literature department. This manual provides additional information on static-dissipative work surfaces, packaging materials, protective clothing, ESD training, and other subjects.

ESD PROTECTION REQUIRES A PARTNERSHIP BETWEEN THE IC SUPPLIER AND THE CUSTOMER

ANALOG DEVICES:

- Circuit Design and Fabrication -
  - Design and manufacture products with the highest level of ESD protection consistent with required analog and digital performance.
- Pack and Ship -
  - Pack in static dissipative material. Mark packages with ESD warning.

CUSTOMERS:

- Incoming Inspection -
  - Inspect at grounded workstation. Minimize handling.
- Inventory Control -
  - Store in original ESD-safe packaging. Minimize Handling.
- Manufacturing -
  - Deliver to work area in original ESD-safe packaging. Open packages only at grounded workstation. Package subassemblies in static dissipative packaging.
- Pack and Ship -
  - Pack in static dissipative material if required. Replacement or optional boards may require special attention.

Figure 7.23

ESD PROTECTION MANUAL

Contact Analog Devices' literature department for a copy of the ESD Prevention Manual, which has further information on:

- Handling Instructions
- Packaging
- Static-Safe Facilities
- Other ESD Issues

Figure 7.24

REFERENCES

1. **Amplifier Applications Guide**, Section XI, pp. 1-10, Analog Devices, Incorporated, Norwood, MA, 1992.

2. **Systems Applications Guide**, Section 1, pp. 56-72, Analog Devices, Incorporated, Norwood, MA, 1993.

3. **Linear Design Seminar**, Section 1, pp. 19-22, Analog Devices, Incorporated, Norwood, MA, 1994.

4. **ESD Prevention Manual**, Analog Devices, Inc.

5. *MIL-STD-883 Method 3015, Electrostatic Discharge Sensitivity Classification.* Available from Standardization Document Order Desk, 700 Robbins Ave., Building #4, Section D, Philadelphia, PA 19111-5094.

6. *EIAJ ED-4701 Test Method C-111, Electrostatic Discharges.* Available from the Japan Electronics Bureau, 250 W 34th St., New York NY 10119, Attn.: Tomoko.

7. *ESD Association Standard S5.2 for Electrostatic Discharge (ESD) Sensitivity Testing -Machine Model (MM)- Component Level.* Available from the ESD Association, Inc., 200 Liberty Plaza, Rome, NY 13440.

8. *ESD Association Draft Standard DS5.3 for Electrostatic Discharge (ESD) Sensitivity Testing - Charged Device Model (CDM) Component Testing.* Available from the ESD Association, Inc., 200 Liberty Plaza, Rome, NY 13440.

9. Niall Lyne, *Electrical Overstress Damage to CMOS Converters*, **Application Note AN-397**, Analog Devices, 1995.

SECTION 8

DISTORTION MEASUREMENTS

- High Speed Op Amp Distortion
- High Frequency Two-Tone Generation
- Using Spectrum Analyzers in High Frequency Low Distortion Measurements
- Measuring ADC Distortion using FFTs
- FFT Testing
- Troubleshooting the FFT Output
- Analyzing the FFT Output

Distortion Measurements

SECTION 8

DISTORTION MEASUREMENTS

HIGH SPEED OP AMP DISTORTION
Walt Kester

Dynamic range of an op amp may be defined in several ways. The most common ways are to specify Harmonic Distortion, Total Harmonic Distortion (THD), or Total Harmonic Distortion Plus Noise (THD + N).

Harmonic distortion is measured by applying a spectrally pure sinewave to an op amp in a defined circuit configuration and observing the output spectrum. The amount of distortion present in the output is usually a function of several parameters: the small- and large-signal nonlinearity of the amplifier being tested, the amplitude and frequency of the input signal, the load applied to the output of the amplifier, the amplifier's power supply voltage, printed circuit board layout, grounding, power supply decoupling, etc. Therefore, any distortion specification is relatively meaningless unless the exact test conditions are specified.

Harmonic distortion may be measured by looking at the output spectrum on a spectrum analyzer and observing the values of the second, third, fourth, etc., harmonics with respect to the amplitude of the fundamental signal. The value is usually expressed as a ratio in %, ppm, dB, or dBc. For instance, 0.0015% distortion corresponds to 15ppm, or –96.5dBc. The unit "dBc" simply means that the harmonic's level is so many dB below the value of the "carrier" frequency, i.e., the fundamental.

Harmonic distortion may be expressed individually for each component (usually only the second and third are specified), or they all may be combined in a root-sum-square (RSS) fashion to give the Total Harmonic Distortion (THD). The distortion component which makes up Total Harmonic Distortion is usually calculated by taking the root sum of the squares of the first five or six harmonics of the fundamental. In many practical situations, however, there is negligible error if only the second and third harmonics are included. This is because the RSS process causes the higher-order terms to have negligible effect on the THD, if they are 3 to 5 times smaller than the largest harmonic. For example,

$$\sqrt{0.10^2 + 0.03^2} = \sqrt{0.0109} = 0.104 \approx 0.1$$

DEFINITIONS OF THD AND THD + N

- V_S = Signal Amplitude (rms Volts)

- V_2 = Second Harmonic Amplitude (rms Volts)

- V_n = nth Harmonic Amplitude (rms Volts)

- V_{noise} = rms value of noise over measurement bandwidth

- $$\text{THD} + \text{N} = \frac{\sqrt{V_2^2 + V_3^2 + V_4^2 + \ldots + V_n^2 + V_{noise}^2}}{V_S}$$

- $$\text{THD} = \frac{\sqrt{V_2^2 + V_3^2 + V_4^2 + \ldots + V_n^2}}{V_S}$$

Figure 8.1

It is important to note that the THD measurement does not include noise terms, while THD + N does. The noise in the THD + N measurement must be integrated over the measurement bandwidth. In audio applications, the bandwidth is normally chosen to be around 100kHz. In narrow-band applications, the level of the noise may be reduced by filtering. On the other hand, harmonics and intermodulation products which fall within the measurement bandwidth cannot be filtered, and therefore may limit the system dynamic range. It should be evident that the THD+N ≈ THD if the rms noise over the measurement bandwidth is several times less than the THD, or even the worst harmonic. It is worth noting that if you know only the THD, you can calculate the THD+N fairly accurately using the amplifier's voltage- and current-noise specifications. (Thermal noise associated with the source resistance and the feedback network may also need to be computed). But if the rms noise level is significantly higher than the level of the harmonics, and you are only given the THD+N specification, you cannot compute the THD.

Special equipment is often used in audio applications for a more sensitive measurement of the noise and distortion. This is done by first using a bandstop filter to remove the fundamental signal (this is to prevent overdrive distortion in the measuring instrument). The total rms value of all the other frequency components (harmonics and noise) is then measured over an appropriate bandwidth. The ratio to the fundamental is the THD+N specification.

DISTORTION MEASUREMENTS

Audio frequency amplifiers (such as the OP-275) are optimized for low noise and low distortion within the audio bandwidth (20Hz to 20kHz). In audio applications, total harmonic distortion plus noise (THD+N) is usually measured with specialized equipment, such as the Audio Precision System One. The output signal amplitude is measured at a given frequency (e.g., 1kHz); then the fundamental signal is removed with a bandstop filter, and the system measures the rms value of the remaining frequency components, which contain both harmonics and noise. The noise and harmonics are measured over a bandwidth that will include the highest harmonics, usually about 100kHz. The measurement is swept over the frequency range for various conditions.

THD+N results for the OP-275 are plotted in Figure 8.2 as a function of frequency. The signal level is 3V rms, and the amplifier is connected as a unity-gain follower. The data is shown for three load conditions: 600Ω, 2kΩ, and 10kΩ. Notice that a THD+N value of 0.0008% corresponds to 8ppm, or –102dBc. The input voltage noise of the OP-275 is typically 6nV/√Hz @ 1kHz, and integrated over an 80kHz noise bandwidth, yields an rms noise level of 1.7μV rms. For a 3V rms signal level, the corresponding signal-to-noise ratio is 125dB. Because the THD is considerably greater than the noise level, the THD component is the primary contributor. Multiple plots with variable bandwidths can be used to help separate noise and distortion.

THD + N FOR THE OP-275 OVER 100kHz BANDWIDTH IS DOMINATED BY DISTORTION

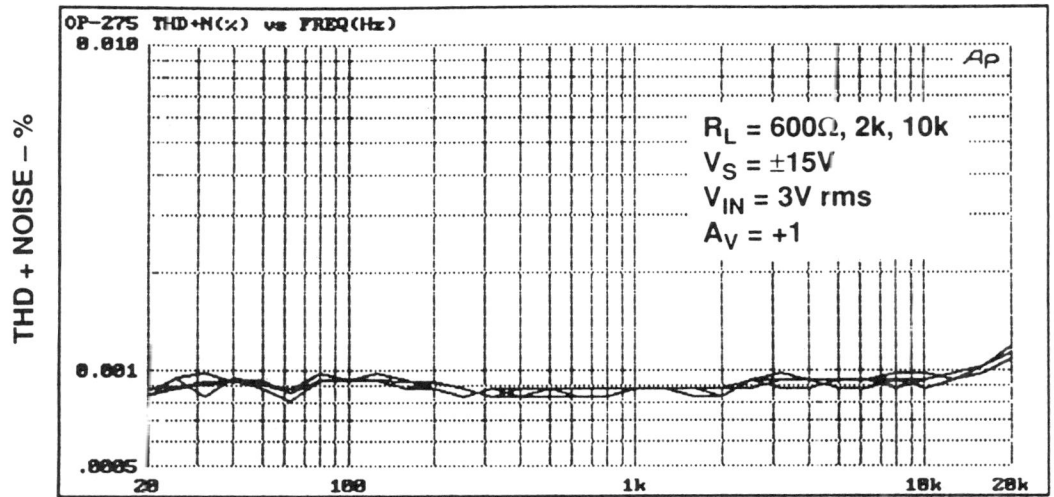

Figure 8.2

DISTORTION MEASUREMENTS

Now, consider the AD797, a low noise amplifier (1nV/√Hz) where measurement equipment distortion, and not the amplifier distortion, limits the measurement. The THD specification for the AD797 is 120dBc @ 20kHz, and a plot is shown in Figure 8.3. The distortion is at the limits of measurement of available equipment, and the actual amplifier noise is even lower by 20dB. The measurement was made with a spectrum analyzer by first filtering out the fundamental sinewave frequency ahead of the analyzer. This is to prevent overdrive distortion in the spectrum analyzer. The first five harmonics were then measured and combined in an RSS fashion to get the THD figure. The legend on the graph indicates that the measurement equipment "floor" is about 120dBc; hence at frequencies below 10kHz, the THD may be even less.

THD OF THE AD797 OP AMP SHOWS MEASUREMENT LIMIT AT –120dBc, WHILE AMPLIFIER NOISE FLOOR IS AT –140dBc (1nV/√Hz INTEGRATED OVER 100kHz BANDWIDTH)

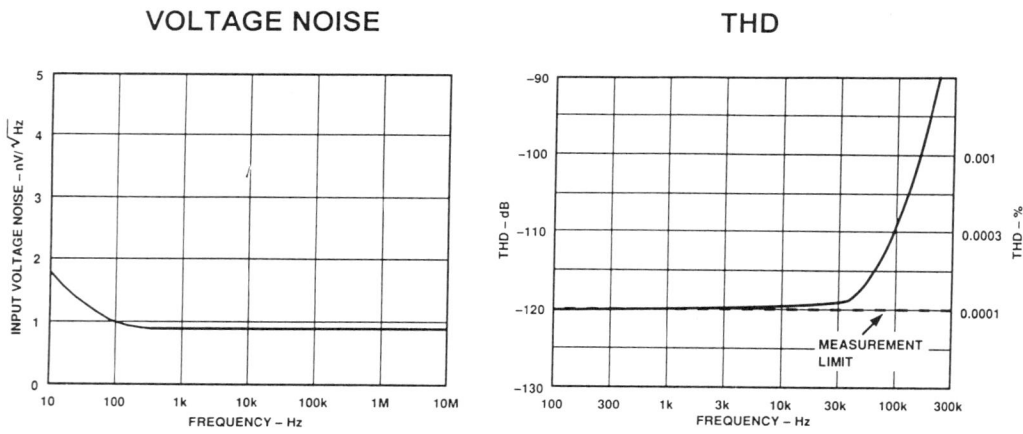

Figure 8.3

To calculate the AD797 noise, multiply the voltage noise spectral density (1nV/√Hz) by the square root of the measurement bandwidth to yield the device's rms noise floor. For a 100kHz bandwidth, the noise floor is 316nV rms, corresponding to a signal-to-noise ratio of about 140dB for a 3V rms output signal.

Rather than simply examining the THD produced by a single tone sinewave input, it is often useful to look at the distortion products produced by two tones. As shown in Figure 8.4, two tones will produce second and third order intermodulation products. The example shows the second and third order products produced by applying two frequencies, f_1 and f_2, to a nonlinear device. The second order products located at $f_2 + f_1$ and $f_2 - f_1$ are located far away from the two tones, and may be removed by filtering. The third order products located at $2f_1 + f_2$ and $2f_2 + f_1$ may likewise be filtered. The third order products located at $2f_1 - f_2$ and $2f_2 - f_1$, however, are close to the original tones, and filtering them is difficult. Third order IMD products are especially troublesome in multi-channel communications systems where the channel separation is constant across the frequency band.

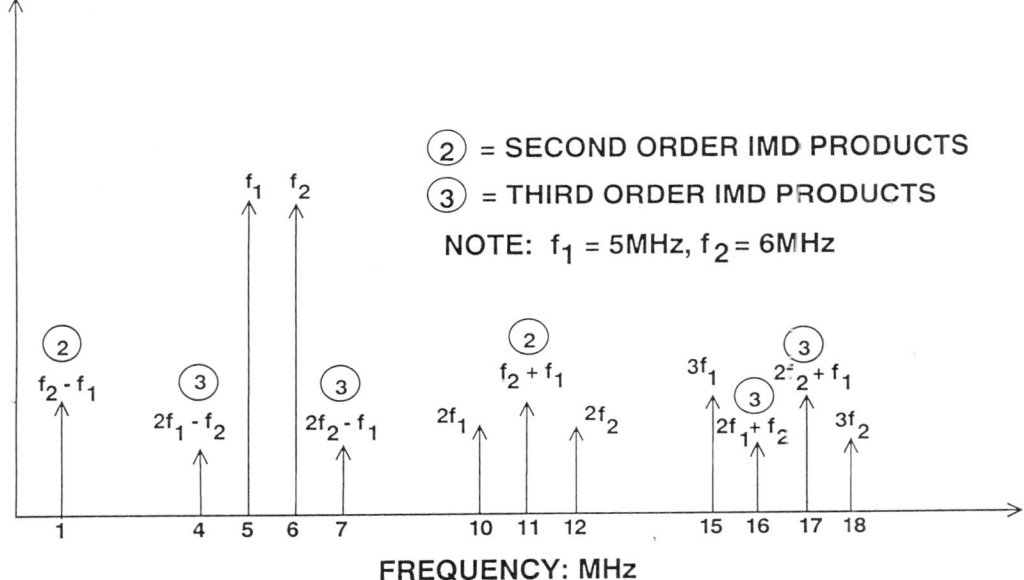

Figure 8.4

Intermodulation distortion products are of special interest in the RF area, and a major concern in the design of radio receivers. Third-order IMD products can mask out small signals in the presence of larger ones. Third order IMD is often specified in terms of the *third order intercept* point as shown in Figure 8.5. Two spectrally pure tones are applied to the system. The output signal power in a single tone (in dBm) as well as the relative amplitude of the third-order products (referenced to a single tone) is plotted as a function of input signal power. If the system non-linearity is approximated by a power series expansion, the second-order IMD amplitudes increase 2dB for every 1dB of signal increase. Similarly, the third-order IMD amplitudes increase 3dB for every 1dB of signal increase. With a low level two-tone input signal, and two data points,

DISTORTION MEASUREMENTS

draw the second and third order IMD lines as are shown in Figure 8.5, because one point and a slope determine each straight line.

Once the input reaches a certain level, however, the output signal begins to soft-limit, or compress. But the second and third-order intercept lines may be extended to intersect the extension of the output signal line. These intersections are called the *second-* and *third order intercept points*, respectively. The values are usually referenced to the output power of the device expressed in dBm. Another parameter which may be of interest is the *1dB compression point*. This is the point at which the output signal is compressed by 1dB from the ideal input/output transfer function. This point is also shown in Figure 8.5.

Knowing the third order intercept point allows calculation of the approximate level of the third-order IMD products as a function of output signal level. Figure 8.6 shows the third order intercept value as a function of frequency for the AD9622 voltage feedback amplifier.

Assume the op amp output signal is 5MHz and 2V peak-to-peak into a 100Ω load (50Ω source and load termination). The voltage into the 50Ω load is therefore 1V peak-to-peak, corresponding to +4dBm. The value of the third order intercept at 5MHz is 36dBm. The difference between +36dBm and +4dBm is 32dB. This value is then multiplied by 2 to yield 64dB (the value of the third-order intermodulation products referenced to the power in a single tone). Therefore, the intermodulation products should be –64dBc (dB below carrier frequency), or at a level of –60dBm. Figure 8.7 shows the graphical analysis for this example.

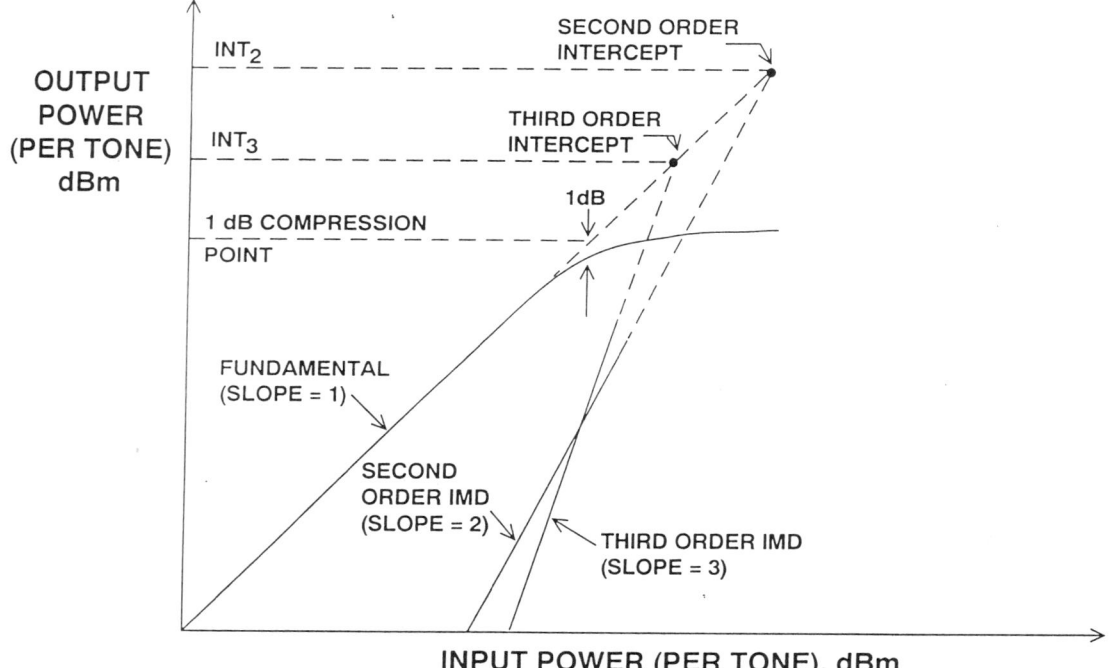

INTERCEPT POINTS, GAIN COMPRESSION, AND IMD

Figure 8.5

DISTORTION MEASUREMENTS

AD9622 THIRD ORDER IMD INTERCEPT VERSUS FREQUENCY

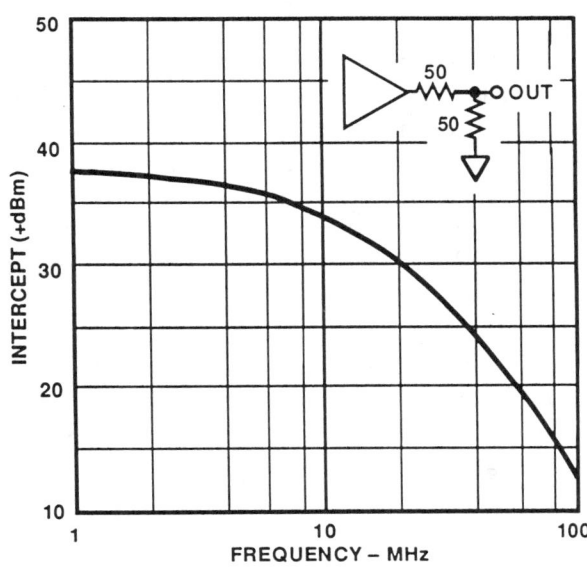

Figure 8.6

USING THE THIRD ORDER INTERCEPT POINT TO CALCULATE IMD PRODUCT FOR THE AD9622 OP AMP

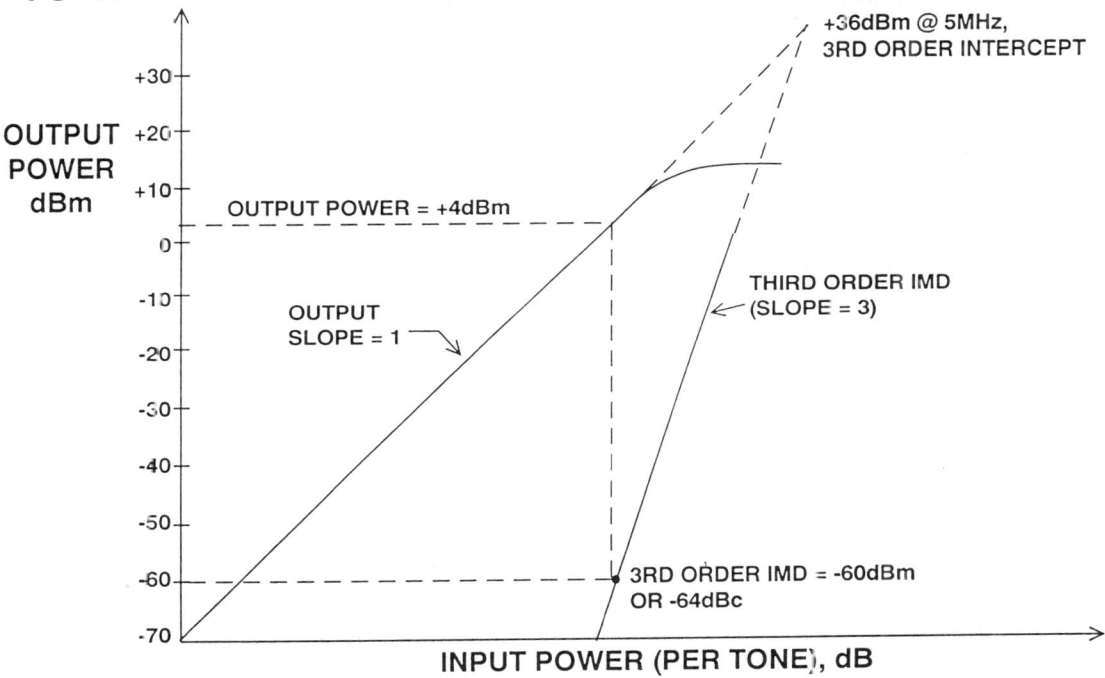

Figure 8.7

DISTORTION MEASUREMENTS

HIGH FREQUENCY TWO-TONE GENERATION

Generating test signals with the spectral purity required to make low distortion high frequency measurements is a challenging task. A test setup for generating a single tone is shown in Figure 8.8. The sinewave oscillator should have low phase noise (e.g., Marconi 2382), especially if the device under test is an ADC, where phase noise increases the ADC noise floor. The output of the oscillator is passed through a bandpass (or lowpass) filter which removes any harmonics present in the oscillator output. The distortion should be 6dB lower than the desired accuracy of the measurement. The 6dB attenuator isolates the DUT from the output of the filter. The impedance at each interface should be maintained at 50Ω for best performance (75Ω components can be used, but 50Ω attenuators and filters are generally more readily available). The termination resistor, R_T, is selected so that the parallel combination of R_T and the input impedance of the DUT is 50Ω.

Before performing the actual distortion measurement, the oscillator output should be set to the correct frequency and amplitude. Measure the distortion at the output of the attenuator with the DUT replaced by a 50Ω termination resistor (generally the 50Ω input of a spectrum analyzer. Next, replace the 50Ω load with R_T and the DUT. Measure the distortion at the DUT input a second time. This allows non-linear DUT loads to be identified. Non-linear DUT loads (such as flash ADCs with signal-dependent input capacitance, or switched-capacitor CMOS ADCs) can introduce distortion at the DUT input.

Generating two tones suitable for IMD measurements is even more difficult. A low-distortion two-tone generator is shown in Figure 8.9. Two bandpass (or lowpass) filters are required as shown. Harmonic suppression of each filter must be better than the desired measurement accuracy by at least 6dB. A 6dB attenuator at the output of each filter serves to isolate the filter outputs from each other and prevent possible cross-modulation. The outputs of the attenuators are combined in a passive 50Ω combining network, and the combiner drives the DUT. The oscillator outputs are set to the required level, and the IMD of the final output of the combiner is measured. The measurement should be made with a single termination resistor, and again with the DUT connected to identify non-linear loads.

DISTORTION MEASUREMENTS

LOW DISTORTION SINGLE-TONE GENERATOR

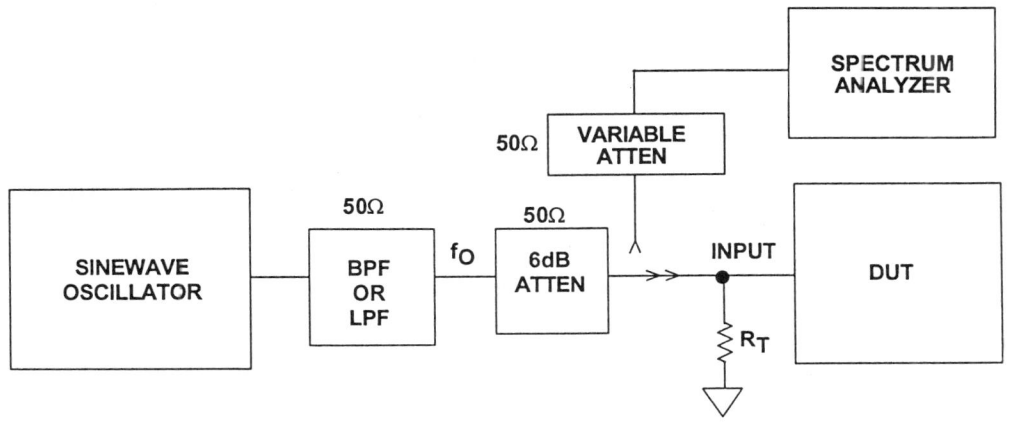

Figure 8.8

LOW DISTORTION TWO TONE GENERATOR

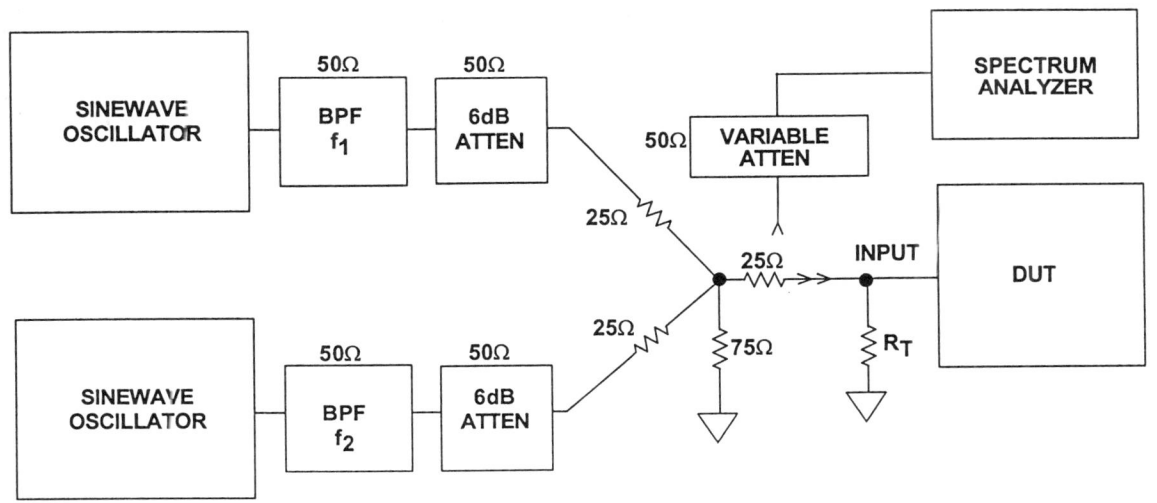

Figure 8.9

Using Spectrum Analyzers in High Frequency Low Distortion Measurements

Analog spectrum analyzers are most often used to measure amplifier distortion. Most have 50Ω inputs, therefore an isolation resistor between the device under test (DUT) and the analyzer is required to simulate DUT loads greater than 50Ω. After adjusting the spectrum analyzer for bandwidth, sweep rate, and sensitivity, check it carefully for input overdrive. The simplest method is to use the variable attenuator to introduce 10dB of attenuation in the analyzer input path. Both the signal and any harmonics should be attenuated by a fixed amount (10dB, for instance) as observed on the screen of the spectrum analyzer. If the harmonics are attenuated by more than 10dB, then the input amplifier of the analyzer is introducing distortion, and the sensitivity should be reduced. Many analyzers have a button on the front panel for introducing a known amount of attenuation when checking for overdrive.

Another method to minimize sensitivity to overdrive is shown in Figure 8.11. The amplitude of the fundamental signal is first measured with the notch filter switched out. The harmonics are measured with the notch filter switched in. The insertion loss of the notch filter, XdB, must be added to the measured level of the harmonics.

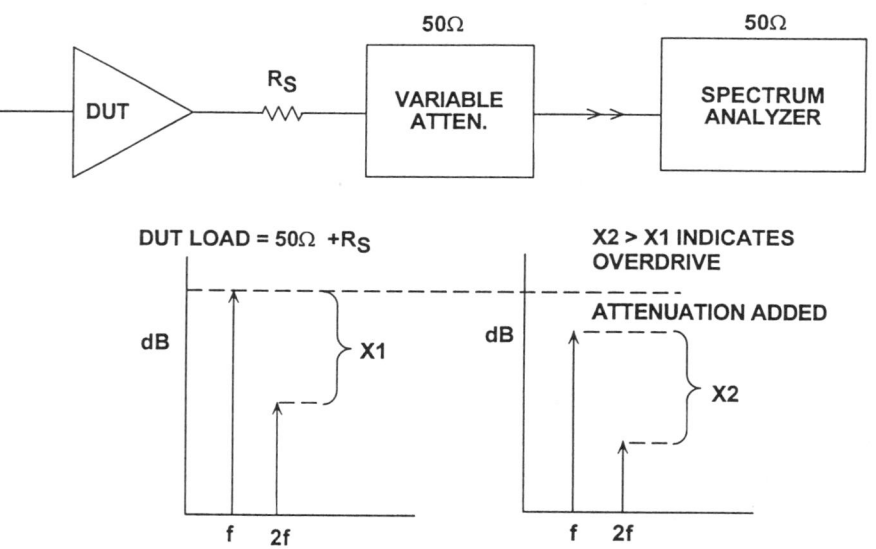

Figure 8.10

NOTCH FILTER REMOVES THE FUNDAMENTAL SIGNAL TO MINIMIZE ANALYZER OVERDRIVE

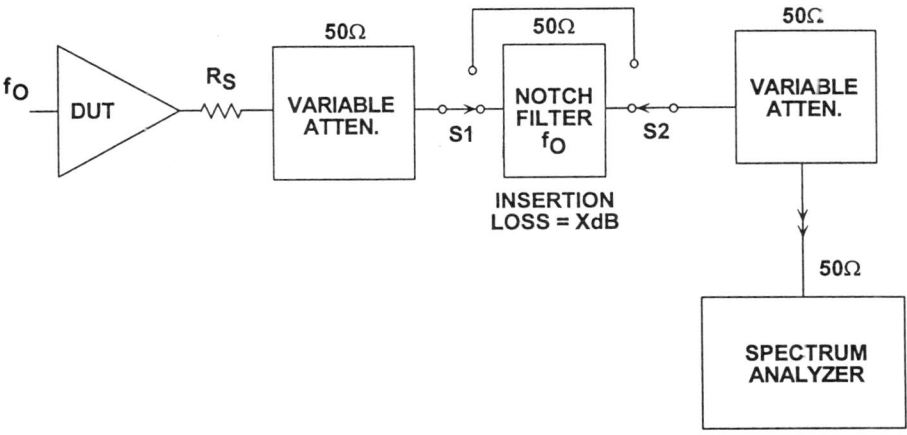

Figure 8.11

Measuring ADC Distortion Using FFTs

The speed of personal computers and the availability of suitable software now makes DSP bench testing of high speed ADCs relatively easy. A block diagram of a typical DSP PC-based test system is shown in Figure 8.12. In order to perform any DSP testing, the first requirement is a high speed buffer memory of sufficient width and depth. High speed logic analyzers make a convenient memory and eliminate the need for designing special hardware. The HP1663A is a 100MHz logic analyzer which has a simple IEEE-488 output port for easy interfacing to a personal computer. The analyzer can be configured as either a 16-bit wide by 8k deep, or a 32-bit wide by 4k deep memory. This is more than sufficient to test a high speed ADC at sample rates up to 100MHz. For higher sample rates, faster logic analyzers are available, but are fairly costly. An alternative to using a high speed logic analyzer is to operate the ADC at the desired sample rate, but only clock the final output register at an even sub-multiple of the sample clock frequency. This is sometimes called *decimation* and is useful for relaxing memory requirements. If an FFT is performed on the decimated output data, the fundamental input signal and its associated harmonics will be present, but translated in frequency. Simple algorithms can be used to find the locations of the signal and its harmonics provided the original signal frequency is known.

DISTORTION MEASUREMENTS

A SIMPLE PC-BASED TEST SYSTEM

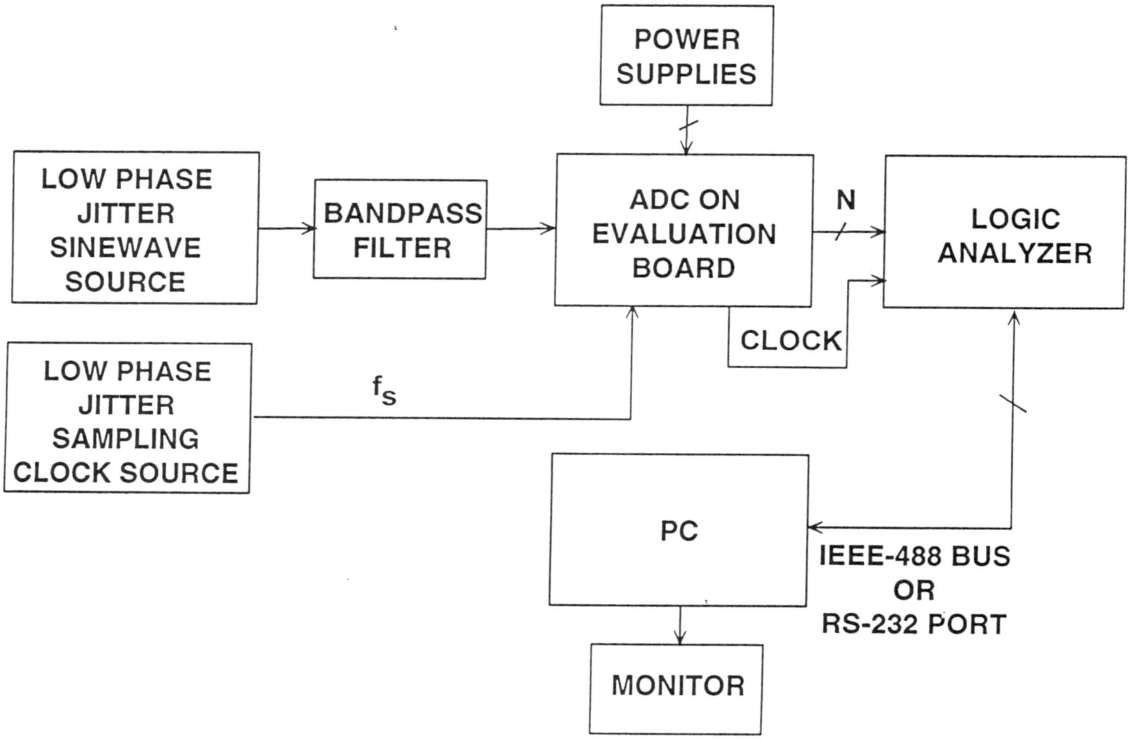

Figure 8.12

FFT Testing

Easy to use mathematical software packages, such as Mathcad™ (available from MathSoft, Inc., 201 Broadway, Cambridge MA, 02139) are available to perform fast FFTs on most 486-based PCs. The use of a co-processor allows a 4096-point FFT to run in a few seconds on a 75MHz, 486 PC. The entire system will run under the Windows environment and provide graphical displays of the FFT output spectrum. It can be programmed to perform SNR, S/(N+D), THD, IMD, and SFDR computations. A simple QuickBasic program transfers the data stored in the logic analyzer into a file in the PC via the IEEE-488 port (Reference 3, Section 16).

Proper understanding of FFT fundamentals is necessary in order to achieve meaningful results. The first step is to determine the number of samples, M, in the FFT record length. In order for the FFT to run properly, M must be a power of 2. The value of M determines the frequency *bin width*, $\Delta f = f_s/M$. The larger M, the more frequency resolution. Figure 8.13 shows the relationship between the average noise floor of the FFT with respect to the broadband quantization noise level. Each time M is doubled, the average noise in the Δf bandwidth decreases by 3dB. Larger values of M also tend to give more repeatable results from run to run (see Figure 8.13).

DISTORTION MEASUREMENTS

RELATIONSHIP BETWEEN AVERAGE NOISE IN FFT BINS AND BROADBAND RMS QUANTIZATION NOISE LEVEL

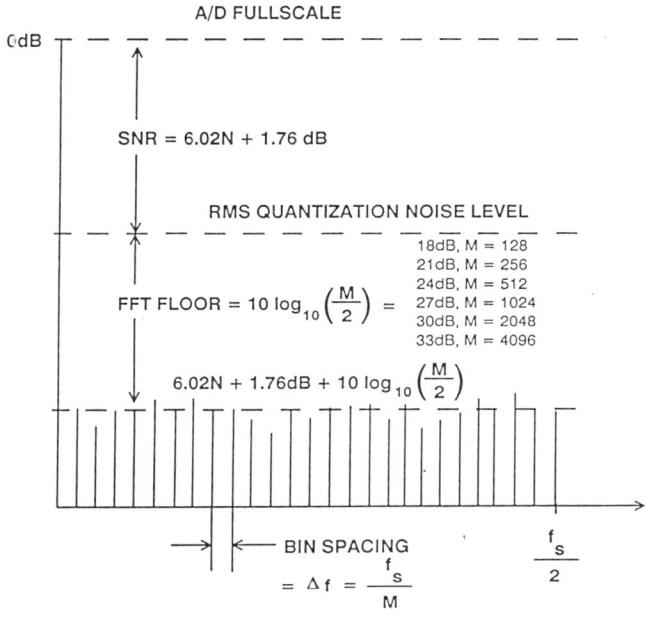

Figure 8.13

M values of 512 (for 8-bit ADCs), 2048 (for 10-bit ADCs), and 4096 (for 12-bit ADCs) have proven to give good accuracy and repeatability. For extremely wide dynamic range applications (such as spectral analysis) M=8192 may be desirable. It should be noted that averaging the results of several FFTs will tend to smooth out the noise floor, but will not change the average value of the floor.

In order to obtain spectrally pure results, the FFT data window must contain an exact integral number of sinewave cycles as shown in Figure 8.14. These frequency ratios must be precisely observed to prevent end-point discontinuity. In addition, it is desirable that the number of sinewave cycles contained within the data window be a prime number. This method of FFT testing is referred to as *coherent* testing because two locked frequency synthesizers are used to insure the proper ratio (coherence) between the sampling clock and the sinewave frequency. The requirements for coherent sampling are summarized in Figure 8.15.

FFT OF SINEWAVE HAVING INTEGRAL NUMBER OF CYCLES IN WINDOW

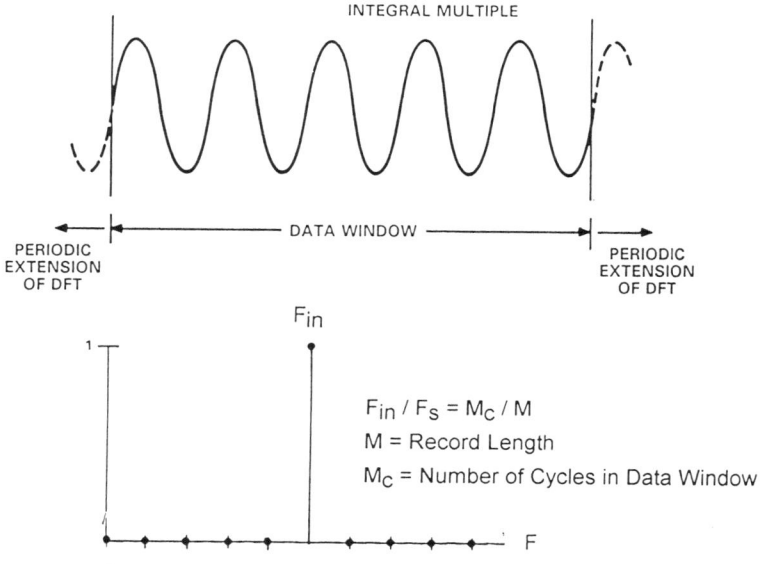

$F_{in} / F_s = M_c / M$
M = Record Length
M_c = Number of Cycles in Data Window

Figure 8.14

REQUIREMENTS FOR COHERENT SAMPLING

- f_s = Sampling Rate

- f_{in} = Input Sinewave Frequency

- M = Number of Samples in Record
  (Integer Power of 2)

- M_c = Prime Integer Number of Cycles of Sinewave
  During Record (Makes All Samples Unique)

- Make $\dfrac{f_{in}}{f_s} = \dfrac{M_c}{M}$

Figure 8.15

CHOOSING A PRIME NUMBER OF CYCLES WITHIN THE FFT RECORD LENGTH ENSURES RANDOMIZATION OF THE QUANTIZATION NOISE (IDEAL 12-BIT ADC)

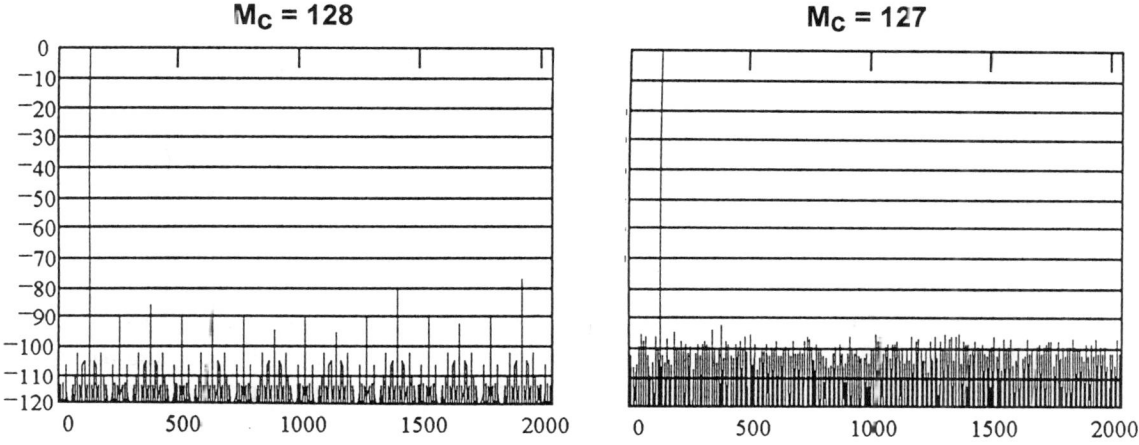

Figure 8.16

FFT OF SINEWAVE HAVING NON-INTEGRAL NUMBER OF CYCLES IN WINDOW

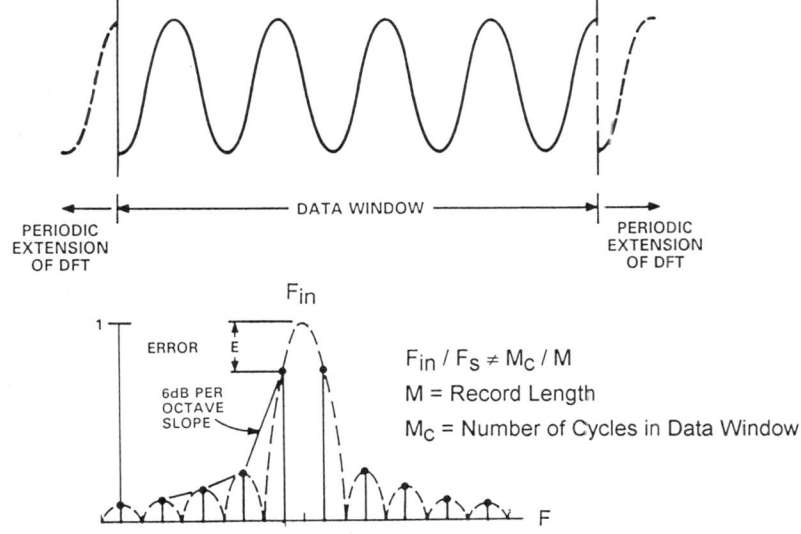

$F_{in} / F_S \neq M_C / M$
M = Record Length
M_C = Number of Cycles in Data Window

Figure 8.17

Distortion Measurements

Making the number of cycles within the record a prime number ensures a unique set of sample points within the data window. An even number of cycles within the record length will cause the quantization noise energy to be concentrated in the harmonics of the fundamental (causing a decrease in SFDR) rather than being randomly distributed over the Nyquist bandwidth. Figure 8.16 shows a 4096-point FFT output for a theoretically perfect 12-bit sinewave. The spectrum on the left was made with exactly 128 samples within the record length, corresponding to a frequency which is 1/32 times f_s. The SFDR is 78dB. The spectrum on the right was made with exactly 127 samples within the record, and the SFDR increases to 92dB.

Coherent FFT testing ensures that the fundamental signal occupies one discrete line in the output spectrum. Any leakage or smearing into adjacent bins is the result of aperture jitter, phase jitter on the sampling clock, or other unwanted noise due to improper layout, grounding, or decoupling.

If the ratio between the sampling clock and the sinewave frequency is such that there is and endpoint discontinuity in the data (shown in Figure 8.17), then spectral leakage will occur. The discontinuities are equivalent to multiplying the sinewave by a rectangular windowing pulse which has a sin(x)/x frequency response. The discontinuities in the time domain result in leakage or smearing in the frequency domain, because many spectral terms are needed to fit the discontinuity. Because of the endpoint discontinuity, the FFT spectral response shows the main lobe of the sinewave being smeared, and a large number of associated sidelobes which have the basic characteristics of the rectangular time pulse. This leakage must be minimized using a technique called *windowing* (or *weighting*) in order to obtain usable results in non-coherent tests.

This situation is exactly what occurs in real-world spectral analysis applications where the exact frequencies being sampled are unknown and uncontrollable. Sidelobe leakage is reduced by choosing a *windowing* (or *weighting*) function other than the rectangular window. The input time samples are multiplied by an appropriate windowing function which brings the signal to zero at the edges of the window. The selection of an appropriate windowing function is primarily a tradeoff between main-lobe spreading and sidelobe rolloff.

The time-domain and frequency-domain characteristics of a simple windowing function (the Hanning Window) are shown in Figure 8.18. A comparison of the frequency response of the Hanning window and the more sophisticated Minimum 4-Term Blackman-Harris window is given in Figures 8.19 and 8.20. For general ADC testing with non-coherent input frequencies, the Hanning window will give satisfactory results. For critical spectral analysis or two-tone IMD testing, the Minimum 4-Term Blackman-Harris window is the better choice because of the increase in spectral resolution.

TIME AND FREQUENCY REPRESENTATION OF THE HANNING WINDOW

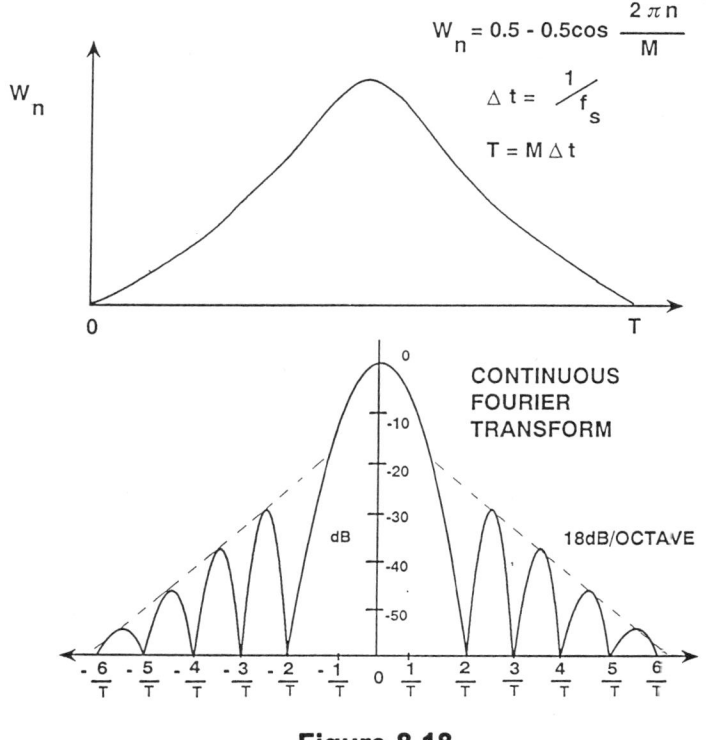

Figure 8.18

FREQUENCY RESPONSE OF THE HANNING WINDOW

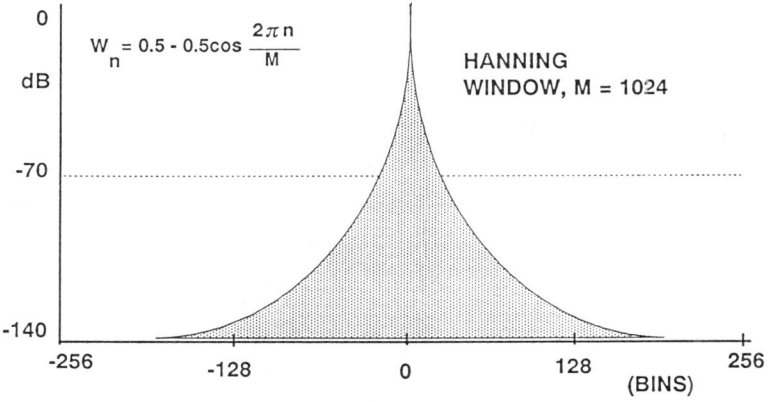

Figure 8.19

DISTORTION MEASUREMENTS

FREQUENCY RESPONSE OF THE MINIMUM 4-TERM BLACKMAN-HARRIS WINDOW

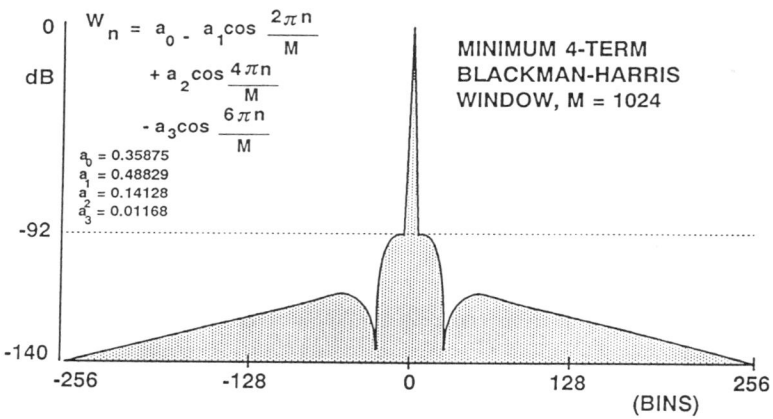

Figure 8.20

The addition of a windowing function to the FFT software involves first calculating the proper coefficient for each time sample within the record. These values are then stored in a memory file. Each time sample is multiplied by its appropriate weighting coefficient before performing the actual FFT. The software routine is easy to implement in QuickBasic.

When analyzing the FFT output resulting from windowing the input data samples, care must be exercised in determining the energy in the fundamental signal and the energy in the various spurious components. For example, sidelobe energy from the fundamental signal should not be included in the rms noise measurement. Consider the case of the Hanning Window function being used to test a 12-bit ADC with a theoretical SNR of 74dB. The sidelobe attenuation of the Hanning Window is as follows:

Bins From Fundamental	Sidelobe Attenuation
2.5	32dB
5.0	50dB
10.0	68dB
20.0	86dB

Therefore, in calculating the rms value of the fundamental signal, you should include at least 20 samples on either side of the fundamental as well as the fundamental itself.

If other weighting functions are used, their particular sidelobe characteristics must be known in order to accurately calculate signal and noise levels.

A typical Mathcad™ FFT output plot is shown in Figure 8.21 for the AD9022 12-bit, 20MSPS ADC using the Hanning Window and a record length of 4096.

MATHCAD™ 4096 POINT FFT OUTPUTS FOR AD9022 12-BIT, 20MSPS ADC (HANNING WEIGHTING)

F_S = 20 MSPS

F_{in} = 0.54 MHz

F_{in} = 9.3 MHz

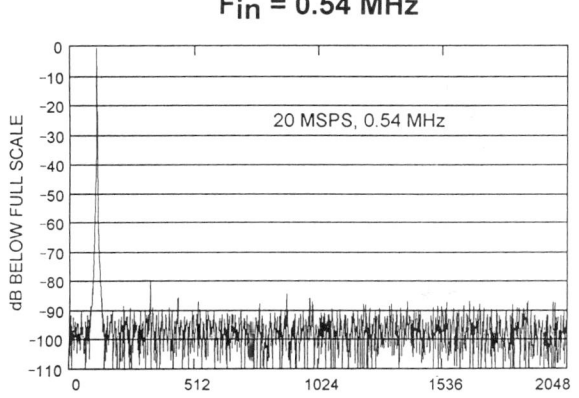
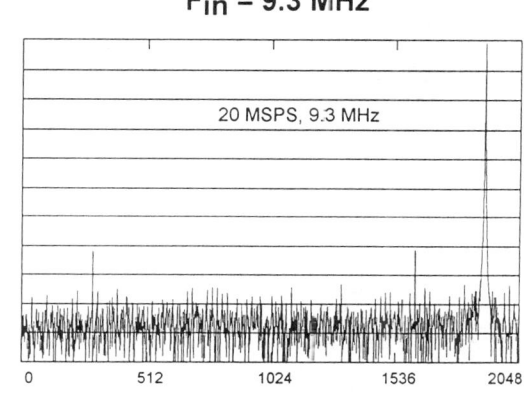

Figure 8.21

The actual QuickBasic routine for transferring the HP analyzer's data to a DOS file in the PC as well as the Mathcad™ routine are given in Reference 3, Section 16.

TROUBLESHOOTING THE FFT OUTPUT

Erroneous results are often obtained the first time an FFT test setup is put together. The most common error is improper timing of the latch strobe to the buffer memory. The HP1663A logic analyzer accepts parallel data and a clock signal. It has an internal DAC which may be used to examine a record of time samples. Large glitches on the stored waveform probably indicate that the timing of the latch strobe with respect to the data should be changed.

After ensuring correct timing, the FFT routine should produce a reasonable spectral output. If there are large values of harmonics, the input signal may be overdriving the ADC at one or both ends of the range. After bringing the signal within the ADC range (usually about 1dB below fullscale), excess harmonic content becomes more difficult to isolate.

Make sure that the sinewave input to the ADC is spectrally pure. Bandpass filters are usually required to clean up the output of most high frequency oscillators, especially if wide dynamic range is expected.

After ensuring the spectral purity of the ADC input, make sure the data output lines are not coupling to either the

DISTORTION MEASUREMENTS

sampling clock or to the ADC analog input. Remember that the glitches produced on the digital lines are signal-dependent and will therefore contribute to harmonic distortion if they couple into either one of these two lines. As has been discussed previously, noise or digital modulation on the sampling clock can also produce harmonic distortion in the FFT output. The use of an evaluation board with separate sampling clock and analog input connectors will usually prevent this. The special ribbon cable used with the logic analyzer to capture the ADC output data has a controlled impedance and should not cause performance degradation.

In addition to the above hardware checks, the FFT software should be verified by applying a theoretically perfect quantized sinewave to the FFT and comparing the results to theoretical SNR, etc. This is easy to do using the "roundoff" function available in most math packages. The effects of windowing non-coherent inputs should also be examined before running actual ADC tests.

In performing calculations with the FFT output, the term at dc and $f_s/2$ should be omitted from any calculations, as they can produce erroneous results. Input frequencies which are integer submultiples of the sampling clock can also produce artificially large harmonics.

TROUBLESHOOTING THE FFT OUTPUT

- Excess harmonic distortion:
    - Distortion on input signal
    - Signal outside ADC input range
    - Digital runs coupling into analog input or sampling clock
    - Poor layout, decoupling, and grounding
    - Buffer memory not clocked at correct time
    - Analog input frequency locked to integer submultiple of sampling clock

- Excess noise floor:
    - Noise or phase jitter on input signal
    - Noise or phase jitter on sampling clock
    - Poor layout, decoupling, and grounding
    - Eliminate dc and $f_s/2$ FFT components from calculations

Figure 8.22

ANALYZING THE FFT OUTPUT

Once the FFT output is obtained, it can be analyzed in a number of ways, similar to that of the display on an analog spectrum analyzer. The spurious free dynamic range (SFDR) is the ratio of the fundamental signal to the worst frequency spur. Total harmonic distortion (THD) is obtained by taking the ratio of the signal to the rms value of the first several harmonics (and then taking the logarithm of the ratio). Because of aliasing, however, locating the harmonics in the frequency spectrum can be difficult. For instance, if a 3MHz signal is sampled at 10MSPS, the second harmonic (6MHz) actually appears in the FFT output at 4MHz (10MHz − 6MHz). The third harmonic (9MHz) appears at 1MHz (10MHz − 9MHz). The fourth harmonic (12MHz) appears at 2MHz (12MHz − 10MHz). Software routines to perform these calculations are easily written.

Two tone intermodulation distortion can be measured by applying two spectrally pure tones to the ADC using the circuit previously shown in Figure 8.9. The concept of second- or third-order intercept point has little meaning when testing ADCs for two reasons. First, the ADC acts as a hard limiter for out-of-range signals, while an amplifier soft limits. Second, as the amplitude of the tones is reduced, the value of the signal-related frequency spurs tends to become somewhat constant because of the discontinuous nature of the ADC transfer function. A "hard distortion" floor is reached beyond which further reduction in signal amplitude has little effect on the spur levels.

Finally, signal-to-noise plus distortion (S/N+D) can be calculated by taking the ratio of the rms signal amplitude to the rms value of all other spectral components (excluding dc and $f_s/2$). From the S/N+D value, the effective number of bits (ENOB) can be calculated. In some applications, the value of S/N+D without the harmonics included is of interest.

Because of the statistical nature of the FFT analysis, there will be some variability in the output from run to run under identical test conditions. The data can be stabilized by averaging the results of several FFT runs. This will not lower the average noise floor of the FFT, but will reduce the varation in the results.

ANALYZING THE FFT OUTPUT

- Signal-to-Noise including Distortion: S/(N+D)
- Effective Number of Bits: (ENOB)
- Signal-to-Noise without distortion: SNR
- Spurious Free Dynamic Range: SFDR
- Harmonic Distortion
- Total Harmonic Distortion: THD
- THD + Noise (Same as S/N +D)
- Two-Tone Intermodulation Distortion

Figure 8.23

REFERENCES

1. Robert A. Witte, *Distortion Measurements Using a Spectrum Analyzer*, **RF Design**, September, 1992, pp. 75-84.

2. Walt Kester, *Confused About Amplifier Distortion Specs?*, **Analog Dialogue**, 27-1, 1993, pp. 27-29.

3. **System Applications Guide**, Analog Devices, 1993, Chapter 16.

4. Frederick J. Harris, *On the Use of Windows for Harmonic Analysis with the Discrete Fourier Transform*, **IEEE Proceedings**, Vol. 66, No. 1, Jan. 1978, pp. 51-83.

5. Joey Doernberg, Hae-Seung Lee, David A. Hodges, *Full Speed Testing of A/D Converters*, **IEEE Journal of Solid State Circuits**, Vol. SC-19, No. 6, Dec. 1984, pp. 820-827.

6. Brendan Coleman, Pat Meehan, John Reidy and Pat Weeks, *Coherent Sampling Helps When Specifying DSP A/D Converters*, **EDN**, October 15, 1987, pp. 145-152.

7. Robert W. Ramierez, **The FFT: Fundamentals and Concepts**, Prentice-Hall, 1985.

8. R. B. Blackman and J. W. Tukey, **The Measurement of Power Spectra**, Dover Publications, New York, 1958.

9. James J. Colotti, "Digital Dynamic Analysis of A/D Conversion Systems Through Evaluation Software Based on FFT/DFT Analysis", **RF Expo East 1987 Proceedings**, Cardiff Publishing Co., pp. 245-272.

10. **HP Journal**, Nov. 1982, Vol. 33, No. 11.

11. **HP Product Note** 5180A-2.

12. **HP Journal**, April 1988, Vol. 39, No. 2.

13. **HP Journal**, June 1988, Vol. 39, No. 3.

14. Dan Sheingold, Editor, **Analog-to-Digital Conversion Handbook, Third Edition**, Prentice-Hall, 1986.

15. W. R. Bennett, "Spectra of Quantized Signals", **Bell System Technical Journal**, No. 27, July 1948, pp. 446-472.

16. Lawrence Rabiner and Bernard Gold, **Theory and Application of Digital Signal Processing**, Prentice-Hall, 1975.

17. Matthew Mahoney, **DSP-Based Testing of Analog and Mixed-Signal Circuits**, IEEE Computer Society Press, Washington, D.C., 1987.

18. **IEEE Trial-Use Standard for Digitizing Waveform Recorders**, No. 1057-1988.

19. Richard J. Higgins, **Digital Signal Processing in VSLI**, Prentice-Hall, 1990.

20. M. S. Ghausi and K. R. Laker, **Modern Filter Design: Active RC and Switched Capacitors**, Prentice Hall, 1981.

21. Mathcad™ 4.0 software package available from MathSoft, Inc., 201 Broadway, Cambridge MA, 02139.

22. **System Applications Guide**, Analog Devices, 1993, Chapter 8 (Audio Applications), Chapter 9.

SECTION 9

HARDWARE DESIGN TECHNIQUES

- Prototyping Analog Circuits
- Evaluation Boards
- Noise Reduction and Filtering for Switching Power Supplies
- Low Dropout References and Regulators
- EMI/RFI Considerations
- Sensors and Cable Shielding

Hardware Design Techniques

SECTION 9

HARDWARE DESIGN TECHNIQUES
*Walt Kester, James Bryant, Walt Jung,
Adolfo Garcia, John McDonald*

PROTOTYPING AND SIMULATING ANALOG CIRCUITS
Walt Kester, James Bryant

While there is no doubt that computer analysis is one of the most valuable tools that the analog designer has acquired in the last decade or so, there is equally no doubt that analog circuit models are not perfect and must be verified with hardware. If the initial test circuit or "breadboard" is not correctly constructed, it may suffer from malfunctions which are not the fault of the design but of the physical structure of the breadboard itself. This section considers the art of successful breadboarding of high performance analog circuits.

Real electronic circuits contain many "components" which were not present in the circuit diagram, but which are there because of the physical properties of conductors, circuit boards, IC packages, etc. These components are difficult, if not impossible, to incorporate into computer modeling software, and yet they have substantial effects on circuit performance at high resolutions, or high frequencies, or both.

It is therefore inadvisable to use SPICE modeling or similar software to predict the ultimate performance of such high performance analog circuits. After modeling is complete, the performance must be verified by experiment.

This is not to say that SPICE modeling is valueless - far from it. Most modern high performance analog circuits could never have been developed without the aid of SPICE and similar programs, but it must be remembered that such simulations are only as good as the models used, and these models are not perfect. We have seen the effects of parasitic components arising from the conductors, insulators and components on the PCB, but it is also necessary to appreciate that the models used within SPICE simulations are not perfect models.

Consider an operational amplifier. It contains some 20-40 transistors, almost as many resistors, and a few capacitors. A complete SPICE model will contain all these components, and probably a few of the more important parasitic capacitances and spurious diodes formed by the diffusions in the op-amp chip. This is the model that the designer will have used to evaluate the device during his design. In simulations, such a model will behave very like the actual op-amp, but not exactly.

SPICE MODELING

- SPICE modeling is a powerful tool for predicting the performance of analog circuits.

- Analog Devices provides macromodels for over 450 ICs

HOWEVER

- Models omit real-life effects

- No model can simulate all the parasitic effects of discrete components and a PCB layout.

THEREFORE

- Prototypes must be built and proven before production.

Figure 9.1

However, this model is not published, as it contains too much information which would be of use to other semiconductor companies who might wish to copy or improve on the design. It would also take far too long for a simulation of a system containing such models of a number of op-amps to reach a useful result. For these, and other reasons, the SPICE models of analog circuits published by manufacturers or software companies are "macro" models, which simulate the major features of the component, but lack some of the fine detail. Consequently, SPICE modeling does not always reproduce the exact performance of a circuit and should always be verified experimentally.

The basic principle of a breadboard is that it is a *temporary* structure, designed to test the performance of a circuit or system, and must therefore be easy to modify.

There are many commercial breadboarding systems, but almost all of them are designed to facilitate the breadboarding of *digital* systems, where noise immunities are hundreds of millivolts or more. (We shall discuss the exception to this generality later.) Non copper-clad Matrix board (Vectorboard, etc.), wire-wrap, and plug-in breadboard systems (Bimboard, etc.) are, without exception, unsuitable for high performance or high frequency analog breadboarding. They have too high resistance, inductance, and capacitance. Even the use of standard IC sockets is inadvisable.

Practical Breadboarding Techniques

The most practical technique for analog breadboarding uses a copper-clad board as a ground plane. The ground pins of the components are soldered directly to the plane and the other components are wired together above it. This allows HF decoupling paths to be very short indeed. All lead lengths should be as short as possible, and signal routing should separate high-level and low-level signals. Ideally the layout should be similar to the layout to be used on the final PCB. This approach is often referred to as "deadbug" because the ICs are often mounted upside down with their leads up in the air (with the exception of the ground pins, which are bent over and soldered directly to the ground plane). The upside-down ICs look liked deceased insects, hence the name.

Figure 9.2 shows a hand-wired breadboard based around two high speed op amps which gives excellent performance in spite of its lack of esthetic appeal. The IC op amps are mounted upside down on the copper board with the leads bent over. The signals are connected with short point-to-point wiring. The characteristic impedance of a wire over a ground plane is about 120Ω, although this may vary as much as ±40% depending on the distance from the plane. The decoupling capacitors are connected directly from the op amp power pins to the copper-clad ground. When working at frequencies of several hundred MHz, it is a good idea to use only one side of the board for ground. Many people drill holes in the board and connect both sides together with short pieces of wire soldered to both sides of the board. If care is not taken, however, this may result in unexpected ground loops between the two sides of the board, especially at RF frequencies.

Pieces of copper-clad may be soldered at right angles to the main ground plane to provide screening, or circuitry may be constructed on both sides of the board (with connections through holes) with the board itself providing screening. In this case, the board will need legs to protect the components on the underside from being crushed.

When the components of a breadboard of this type are wired point-to-point in the air (a type of construction strongly advocated by Robert A. Pease of National Semiconductor (Reference 1) and sometimes known as "bird's nest" construction) there is always the risk of the circuitry being crushed and resulting short-circuits. Also if the circuitry rises high above the ground plane, the screening effect of the ground plane is diminished, and interaction between different parts of the circuit is more likely. Nevertheless the technique is very practical and widely used because the circuit may so easily be modified.

Another "deadbug" prototype is shown in Figure 9.3. The board is single-sided copper clad with holes pre-drilled on 0.1" centers. Power busses are at the top and bottom of the board. The decoupling capacitors are used on the power pins of each IC. Because of the loss of copper area due to the pre-drilled holes, this technique does not provide as low a ground impedance as a completely covered copper-clad board.

HARDWARE DESIGN TECHNIQUES

"DEADBUG" PROTOTYPE TECHNIQUE

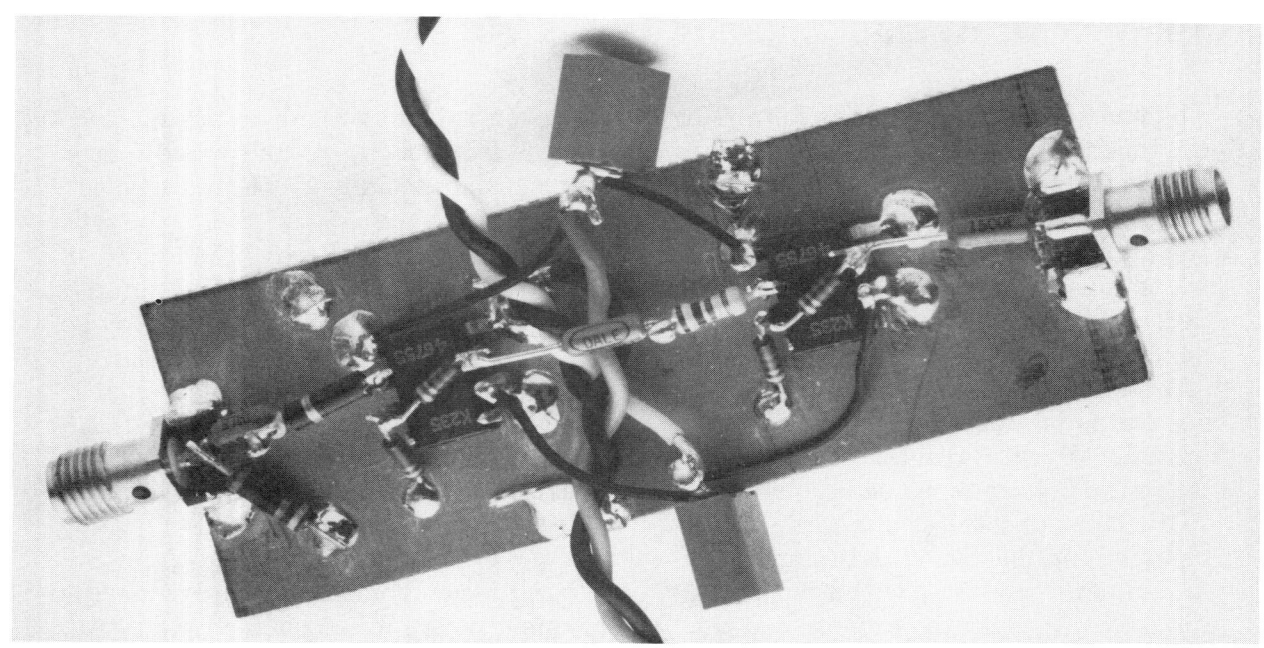

Figure 9.2

"DEADBUG" PROTOTYPE USING PRE-DRILLED SINGLE-SIDED COPPER-CLAD BOARD

Figure 9.3

Hardware Design Techniques

In a variation of this technique, the ICs and other components are mounted on the non-copper-clad side of the board. The holes are used as vias, and the point-to-point wiring is done on the copper-clad side of the board. The copper surrounding each hole used for a via must be drilled out so as to prevent shorting. This approach requires that all IC pins be on 0.1" centers. Low profile sockets can be used for low frequency circuits, and the socket pins allow easy point-to-point wiring.

IC sockets can degrade the performance of high speed or high precision analog ICs. Even "low-profile" sockets often introduce enough parasitic capacitance and inductance to degrade the performance of the circuit. If sockets must be used, an IC socket made of individual "pin sockets" (sometimes called "cage jacks") mounted in the ground plane board may be acceptable (clear the copper, on both sides of the board, for about 0.5mm around each ungrounded pin socket and solder the grounded ones to ground on both sides of the board. Both capped and uncapped versions of these pin sockets are available (AMP part numbers 5-330808-3, and 5-330808-6, respectively).

PIN SOCKETS (CAGE JACKS) HAVE MINIMUM PARASITIC RESISTANCE, INDUCTANCE, AND CAPACITANCE

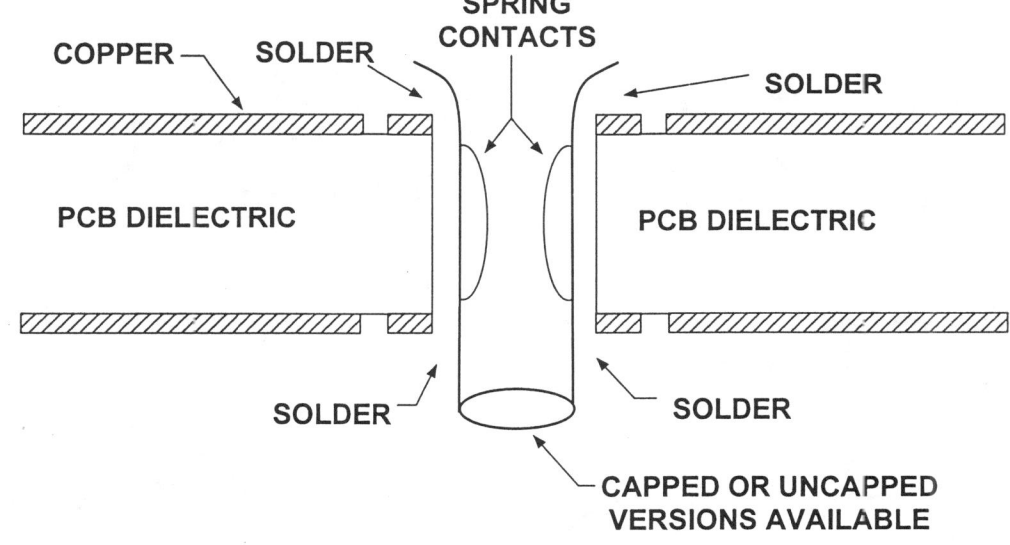

Figure 9.4

There is a commercial breadboarding system which has most of the advantages of "bird's nest over a ground plane, or deadbug" (robust ground, screening, ease of circuit alteration, low capacitance and low inductance) and several additional advantages:- it is rigid, components are close to the ground plane, and where necessary node capacitances and line impedances can be calculated easily. This system is made by Wainwright Instruments and is available in Europe as "Mini-Mount" and in the USA (where the trademark "Mini-Mount" is the property of another company) as "Solder-Mount"[Reference 2].

Solder-Mount consists of small pieces of PCB with etched patterns on one side and contact adhesive on the other. They are stuck to the ground plane, and components are soldered to them. They are available in a wide variety of patterns, including ready-made pads for IC packages of all sizes from 8-pin SOICs to 64-pin DILs, strips with solder pads at intervals (which intervals range from 0.040" to 0.25", the range includes strips with 0.1" pad spacing which may be used to mount DIL devices), strips with conductors of the correct width to form microstrip transmission lines (50Ω, 60Ω, 75Ω or 100Ω) when mounted on the ground plane, and a variety of pads for mounting various other components. A few of the many types of Solder-Mount building-block components are shown in Figure 9.5.

SAMPLES OF "SOLDER-MOUNT" COMPONENTS

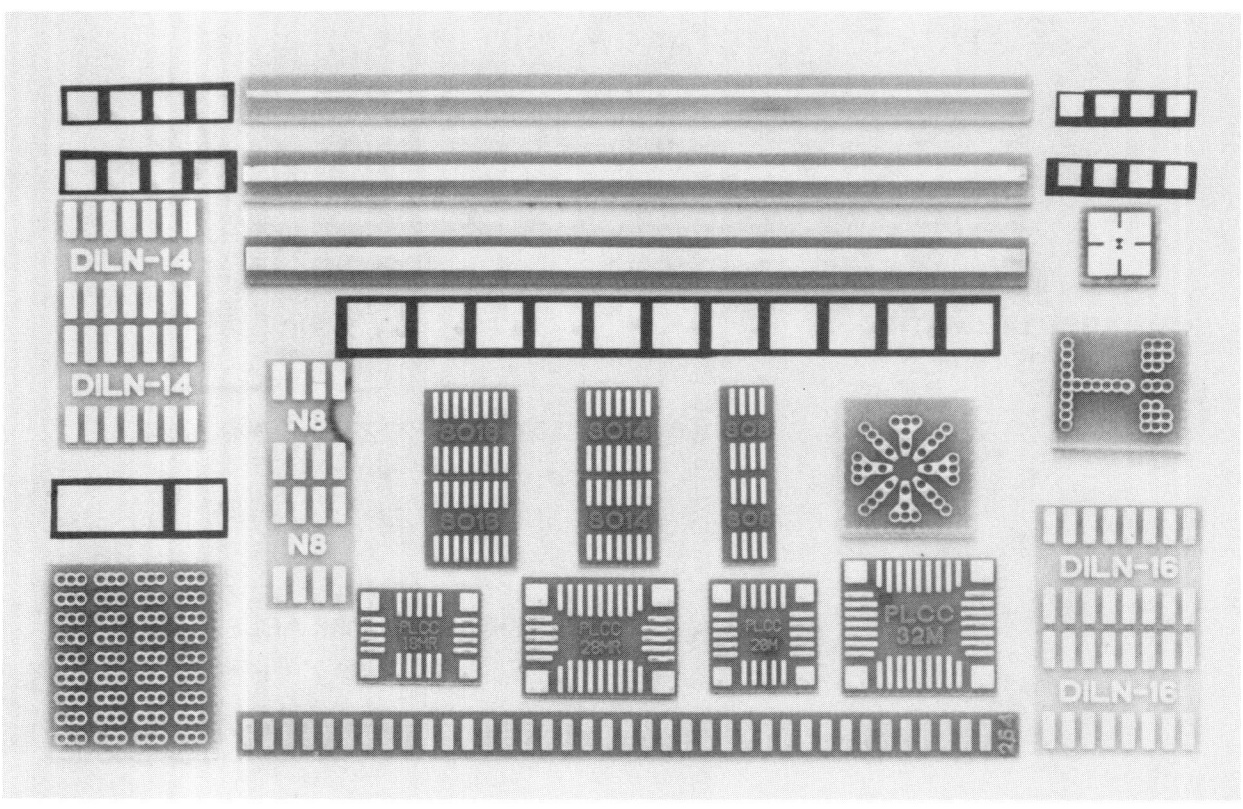

Figure 9.5

Hardware Design Techniques

The main advantage of Solder-Mount construction over "bird's nest" or "deadbug" is that the resulting circuit is far more rigid, and, if desired, may be made far smaller (the latest Solder-Mounts are for surface-mount devices and allow the construction of breadboards scarcely larger than the final PCB, although it is generally more convenient if the prototype is somewhat larger). Solder-Mount is sufficiently durable that it may be used for small quantity production as well as prototyping.

Figure 9.6 shows an example of a 2.5GHz phase-locked-loop prototype built with Solder-Mount. This is a high speed circuit, but the technique is equally suitable for the construction of high resolution low frequency analog circuitry. A particularly convenient feature of Solder-Mount at VHF is the ease with which it is possible to make a transmission line.

"SOLDER-MOUNT" PROTOTYPE

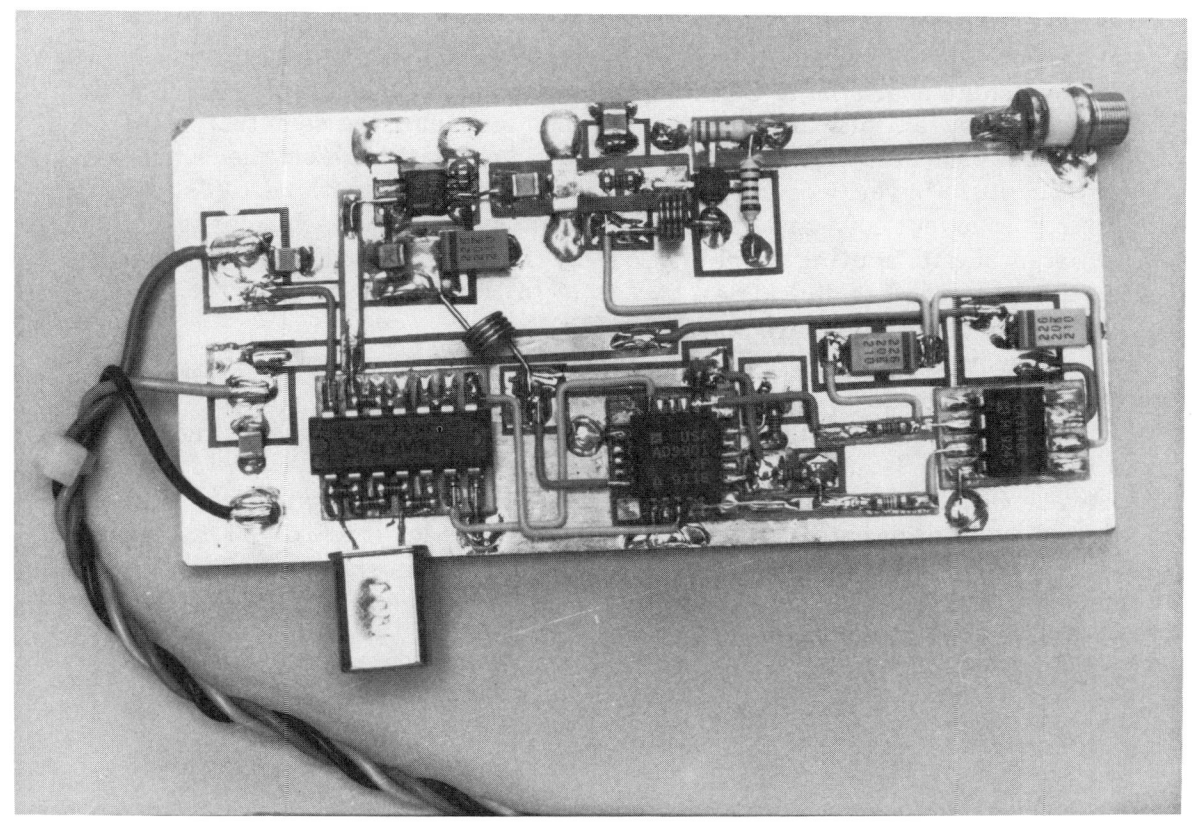

Figure 9.6

Hardware Design Techniques

If a conductor runs over a ground plane it forms a microstrip transmission line. Solder-Mount has strips which form microstrip lines when mounted on a ground plane (they are available with impedances of 50Ω, 60Ω, 75Ω, and 100Ω). These strips may be used as transmission lines, for impedance matching, or simply as power buses. (Glass fiber/epoxy PCB is somewhat lossy at VHF and UHF, but the losses will probably be tolerable if microstrip runs are short.)

Both the "deadbug" and the "Solder-Mount" breadboarding techniques become tedious for complex circuits. Larger circuits are often better prototyped using more formal layout techniques.

An approach to prototyping more complex analog circuits is to actually lay out a double-sided board using CAD techniques. PC-based software layout packages offer ease of layout as well as schematic capture to verify connections. Although most layout software has some amount of auto-routing capability, this feature is best left to digital designs. After the components are placed in their approximate position, the interconnections should be routed manually following good analog layout guidelines. After the layout is complete, the software verifies the connections per the schematic diagram net list.

Many design engineers find that they can use CAD techniques (Reference 3) to lay out simple boards themselves, or work closely with a layout person who has experience in analog circuit boards. The result is a pattern-generation tape (or Gerber file) which would normally be sent to a PCB manufacturing facility where the final board is made. Rather than use a PC board manufacturer, however, automatic drilling and milling machines are available which accept the PG tape (Reference 4). These systems produce single and double-sided circuit boards directly by drilling all holes and using a milling technique to remove copper and create insulation paths and finally, the finished board. The result is a board very similar to the final manufactured double-sided PC board, the chief exception being that there is no "plated-through" hole capability, and any "vias" between the two layers of the board must be wired and soldered on both sides. Minimum trace widths of 25 mils (1 mil = 0.001") and 12 mil spacing between traces are standard, although smaller trace widths can be achieved with care. The minimum spacing between lines is dictated by the size of the milling bit, typically 10 to 12 mils.

An example of such a prototype board is shown in Figure 9.7. This is a "daughter" board designed to interface an AD9562 Dual Pulse-Width Modulator in a 44-pin PLCC package to a "mother" board. The leads are on 50 mil centers, and the traces are approximately 25 mils wide.

"MILLED" PROTOTYPE PC BOARD

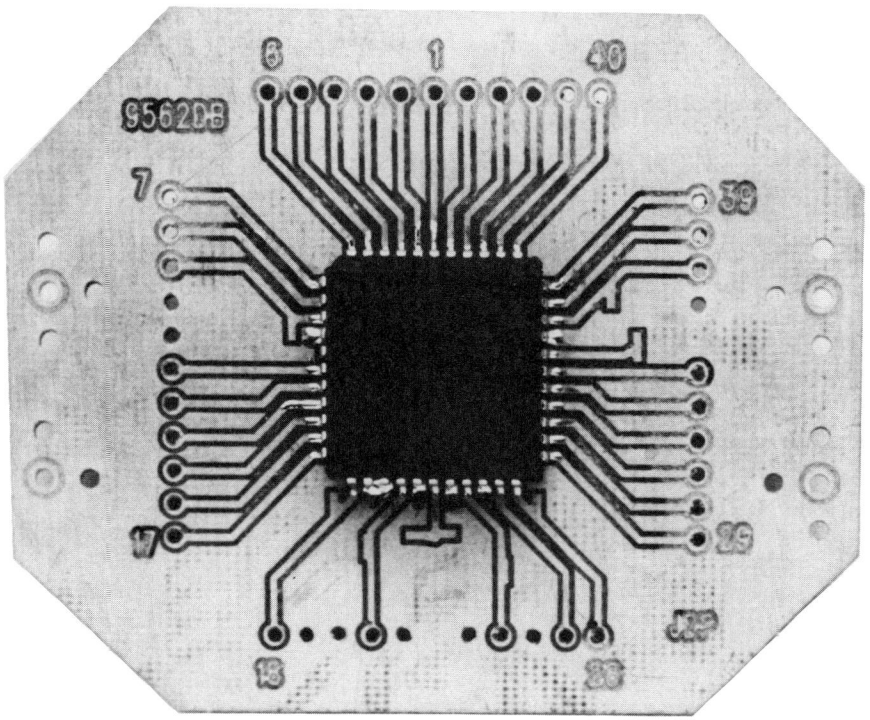

Figure 9.7

Multilayer PC boards do not easily lend themselves to standard prototyping techniques. One side of a double-sided board can be used for ground and the other side for power and signals. Point-to-point wiring can be used for additional runs which would normally be placed on the additional layers provided by a multi-layer board. However, it is difficult to control the impedance of the point-to-point wiring runs, and the high frequency performance of a circuit prototyped in this manner may differ significantly from the final multilayer board.

SUCCESSFUL PROTOTYPING

- Always use a ground plane for precision or high frequency circuits
- Minimize parasitic resistance, capacitance, and inductance
- If sockets are required, use "pin sockets" ("cage jacks")
- Pay equal attention to signal routing, component placement, grounding, and decoupling in both the prototype and the final design
- Popular prototyping techniques:
    - Freehand "deadbug" using point-to-point wiring
    - "Solder-Mount"
    - Milled PC board from CAD layout
    - Multilayer boards: Double-sided with additional point-to-point wiring

Figure 9.8

EVALUATION BOARDS

Manufacturer's evaluation boards provide a convenient way of evaluating high-performance ICs without the need for constructing labor-intensive prototype boards. Analog Devices provides evaluation boards for almost all new high speed and precision products. The boards are designed with good layout, grounding, and decoupling techniques. They are completely tested, and artwork (including PG tape) is available to customers.

Because of the popularity of dual precision op amps in 8-pin DIPs, a universal evaluation board has been developed (see Figure 9.9). This board makes extensive use of pin sockets to allow resistors or jumpers to configure the two op amps in just about any conceivable feedback, input/output, and load condition. The inputs and outputs are convenient right-angle BNC connectors. Because of the use of sockets and the less-than-compact layout, this board is not useful for op amps having gain-bandwidth products much greater than 10MHz.

UNIVERSAL EVALUATION BOARD FOR DUAL PRECISION OP AMPS IN 8-PIN DIPs

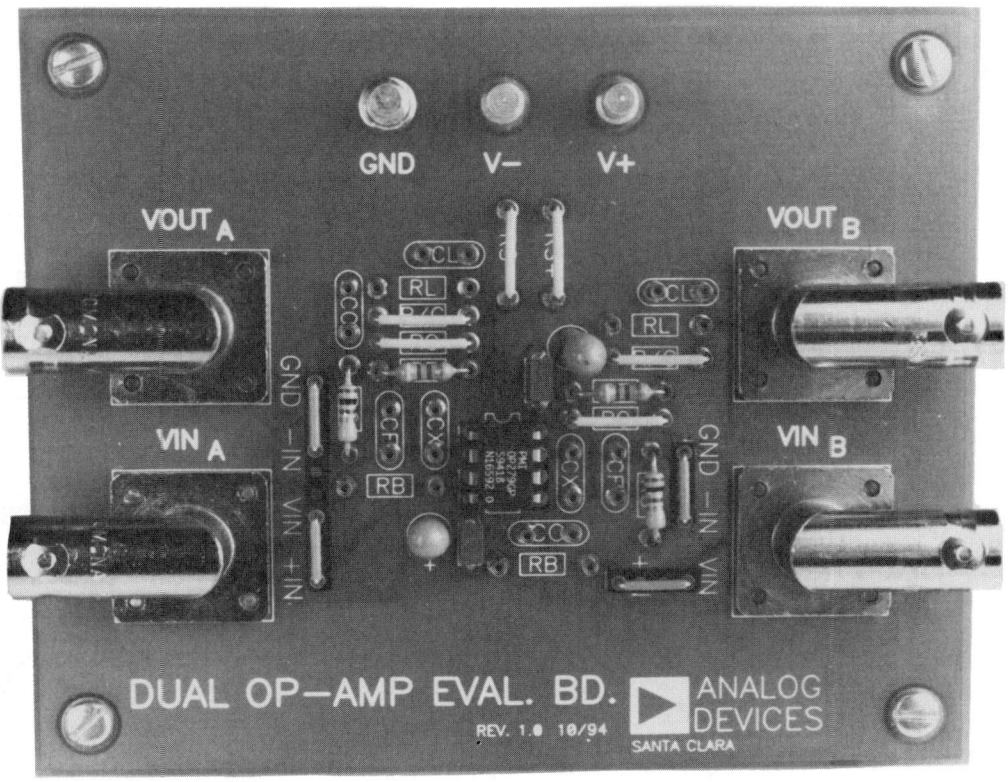

Figure 9.9

A schematic of the AD8001 800MHz 50mW current feedback op amp evaluation board for the 8-lead SOIC package is shown in Figure 9.10. The board is designed for a non-inverting gain of 2. (Boards for inverting and non-inverting modes are available for both the 8-lead SOIC and the DIP package).

Decoupling on both the + and − supplies consists of 1000pF and 0.01µF surface mount chip ceramic capacitors in addition to a 10µF/25V surface mount tantalum electrolytic. The top view of the PC board is shown in Figure 9.11, and the bottom view in Figure 9.12.

HARDWARE DESIGN TECHNIQUES

AD8001AR (SOIC) 800MHz OP AMP: NON-INVERTING MODE EVALUATION BOARD SCHEMATIC

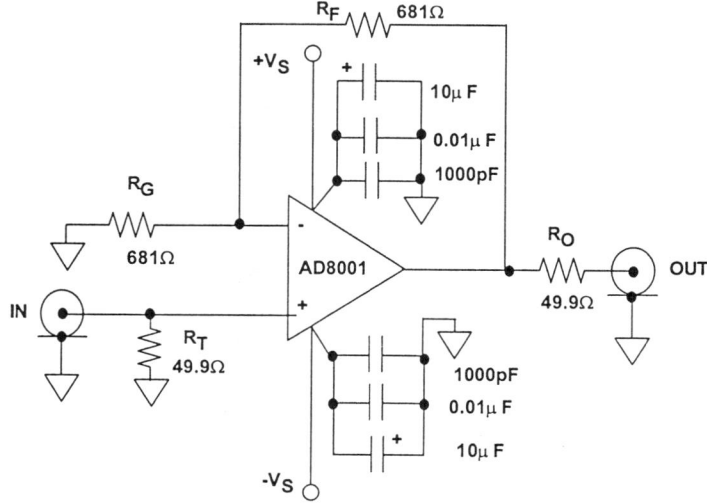

Figure 9.10

AD8001AR (SOIC) EVALUATION BOARD - TOP VIEW

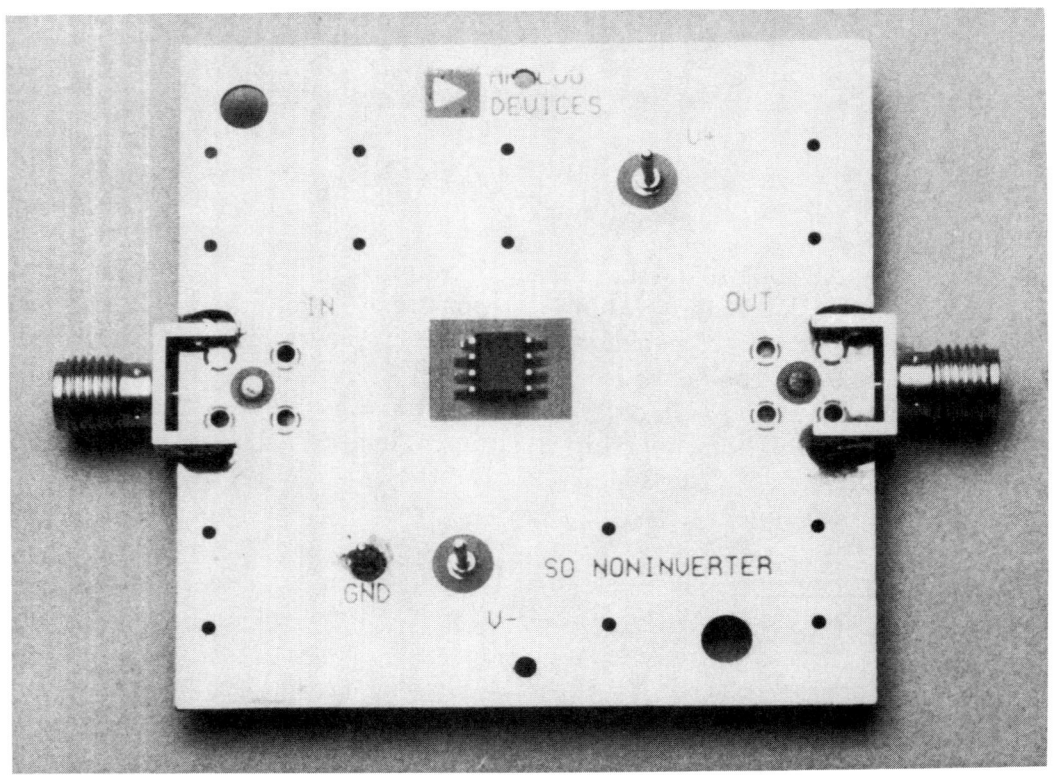

Figure 9.11

HARDWARE DESIGN TECHNIQUES

AD8001AR (SOIC) EVALUATION BOARD - BOTTOM VIEW

Figure 9.12

Figure 9.12 (bottom side of board) shows the surface mount resistors and capacitors. Notice that the ceramic chip capacitors are mounted as close as possible to the power pins as possible. The input and output runs are 50Ω transmission lines. The input from the SMA connector is terminated in a 50Ω chip resistor at the op amp, and the output has a 50Ω source termination for driving a 50Ω cable through the output SMA connector.

All resistors are surface mount film resistors. Notice that the ground plane is etched away from the area immediately surrounding the inputs of the op amp to minimize stray capacitance.

Slightly different resistor values are required to achieve optimum performance in the SOIC and the DIP packages (see Figure 9.13), because the SOIC package has slightly lower parasitic capacitance and inductance than the DIP. The criteria for selection of the components was maximum 0.1dB bandwidth.

Hardware Design Techniques

OPTIMUM VALUES OF R_F AND R_G FOR AD8001 DIP AND SOIC PACKAGES (MAXIMUM 0.1dB BANDWIDTH)

AD8001AN (DIP) GAIN					
Component	–1	+1	+2	+10	+100
R_F	649Ω	1050Ω	750Ω	470Ω	1000Ω
R_G	649Ω	-	750Ω	51Ω	10Ω
Small Signal BW	340MHz	880MHz	460MHz	260MHz	20MHz
0.1dB Flatness	105MHz	70MHz	105MHz	-	-

AD8001AR (SOIC) GAIN					
Component	–1	+1	+2	+10	+100
R_F	604Ω	953Ω	681Ω	470Ω	1000Ω
R_G	604Ω	-	681Ω	51Ω	10Ω
Small Signal BW	370MHz	710MHz	440MHz	260MHz	20MHz
0.1dB Flatness	130MHz	100MHz	120MHz	-	-

Figure 9.13

ADC evaluation boards include more support circuitry than op amp boards. An example is the AD7714 (22-bit precision measurement sigma-delta ADC) evaluation board (see Figure 9.14 for a simplified block diagram). Included on the evaluation board are an AD780 precision reference, a 2.4576MHz crystal and digital buffers to buffer the signal to and from the edge connectors.

AD7714 ADC EVALUATION BOARD

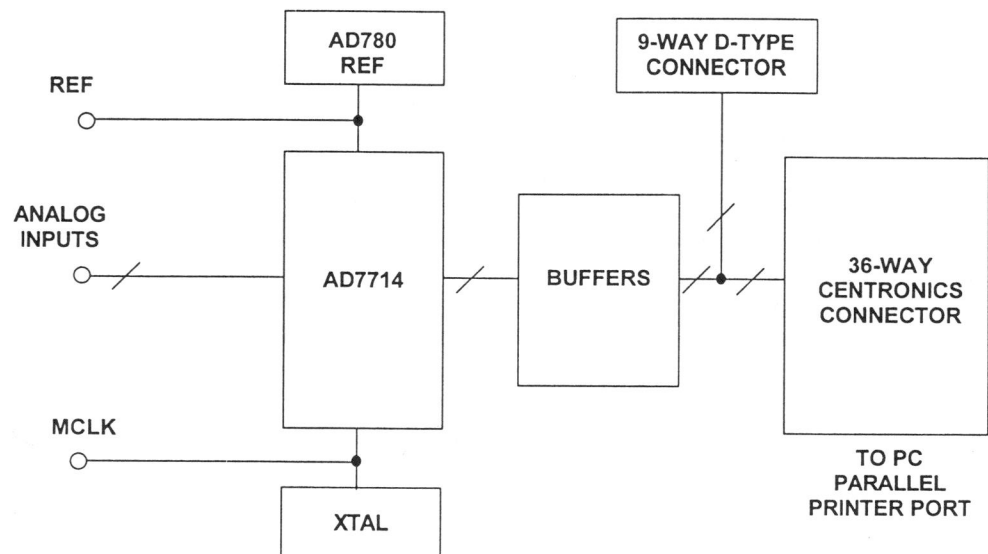

Figure 9.14

HARDWARE DESIGN TECHNIQUES

Interfacing to this board is provided either through a 36-Way Centronics connector or through a 9-Way D-type connector. The Centronics connector is intended for connection to the printer port of a PC. External sockets are provided for the analog inputs, an external reference input option, and an external master clock option.

Included in the evaluation board package is a PC-compatible DOS-based disk which contains software for controlling and evaluating the performance of the AD7714 using the printer port. There are two files on the disk, an executable file, and a "readme" text file which gives details of the functions available in the executable program. The evaluation software is run by running the executable file. The program provides a number of different menu-type screens, each screen containing several function options.

The first menu gives options on the type of PC being used. The next menu in the sequence is the Main Menu which contains various options. These allow reading from the AD7714 Data Register, configuration of the Communications Register, file options (read and write data to files), noise analysis, printer port setup, and resetting the AD7714.

The Noise Menu allows the user to get statistical results from the data, to plot the raw data, plot a histogram of the data on the screen, or perform a rolling average of the data. A photograph of the AD7714 evaluation board is shown in Figure 9.15. Notice the parallel printer connector on the left and the use of low-profile sockets for convenience. It should be noted that although the use of sockets is discouraged, any sockets used on evaluation boards have been proven to cause minimal performance degradation.

Evaluation boards for high speed sampling ADCs contain the required support circuitry for proper evaluation of the converter. Figure 9.16 shows a block diagram of the AD9026 (12-bit, 31MSPS) evaluation board. The analog input is connected directly to the ADC input via an SMA connector. The sampling clock (Encode Input) is conditioned by the low-jitter high-speed AD9698 dual comparator. The parallel digital outputs of the AD9026 are buffered by latches which drive the output connector as well as the AD9713 12-bit DAC. The DAC output is connected to an output SMA connector.

Hardware Design Techniques

AD7714 EVALUATION BOARD PHOTO

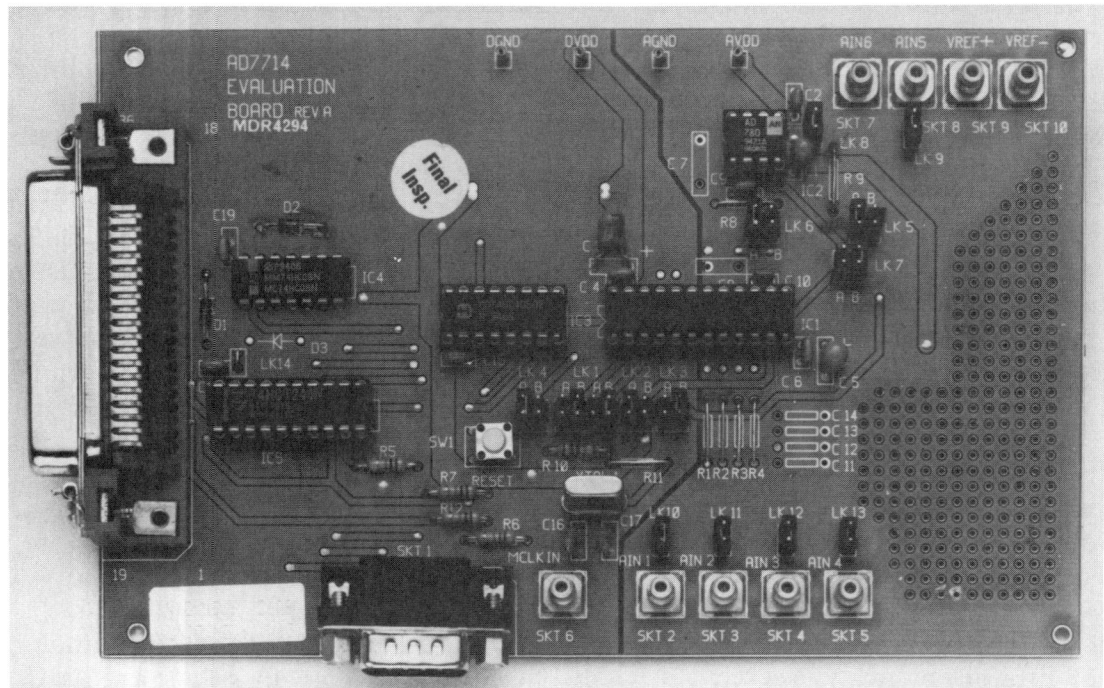

Figure 9.15

AD9026 12-BIT, 31MSPS ADC EVALUATION BOARD

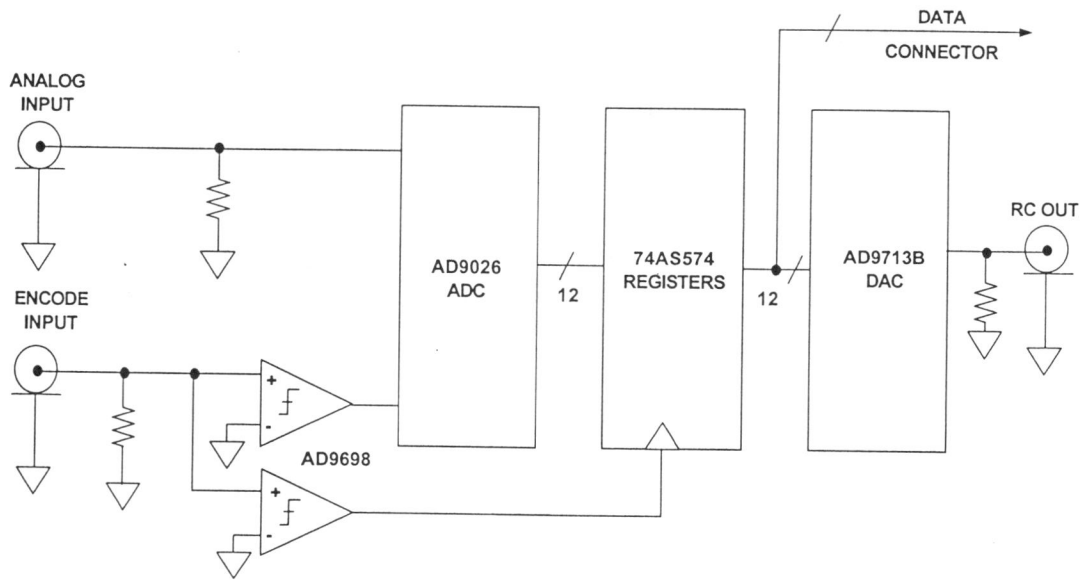

Figure 9.16

Hardware Design Techniques

The output connector is designed for convenient interfacing to an external buffer memory or to a logic analyzer input (a very convenient high speed buffer memory).

The top side of the board is shown in Figure 9.17. The board is a 3-layer board consisting of one ground plane (outer layer), one power/signal plane (inner layer), and an additional signal plane (outer layer). Pin sockets are used to mount the AD9026. Figure 9.18 shows the bottom side of the board and the surface mounted AD9698 SOIC comparator and the AD9713B PLCC DAC.

AD9026 EVALUATION BOARD - TOP VIEW

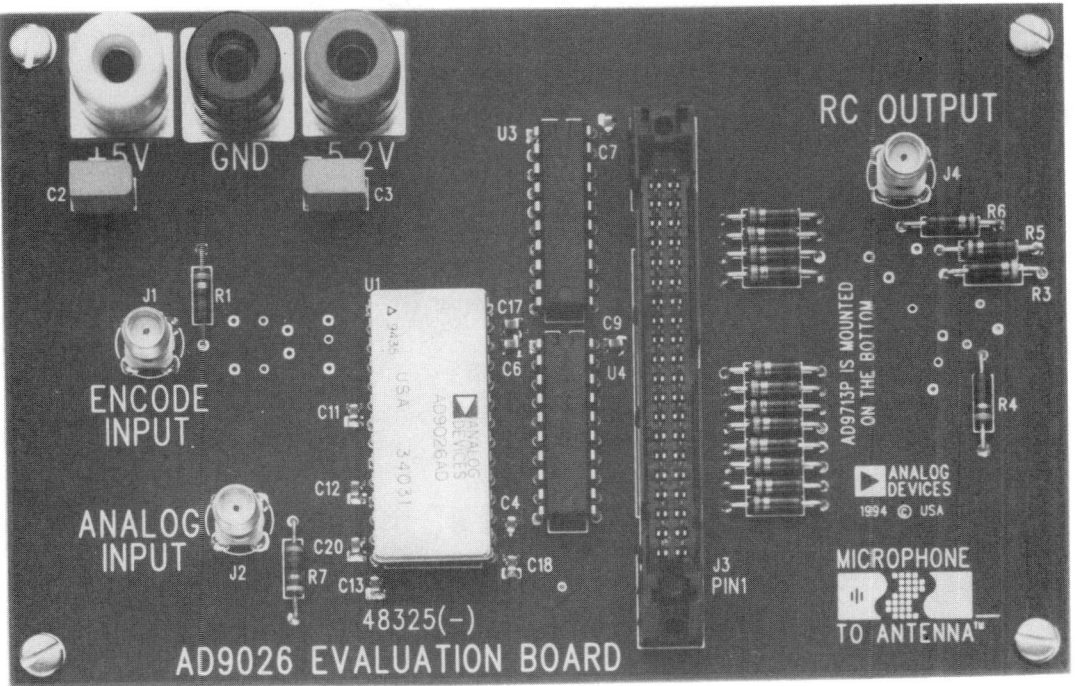

Figure 9.17

AD9026 EVALUATION BOARD - BOTTOM VIEW

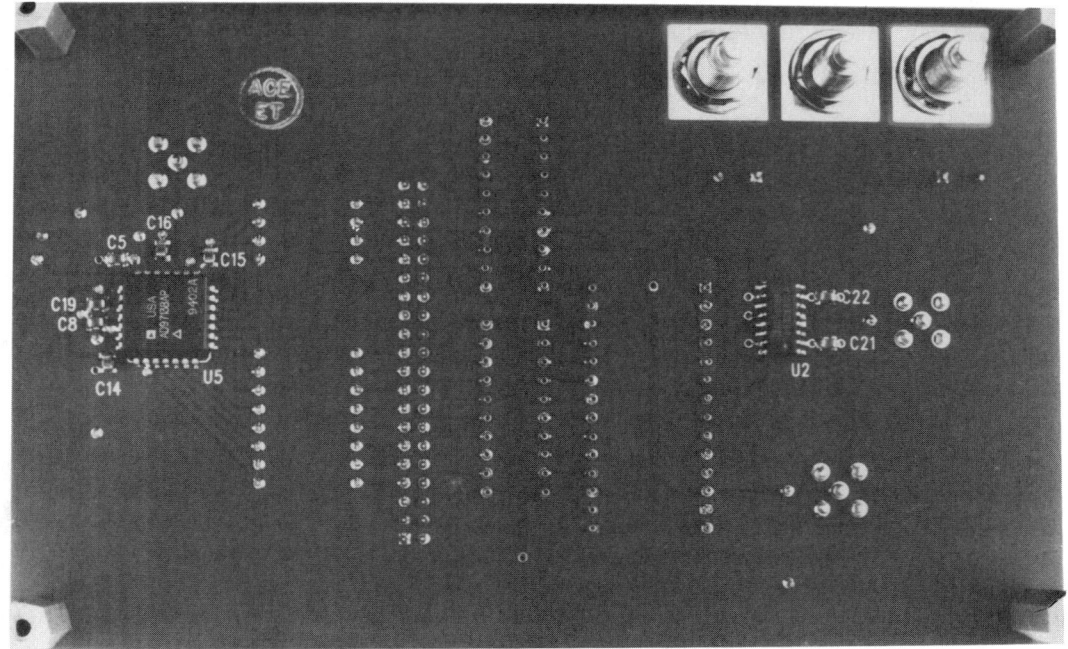

Figure 9.18

REFERENCES:
PROTOTYPING AND EVALUATION BOARDS

1. Robert A. Pease, **Troubleshooting Analog Circuits**, Butterworth-Heinemann, 1991.

2. Wainwright Instruments Inc., 7770 Regents Rd., #113, Suite 371, San Diego, CA 92122, Tel. 619-558-1057, Fax. 619-558-1019.

   Wainwright Instruments GmbH, Widdersberger Strasse 14, DW-8138 Andechs-Frieding, Germany. Tel: +49-8152-3162, Fax: +49-8152-40525.

3. Schematic Capture and Layout Software:

   PADS Software, INC, 165 Forest St., Marlboro, MA, 01752

   ACCEL Technologies, Inc., 6825 Flanders Dr., San Diego, CA, 92121

4. Prototype Board Cutters:

   LPKF CAD/CAM Systems, Inc., 1800 NW 169th Place, Beaverton, OR, 97006

   T-Tech, Inc., 5591-B New Peachtree Road, Atlanta, GA, 34341

Noise Reduction and Filtering for Switching Power Supplies
Walt Jung and John McDonald

Precision analog circuitry has traditionally been powered from well regulated, low noise linear power supplies. During the last decade however, switching power supplies have become much more common in electronic systems. As a consequence, they also are being used for analog supplies. There are several good reasons for their popularity, including high efficiency, low temperature rise, small size, and light weight.

Switching power supplies, or more simply *switchers*, a category including switching regulators and switching converters, are by their nature efficient. Often this can be above 90%, and as a result, these power supply types use less power and generate less heat than do equivalent linear supplies.

A switcher can be as much as one third the size and weight of a linear supply delivering the same voltage and current. Switching frequencies can range from 20kHz to 1MHz, and as a result, relatively small components can be used in their design.

In spite of these benefits, switchers have their drawbacks, most notably high output noise. This noise generally extends over a broad band of frequencies, and occurs as both conducted and radiated noise, and unwanted electric and magnetic fields. Voltage output noise of switching supplies are short-duration voltage transients, or spikes. Although the fundamental switching frequency can range from 20kHz to 1MHz, the voltage spikes will contain frequency components extending easily to 100MHz or more.

Because of wide variation in the noise output characteristics of commercial switchers, they should always be purchased in accordance with a specification-control drawing. Although specifying switching supplies in terms of RMS noise is common vendor practice, as a user you should also specify the *peak* (or p-p) amplitudes of the switching spikes, with the output loading of your system. You should also insist that the switching-supply manufacturer inform you of any internal supply design changes that may alter the spike amplitudes, duration, or switching frequency. These changes may require corresponding changes in external filtering networks.

HARDWARE DESIGN TECHNIQUES

SWITCHING POWER SUPPLY CHARACTERISTICS

ADVANTAGES:

- Efficient
- Small Size, Light Weight
- Low Operating Temperature Rise
- Isolation from Line Transients
- Wide Input/Output Range

DISADVANTAGES:

- Noise: LF, HF, Electric Field, Magnetic Field
  Conducted, Radiated
- DC regulation and accuracy can be poor

Figure 9.19

This section discusses filter techniques for rendering a noisy switcher output *analog ready*, that is sufficiently quiet to power precision analog circuitry with relatively small loss of DC terminal voltage. These techniques include characterization of switcher output noise, identification of the frequency range of interference produced by the switching power supply, evaluation of passive components commonly used in external power supply filters, and the design and construction of a switching power supply filter. The filter solutions presented are generally applicable to all power supply types incorporating a switch element in their energy path. This includes various DC-DC converters, as well as the 5V PC type supply used in the example.

A typical 5V PC type switcher is shown in Figure 9.20, and typifies the style. A display of the 5V power buss of an operating desktop PC (Dell Dimension XPS P-90) using a similar (but not identical) supply is shown in Figure 9.21. The unfiltered output shown from this switcher exhibits a ~60mV p-p transient component at an 80kHz (roughly) switching frequency, as seen in the (A) left photo with the 5s time base. The expanded scale photo for the same operating conditions of (B) right shows the detail of the switching glitches on a 100ns time base. The fast voltage spikes produce significant harmonics well into the high MHz range.

HARDWARE DESIGN TECHNIQUES

A TYPICAL 5V, 150W PC SWITCHING POWER SUPPLY

Figure 9.20

OUTPUT OF 5V PC SWITCHING POWER SUPPLY (UNFILTERED)

(A)

(B)

Vertical Scale: 20mV/div
Horizontal Scale: 5µs/div

Vertical Scale: 20mV/div
Horizontal Scale: 100ns/div

Figure 9.21

It is clear that this switcher is *not* analog ready, just as it is shown. Since many analog ICs show degraded power supply rejection at frequencies above a few kHz, some filtering is necessary. This will be particularly true for use with low power op amps, which can show PSRR degradation above a few hundred Hz. In general, all op amps, voltage references, DACs, and ADCs require clean supplies to meet their design accuracy. Switcher noise can prevent this happening, if left unchecked.

Before offering techniques to reduce switcher noise, a brief examination of switching supply architectures is helpful, and two popular ones are shown in Figure 9.22. Higher efficiency designs use a pulse width modulation technique for voltage regulation, while lower efficiency, lower noise designs use linear post regulators.

BASIC SWITCHING POWER SUPPLY TOPOLOGY

BETTER REGULATION

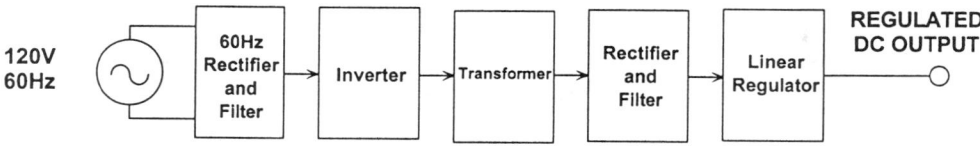

HIGHER EFFICIENCY

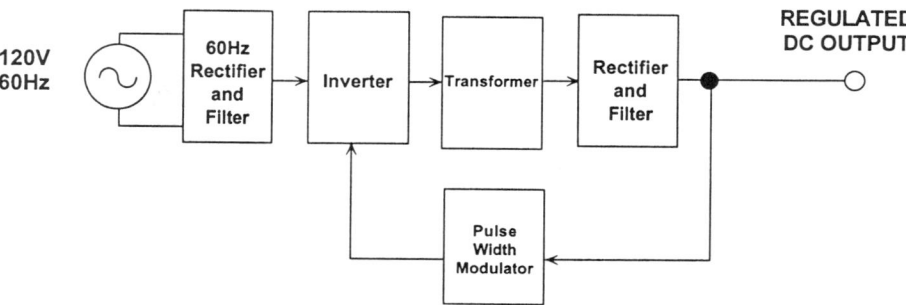

Figure 9.22

The raw AC line voltage is first rectified and filtered to a high DC level, then converted to a 30kHz (or higher) frequency square wave, which drives a transformer. The signal is again rectified, and filtered at the transformer output. Some switchers may use a linear regulator to form the final output voltage, as in the upper diagram. Others use pulse width regulation techniques to control the duty cycle of the transformer drive (lower diagram). Although more efficient, this results in more noise at the output than linear post-regulation.

It is important to understand that noise is generated in every stage of the switcher. The first stage AC line rectification creates current spikes, which produce line-related harmonic noise. In the inverter stage, fast pulse edges generate harmonics extending well beyond 5MHz. Furthermore, the parasitic capacitance within the primary and secondary windings of the transformer provide an additional path through which high-frequency noise can corrupt the DC output voltage. Finally, high-frequency noise is also generated by both rectifier stages.

An understanding of the EMI process is necessary to understand the effects of supply noise on analog circuits and systems. Every interference problem has a *source*, a *path*, and a *receptor* [Reference 1]. In general, there are three methods for dealing with interference. First, source emissions can be minimized by proper layout, pulse-edge rise time control/reduction, filtering, and proper grounding. Second, radiation and conduction paths should be reduced through shielding and physical separation. Third, receptor immunity to interference can be improved, via supply and signal line filtering, impedance level control, impedance balancing, and utilizing differential techniques to reject undesired common-mode signals. From this array of general noise immunity approaches, this section focuses on reducing switching power supply noise with external post filters.

THE INTERFERENCE PROCESS

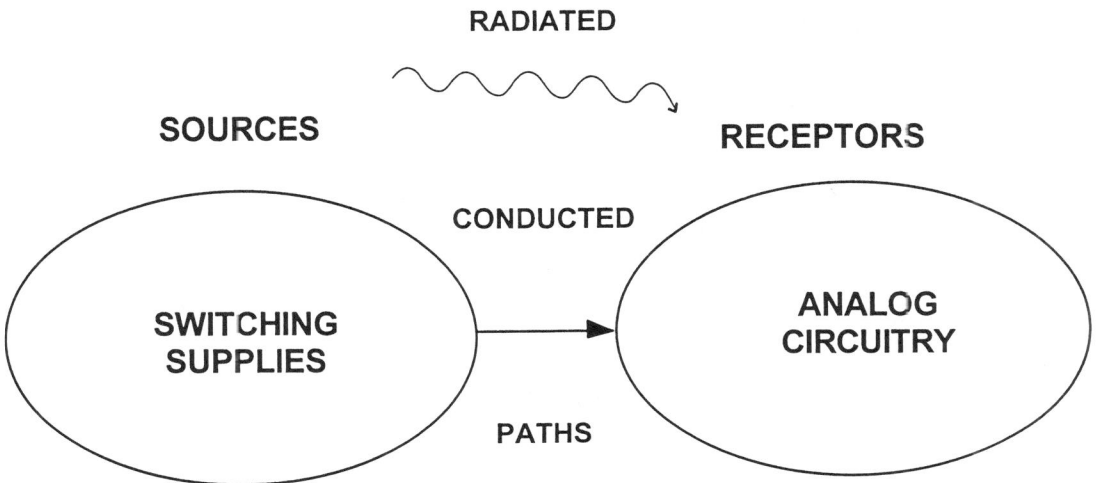

Figure 9.23

Before designing a switching supply filter, it is helpful to determine whether or not the supply noise is actually affecting the circuit performance. If critical node voltages in the circuit have transients synchronous with the switcher's operating frequency, then the supply is the likely culprit. A highly recommended method for determining if the supply is the noise source is to temporarily operate the circuit from a clean linear power supply or battery. If the interfering noise level drops dramatically, the switcher is guilty as charged. Note that lowering the power supply noise level may also help identify other noise sources which were masked by the higher switcher noise. Once the noise source is quantified and its path (radiated or conducted) identified, the process of reducing or eliminating it can begin.

Tools which can be used to combat switcher noise are highlighted by Figure 9.24. These tools differ in their electrical characteristics as well as their practicality towards noise reduction. For this reason they are listed roughly in a suggested order of priorities. Of the tools, inductance and capacitance are the most powerful filtering elements, and can also be the most cost-effective and small in size.

NOISE REDUCTION TOOLS

- Capacitors
- Inductors
- Ferrites
- Resistors
- Linear Post Regulation
- PHYSICAL SEPARATION FROM SENSITIVE ANALOG CIRCUITS !!

Figure 9.24

Capacitors are probably the single most important filter component for switchers. There are many different types of capacitors, and an understanding of their individual characteristics is absolutely mandatory to the design of practical, effective power supply filters. There are generally three classes of capacitors useful in filters in the 10kHz-100MHz frequency range suitable for switchers. These are broadly distinguished by their generic dielectric types; *electrolytic*, *film*, and *ceramic*. These dielectrics can in turn can be further sub-divided as discussed below. A thumbnail sketch of capacitor characteristics is shown in the chart of Figure 9.25. Background and tutorial information on capacitors can be found in Reference 2 and many vendor catalogs.

CAPACITOR SELECTION

	Aluminum Electrolytic (General Purpose)	Aluminum Electrolytic (Switching Type)	Tantalum Electrolytic	Polyester (Stacked Film)	Ceramic (Multilayer)
Size	100 µF (1)	120 µF (1)	100 µF (1)	1 µF	0.1 µF
Rated Voltage	25 V	25 V	20 V	400 V	50 V
ESR	0.6 Ω @ 100 kHz	0.18 Ω @ 100 kHz	0.12 Ω @ 100 kHz	0.11 Ω @ 1 MHz	0.12 Ω @ 1 MHz
Operating Frequency (2)	≅ 100 kHz	≅ 500 kHz	≅ 1 MHz	≅ 10 MHz	≅ 1 GHz

(1) Types shown in Figure 9.26 data
(2) Upper frequency limit is strongly size and package dependent

Figure 9.25

With any dielectric, a major potential filter loss element is ESR (equivalent series resistance), the net parasitic resistance of the capacitor. ESR provides an ultimate limit to filter performance, and requires more than casual consideration, because it can vary both with frequency and temperature in some types. Another capacitor loss element is ESL (equivalent series inductance). ESL determines the frequency where the net impedance of the capacitor switches from a capacitive to inductive characteristic. This varies from as low as 10kHz in some electrolytics to as high as 100MHz or more in chip ceramic types. Both ESR and ESL are minimized when a leadless package is used, and all capacitor types discussed here are available in surface mount packages, which are preferable for high speed uses.

The *electrolytic* family provides an excellent, cost-effective low-frequency filter component, because of the wide range of values, a high capacitance-to-volume ratio, and a broad range of working voltages. It includes *general purpose aluminum electrolytic* types, available in working voltages from below 10V up to about 500V, and in size from 1 to several thousand µF (with proportional case sizes). All electrolytic capacitors are polarized, and thus cannot withstand more than a volt or so of reverse bias without damage. They have relatively high leakage currents (this can be tens of µA, but is strongly dependent upon specific family design, electrical size and voltage rating vs. applied voltage). However, this is not likely to be a major factor for basic filtering applications.

Also included in the electrolytic family are *tantalum* types, which are generally limited to voltages of 100V or less, with capacitance of 500µF or less[Reference 3]. In a given size, tantalums exhibit a higher capacitance-to-volume ratios than do the general purpose electrolytics, and have both a higher frequency range and lower ESR. They are generally more expensive than standard electrolytics, and must be carefully applied with respect to surge and ripple currents.

A subset of aluminum electrolytic capacitors is the *switching* type, which is designed and specified for handling high pulse currents at frequencies up to several hundred kHz with low losses [Reference 4]. This type of capacitor competes directly with the tantalum type in high frequency filtering applications, and has the advantage of a much broader range of available values.

More recently, high performance aluminum electrolytic capacitors using an organic semiconductor electrolyte have appeared [Reference 5]. These *OS-CON* families of capacitors feature appreciably lower ESR and higher frequency range than do the other electrolytic types, with an additional feature of low low-temperature ESR degradation.

Film capacitors are available in very broad ranges of values and an array of dielectrics, including polyester, polycarbonate, polypropylene, and polystyrene. Because of the low dielectric constant of these films, their volumetric efficiency is quite low, and a 10µF/50V polyester capacitor (for example) is actually a handful. Metalized (as opposed to foil) electrodes does help to reduce size, but even the highest dielectric constant units among film types (polyester, polycarbonate) are still larger than any electrolytic, even using the thinnest films with the lowest voltage ratings (50V). Where film types excel is in their low dielectric losses, a factor which may not necessarily be a practical advantage for filtering switchers. For example, ESR in film capacitors can be as low as 10mΩ or less, and the behavior of films generally is very high in terms of Q. In fact, this can cause problems of spurious resonance in filters, requiring damping components.

Typically using a wound layer-type construction, film capacitors can be inductive, which can limit their effectiveness for high frequency filtering. Obviously, only non-inductively made film caps are useful for switching regulator filters. One specific style which is non-inductive is the *stacked-film* type, where the capacitor plates are cut as small overlapping linear sheet sections from a much larger wound drum of dielectric/plate material. This technique offers the low inductance attractiveness of a plate sheet style capacitor with conventional leads [see type "V" of

HARDWARE DESIGN TECHNIQUES

Reference 4, plus Reference 6]. Obviously, minimal lead length should be used for best high frequency effectiveness. Very high current polycarbonate film types are also available, specifically designed for switching power supplies, with a variety of low inductance terminations to minimize ESL [Reference 7].

Dependent upon their electrical and physical size, film capacitors can be useful at frequencies to well above 10MHz. At the very high frequencies, stacked film types only should be considered. Some manufacturers are also supplying film types in leadless surface mount packages, which eliminates the lead length inductance.

Ceramic is often the capacitor material of choice above a few MHz, due to its compact size, low loss, and availability in values up to several µF in the high-K dielectric formulations of X7R and Z5U, at voltage ratings up to 200V [see ceramic families of Reference 3]. NP0 (also called COG) types use a lower dielectric constant formulation, and have nominally zero TC, plus a low voltage coefficient (unlike the less stable high-K types). The NP0 types are limited in available values to 0.1µF or less, with 0.01µF representing a more practical upper limit.

Multilayer ceramic "chip caps" are increasingly popular for bypassing and filtering at 10MHz or more, because their very low inductance design allows near optimum RF bypassing. In smaller values, ceramic chip caps have an operating frequency range to 1GHz. For these and other capacitors for high frequency applications, a useful value can be ensured by selecting a value which has a self-resonant frequency *above* the highest frequency of interest.

All capacitors have some finite ESR. In some cases, the ESR may actually be helpful in reducing resonance peaks in filters, by supplying "free" damping. For example, in general purpose, tantalum and switching type electrolytics, a broad series resonance region is noted in an impedance vs. frequency plot. This occurs where |Z| falls to a minimum level, which is nominally equal to the capacitor's ESR at that frequency. In an example below, this low Q resonance is noted to encompass quite a wide frequency range, several octaves in fact. Contrasted to the very high Q sharp resonances of film and ceramic caps, this low Q behavior can be useful in controlling resonant peaks.

In most electrolytic capacitors, ESR degrades noticeably at low temperature, by as much as a factor of 4-6 times at –55°C vs. the room temperature value. For circuits where a high level of ESR is critical, this can lead to problems. Some specific electrolytic types do address this problem, for example within the HFQ switching types, the –10°C ESR at 100kHz is no more than 2× that at room temperature. The OSCON electrolytics have a ESR vs. temperature characteristic which is relatively flat.

Figure 9.26 illustrates the high frequency impedance characteristics of a number of electrolytic capacitor types, using nominal 100µF/20V samples. In these plots, the impedance, |Z|, vs. frequency over the 20Hz-200kHz range is displayed using a high resolution 4-terminal setup [Reference 8]. Shown in this display are performance samples for a 100µF/25V general purpose aluminum unit (top curve @ right), a 120µF/25V HFQ unit (next curve down @ right), a 100µF/20V tantalum bead type (next curve down @ right), and a

HARDWARE DESIGN TECHNIQUES

100µF/20V OS-CON unit (lowest curve @ right). While the HFQ and tantalum samples are close in 100kHz impedance, the general purpose unit is about 4 times worse. The OS-CON unit is nearly an order of magnitude lower in 100kHz impedance than the tantalum and switching electrolytic types.

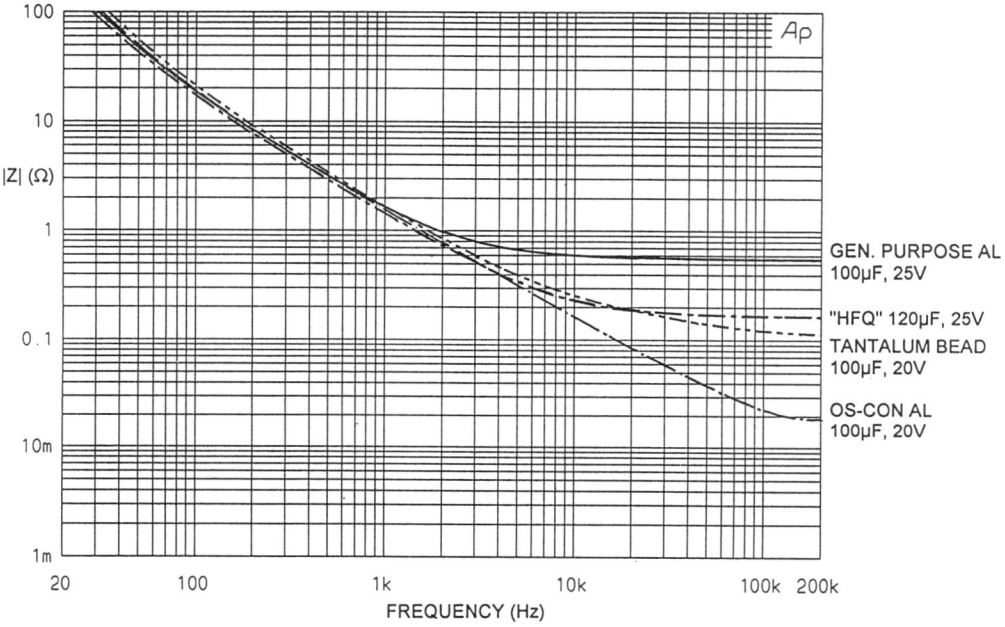

Figure 9.26

As noted above, all real capacitors have parasitic elements which limit their performance. As an insight into why the impedance curves of Figure 9.26 appear the way they do, a (simplified) model for a 100µF/20V tantalum capacitor is shown in Figure 9.27.

The electrical network representing this capacitor is shown, and it models the ESR and ESL components with simple R and L elements, plus a 1MΩ shunt resistance. While this simple model ignores temperature and dielectric absorption effects which occur in the real capacitor, it is still sufficient for this discussion.

When driven with a constant level of AC current swept from 10Hz to 100MHz, the voltage across this capacitor model is proportional to its net impedance, which is shown in Figure 9.28.

Hardware Design Techniques

SIMPLIFIED SPICE MODEL FOR LEADED 100µF/20V TANTALUM ELECTROLYTIC CAPACITOR

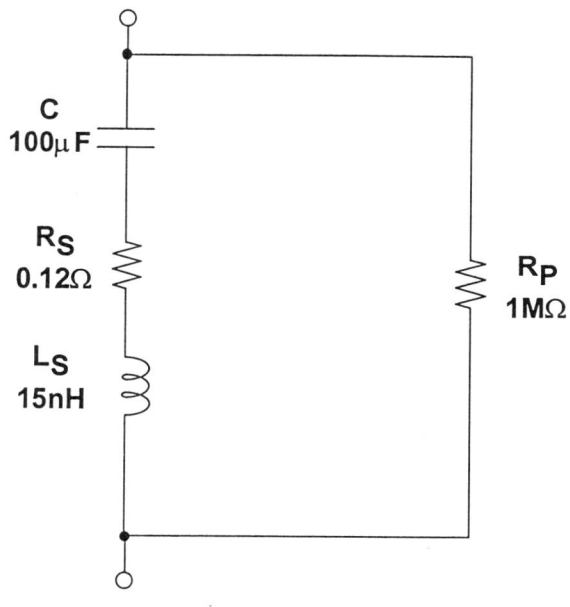

Figure 9.27

100µF / 20V TANTALUM CAPACITOR SIMPLIFIED MODEL IMPEDANCE (Ω) VS. FREQUENCY (Hz)

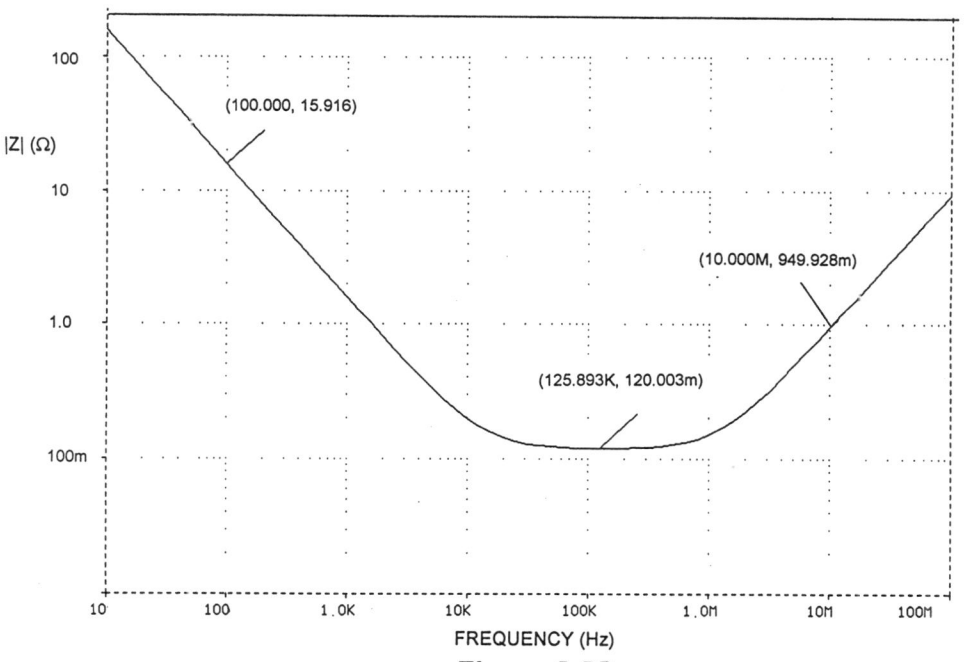

Figure 9.28

Hardware Design Techniques

At low frequencies the net impedance is almost purely capacitive, as noted by the 100Hz impedance of 15.9Ω. At the bottom of this "bathtub" curve, the net impedance is determined by ESR, which is shown to be 0.12Ω at 125kHz. Above about 1MHz this capacitor becomes inductive, and impedance is dominated by the effect of ESL. While this particular combination of capacitor characteristics have been chosen purposely to correspond to the tantalum sample used with Figure 9.26, it is also true that all electrolytics will display impedance curves which are similar in general shape. The minimum impedance will vary with the ESR, and the inductive region will vary with ESL (which in turn is strongly effected by package style). The simulation curve of Figure 9.28 can be considered as an extension of the 100μF/20V tantalum capacitor curve from Figure 9.26.

Ferrites (non-conductive ceramics manufactured from the oxides of nickel, zinc, manganese, or other compounds) are extremely useful for decoupling in power supply filters [Reference 9]. At low frequencies (<100kHz), ferrites are inductive; thus they are useful in low-pass LC filters. Above 100kHz, ferrites become resistive, an important characteristic in high-frequency filter designs. Ferrite impedance is a function of material, operating frequency range, DC bias current, number of turns, size, shape, and temperature. Figure 9.29 summarize a number ferrite characteristics.

CHARACTERISTICS OF FERRITES

- Good for frequencies above 25kHz
- Many sizes and shapes available including leaded "resistor style"
- Ferrite impedance at high frequencies is primarily resistive -- Ideal for HF filtering
- Low DC loss: Resistance of wire passing through ferrite is very low
- High saturation current
- Low cost

Figure 9.29

Hardware Design Techniques

Several ferrite manufacturers offer a wide selection of ferrite materials from which to choose, as well as a variety of packaging styles for the finished network (see References 10 and 11). The most simple form is the *bead* of ferrite material, a cylinder of the ferrite which is simply slipped over the power supply lead to the stage being decoupled.

Alternately, the *leaded ferrite bead* is the same bead, mounted by adhesive on a length of wire, and used simply as a component (Reference 11 typifies these two styles). More complex beads offer multiple holes through the cylinder for increased decoupling, plus other variations. Surface mount bead styles are also available.

FERRITE IMPEDANCE DEPENDS ON

- Material
- Permeability
- Frequency
- Number of Turns
- Size
- Shape
- Temperature
- Field Strength (generated by current flowing through wire)

Figure 9.30

Recently, PSpice ferrite models for Fair-Rite materials have become available that allow ferrite impedance to be estimated [Reference 12]. The models of Fair-Rite materials #43 and #73 can be downloaded from the MicroSim bulletin board (714-830-1550). These models have been designed to match measured impedances rather than theoretical impedances.

A ferrite's impedance is dependent upon a number of inter-dependent variables, and is difficult to quantify analytically, thus selecting the proper ferrite is not straightforward. However, knowing the following system characteristics will make selection easier. First, determine the frequency range of the noise to be filtered. A spectrum analyzer is useful here. Second, the expected temperature range of the filter should be known, because ferrite impedance varies with temperature. Third, the DC bias current flowing through the ferrite must be known, to ensure that the ferrite does

HARDWARE DESIGN TECHNIQUES

not saturate. Although models and other analytical tools may prove useful, the general guidelines given above, coupled with some experimentation with the actual filter connected to the supply output under system load conditions, should ultimately lead to the selection of the proper ferrite.

CHOOSING THE RIGHT FERRITE DEPENDS ON

- Source of Interference
- Interference Frequency Range
- Impedance Required at Interference Frequency
- Environmental Conditions:

    Temperature, AC and DC Field Strength, Size / Space Available

- Don't fail to Test the Design -------

    EXPERIMENT! EXPERIMENT!

Figure 9.31

Maintaining high power supply efficiency requires the intelligent limiting of series resistors and linear post regulators in the switching supply's output. However, small resistors (generally less than 10Ω) can be used in applications where load currents are low and load regulation is not highly critical.

Higher performance linear post regulators can provide 60dB and more of power supply rejection up to 100kHz, for example see the designs of [Reference 8]. When used with effective input filtering, their PSRR can be extended above 1MHz or higher. Linear post regulation will generally result in a net efficiency decrease, which can be serious if the regulator requires several volts of headroom. The PSRR vs. frequency performance of a linear post regulator should also be carefully considered. For example, some low dropout linear regulators offer very little power supply rejection at frequencies above a few kHz, a performance fact of life which must be traded off against the efficiency advantages of the <100mV level dropouts they can boast.

Using the component selection choices mentioned above, low and high frequency band filters can be designed to smooth the noisy switcher's DC output so as to produce an *analog ready* 5V supply. It is most practical to do this over two (and sometimes more) stages, with each stage optimized for a range of frequencies. One basic stage can be used to carry all of the DC load current, and filter noise by 60dB or more up to about 10MHz. This larger filter is used as a *card entry filter* providing broadband filtering for all analog power entering the PC board. Thereafter, smaller and more simple local filter stages can be used to provide very high frequency decoupling, right at the power pins of the individual ICs.

Figure 9.32 illustrates a card entry filter suitable for use with switching supplies. Because of the low rolloff point of 1.5kHz and mV level DC errors, it will be effective for a wide variety of filter applications just as it is shown. This filter is a single stage LC low-pass filter covering the 1kHz to 1MHz range, using carefully chosen parts. Because of component losses, it begins to lose effectiveness above a few MHz, but is still able to achieve an attenuation approaching 60dB at 1MHz.

The key to low DC losses is the use of input choke, L1, a ferrite-core unit selected for a low DC resistance (DCR) of <0.25Ω at the 100µH inductance. The prototype was tested with an axial lead type 5250 choke, but the radial style 6000-101K should give comparable results [Reference 13]. Both chokes have a low inductance shift due to the 300mA load current. The low DCR allows the 300mA to be passed with no more than 75mV of DC error at full load. Alternately, resistive filtering might be used in place of L1, but the basic tradeoff here is that load current capacity will be compromised for comparable DC errors. For example, a 1Ω resistor with a 75mV DC allowable error can pass only a 75mA current.

C1, a 100µF/20V tantalum type, provides the bulk of the capacitive filtering, shunted by a 1F multilayer ceramic. The remaining part of the filter is R1, a damping resistor used to control resonant peaks.

"CARD-ENTRY" SWITCHING SUPPLY FILTER

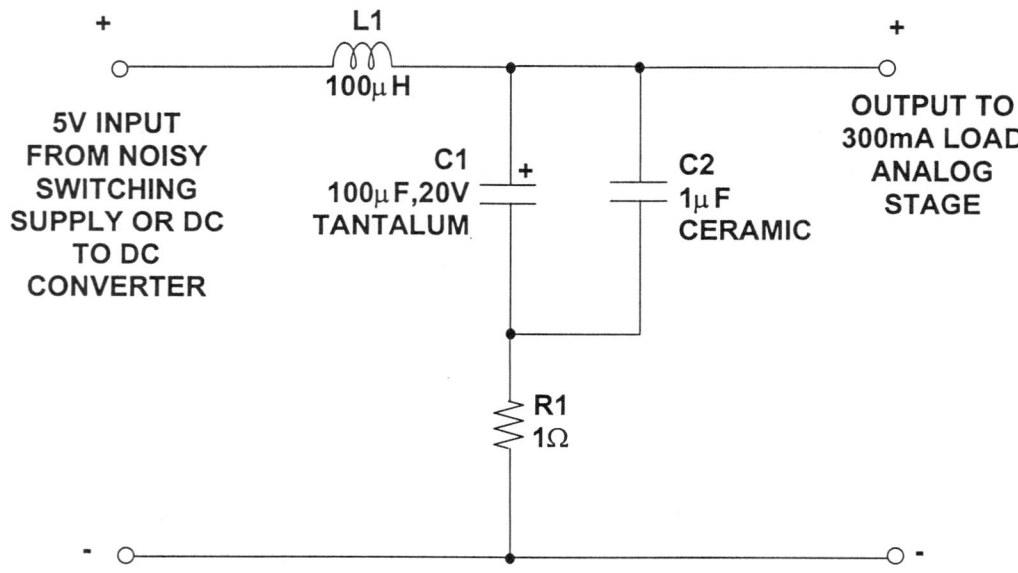

Figure 9.32

Figure 9.33 shows the frequency response of this filter, both in terms of a SPICE simulation as well as lab measurements. There is good agreement between the simulation and the measurements for the common range below 1MHz. Measurements were not made above 1MHz, since higher frequencies are attenuated by second stage localized high frequency filters.

This filter has some potential pitfalls, and one of them is the control of resonances. If the LCR circuit formed does not have sufficiently high resistance at the resonant frequency, amplitude peaking will result. This peaking can be minimized with resistance at two locations: in series with L1, or in series with C1+C2. Obviously, limited resistance is usable in series with L1, as this increases the DC errors. Thus the use of the C1+C2 series damping resistor R1, which should not be eliminated. The 1Ω value used actually provides a slightly underdamped response, with peaking on the order of 1dB. An alternate value of 1.5Ω can also be used for less peaking, if this is desired, but the tradeoff here is that the attenuation below 1MHz will then suffer.

Note that if the damping resistor were to be eliminated or an excessively low value used, it is possible that a transient at the frequency of L1-(C1+C2) resonance could cause the filter to ring, actually exacerbating whatever peak amplitude occurs at the input. Of course, keeping the basic filter corner frequency well below the lowest commonly used switcher frequency of 20kHz also helps to minimize this possibility. A side benefit of R1 is that it buffers variations in parasitic resistance in L1 and C1, by making them smaller percentage of the total resistance, and thus less likely to effect overall performance. For wide temperature applications however, temperature changes of the filter characteristics will still need consideration.

Hardware Design Techniques

OUTPUT RESPONSE OF "CARD-ENTRY" FILTER LAB VS. SIMULATION

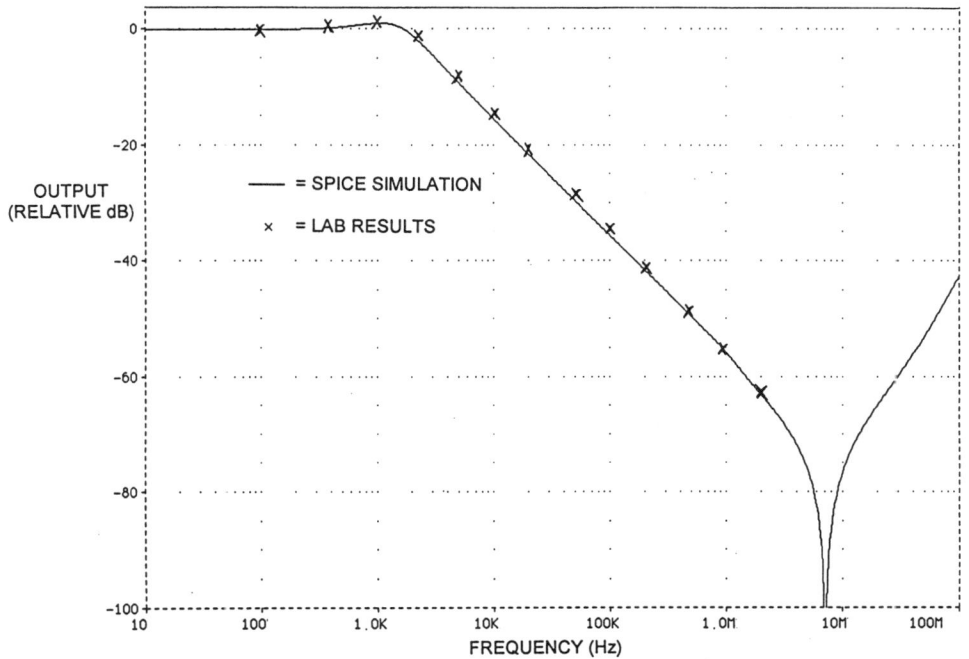

Figure 9.33

Because of the high sensitivity to source series resistance of this filter, measuring its frequency response is not a trivial task. The low output impedance high current unity gain buffer of Figure 9.34 was used for the data of Figure 9.33. Note that the filter presents a load of ~1Ω to the source at resonance, so the buffer drive level is kept ≤100mV RMS, to prevent buffer current limiting.

TEST SETUP FOR MEASURING FILTER FREQUENCY RESPONSE

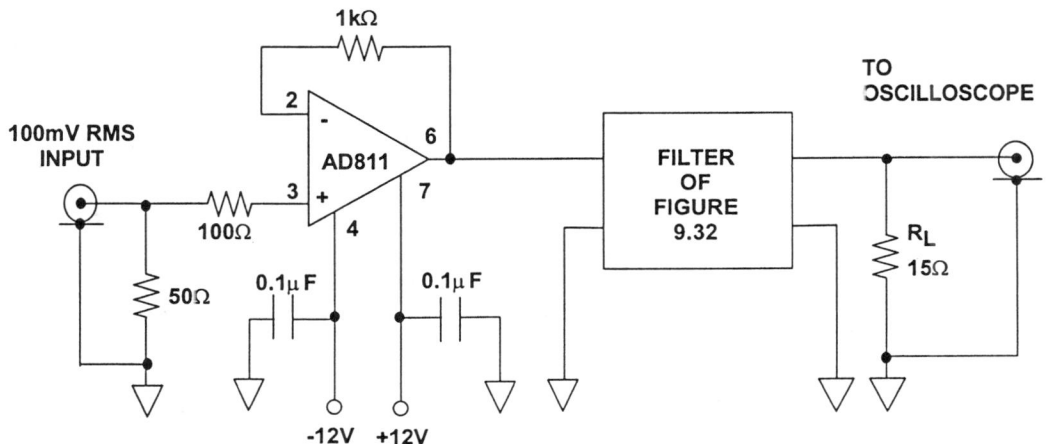

Figure 9.34

Hardware Design Techniques

A local high frequency filter which can be used in conjunction with the card entry filter is shown in Figure 9.35. This simple filter can be considered an option, one which is exercised dependent upon the high frequency characteristics of the associated IC and the relative attenuation desired. It is composed of Z1, a leaded ferrite bead such as the Panasonic EXCELSA39, which provides a resistance of more than 80Ω at 10MHz, increasing to over 100Ω at 100MHz. The ferrite bead is best used with a local high frequency decoupling cap right at the IC power pins, such as a 0.1μF ceramic unit shown.

HIGH FREQUENCY LOCALIZED DECOUPLING

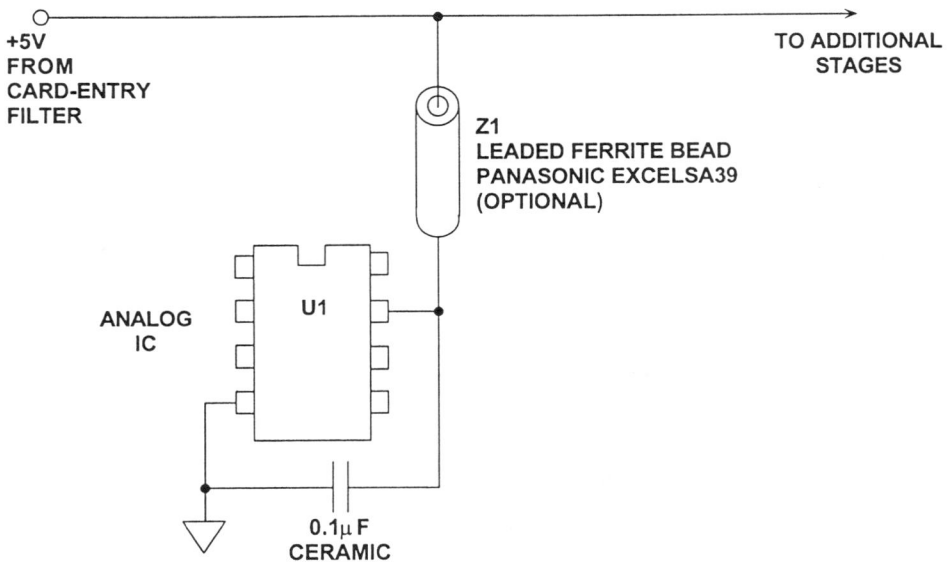

Figure 9.35

Both the card entry filter and the local high frequency decoupling filters are designed to filter differential-mode noise only. They use components commonly available off the shelf from national distributors [Reference 14].

The following is a list of switching power supply filter layout and construction guidelines which will help guarantee that the filter does the best job possible:

(1) *Pick the highest electrical value and voltage rating for filter capacitors which is consistent with budget and space limits. This minimizes ESR, and maximizes filter performance. Pick chokes for low ΔL at the rated DC current, as well as low DCR.*

(2) *Use short and wide PCB tracks to decrease voltage drops and minimize inductance. Make track widths at least 200 mils for every inch of track length*

HARDWARE DESIGN TECHNIQUES

for lowest DCR, and use 1 oz or 2 oz copper PCB traces to further reduce IR drops and inductance.

(3) *Use short leads or leadless components, to minimize lead inductance. This will minimize the tendency to add excessive ESL and/or ESR. Surface mount packages are preferred.*

(4) *Use a large-area ground plane for minimum impedance.*

(5) *Know what your components do over frequency, current and temperature variations! Make use of vendor component models for the simulation of prototype designs, and make sure that lab measurements correspond reasonably with the simulation. While simulation is not absolutely necessary, it does instill confidence in a design when correlation is achieved* (see Reference 15).

The discussion of switching power supplies so far has focused on filtering the output of the supply. This assumes that the incoming AC power is relatively clean, an assumption not always valid. However, the AC power can also be an EMI path, both entering and exiting the equipment. To remove this path and reduce emissions caused by the switching power supply and other circuits in the instrument, a power line filter is required. *Remember that the AC line power can be lethal! Do not experiment without proper equipment and training!*

All components used in power line filters should be UL approved, and the best way to provide this for your equipment is to specify only a complete, packaged UL approved filter. It should be installed in such a manner that it is the first thing the AC line sees upon entering the equipment. Standard three wire IEC style line cords are designed to mate with three terminal male connectors, which are integral to many line filters. This is the best way to do this function, as this automatically grounds the third wire to the shell of the filter and equipment chassis via a low inductance path.

Commercial power line filters, such as the one shown schematically in Figure 9.37, can be quite effective in reducing noise. AC power-line noise is generally has both common-mode and differential-mode components. Common-mode noise is noise that is found on any two of the three power connections (black, white, or green) with the same amplitude and polarity. In contrast, differential-mode noise is noise found only between two lines.

Hardware Design Techniques

POWER LINE FILTERING IS ALSO IMPORTANT

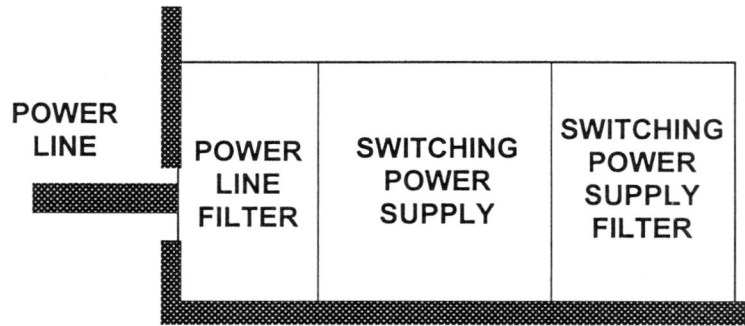

Power Line Filter Blocks EMI from
Entering or Exiting Box Via Power Lines

Figure 9.36

TYPICAL COMMERCIAL POWER LINE FILTER

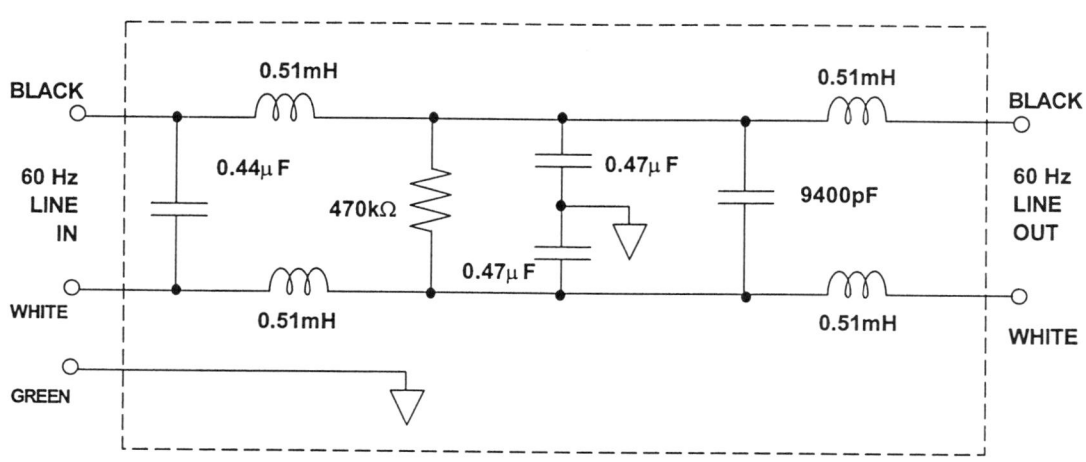

EICHHOFF
P/N: F033-106/500
6A current rated device

Figure 9.37

HARDWARE DESIGN TECHNIQUES

Common-mode noise is dominant at frequencies over 1MHz and is generally introduced into the AC lines through capacitive coupling. Ferrites provide effective filtering of the high frequency common-mode noise when used as common-mode chokes. For example, to create an effective common-mode choke, a few turns of the input power leads can be wound around a large ferrite. This provides a simple and effective solution for common-mode noise, but is ineffective against differential-mode noise. Differential-mode noise can be minimized by using proper LC filtering techniques as described earlier in this section, using proper UL approved across-the-line rated components. A power-line filter must be designed to minimize both common- and differential-mode noise to keep EMI from entering and leaving the system.

POWER LINE NOISE MODES

- **Common Mode:**
    - ◆ Dominates above 1MHz, Primarily Capacitive Coupled
    - ◆ 2200 to 4700pF typical shunt capacitor values
    - ◆ 0.5 to 10mH typical series ferrite values

- **Differential Mode:**
    - ◆ Dominates below 1MHz
    - ◆ 0.1 to 2.2µF typical shunt capacitor values
    - ◆ Molyperm or powered iron inductors, 100 to 200µH, typical series values

Figure 9.38

Notice the common-mode choke formed by the inductors on both sides of the filter in Figure 9.37. These inductors have a dual role, because they are also part of the differential-mode LC filters. This filter, using multiple stages, is an example of a higher quality type.

As noted, the AC power line filter should be in good electrical contact with the chassis of the instrument. Furthermore, connecting the AC power line directly into the power line filter reduces the possibility of EMI entering or exiting the instrument. If this is not possible, routing the filter AC input power lines close to the chassis and twisting them will help minimize loop areas and minimize LF magnetic coupling.

The power supply filter should be located as close as possible to the switching power supply; i.e., in commercial units it is always integral to the supply. Many switching power supplies

Hardware Design Techniques

have steel enclosures which can be used for electric and LF magnetic shielding (however, the enclosure must be connected to chassis ground to act as a Faraday shield).

If digital circuitry is present, the digital power pick-off point should occur *before* the switching power supply filter. This minimizes the digital noise in the output of the switching supply filter, plus minimizes any potential DC error form the higher currents.

POWER LINE FILTER PLACEMENT IS IMPORTANT

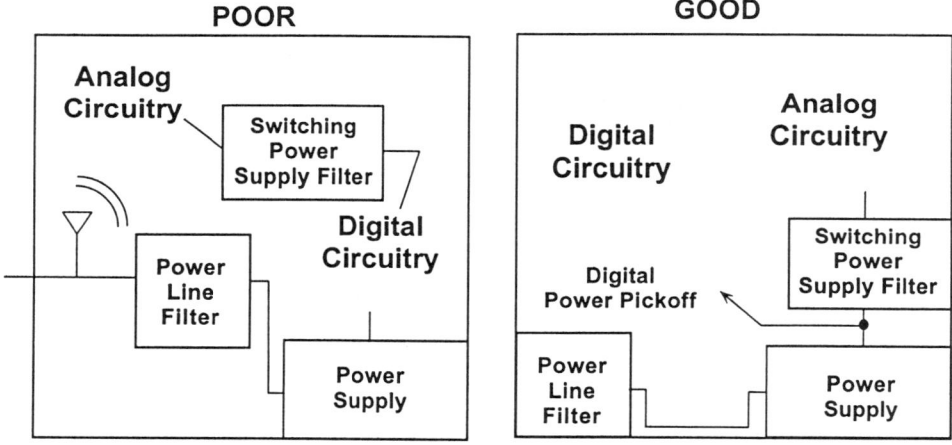

Figure 9.39

REFERENCES: NOISE REDUCTION AND FILTERING FOR SWITCHING POWER SUPPLIES

1. **EMC Design Workshop Notes**, Kimmel-Gerke Associates, Ltd., St. Paul, MN. 55108, (612) 330-3728.

2. Walt Jung, Dick Marsh, *Picking Capacitors, Parts 1 & 2*, **Audio**, February, March, 1980.

3. Tantalum Electrolytic and Ceramic Capacitor Families, Kemet Electronics, Box 5928, Greenville, SC, 29606, (803) 963-6300.

4. Type HFQ Aluminum Electrolytic Capacitor and type V Stacked Polyester Film Capacitor, Panasonic, 2 Panasonic Way, Secaucus, NJ, 07094, (201) 348-7000.

5. OS-CON Aluminum Electrolytic Capacitor 93/94 Technical Book, Sanyo, 3333 Sanyo Road, Forrest City, AK, 72335, (501) 633-6634.

6. Ian Clelland, *Metalized Polyester Film Capacitor Fills High Frequency Switcher Needs*, **PCIM**, June 1992.

7. Type 5MC Metallized Polycarbonate Capacitor, Electronic Concepts, Inc., Box 1278, Eatontown, NJ, 07724, (908) 542-7880.

8. Walt Jung, *Regulators for High-Performance Audio, Parts 1 and 2*, **The Audio Amateur**, issues 1 and 2, 1995.

9. Henry Ott, **Noise Reduction Techniques in Electronic Systems, 2d Ed.,** 1988, Wiley.

10. Fair-Rite Linear Ferrites Catalog, Fair-Rite Products, Box J, Wallkill, NY, 12886, (914) 895-2055.

11. Type EXCEL leaded ferrite bead EMI filter, and type EXC L leadless ferrite bead, Panasonic, 2 Panasonic Way, Secaucus, NJ, 07094, (201) 348-7000.

12. Steve Hageman, *Use Ferrite Bead Models to Analyze EMI Suppression*, **The Design Center Source**, MicroSim Newsletter, January, 1995.

13. Type 5250 and 6000-101K chokes, J. W. Miller, 306 E. Alondra Blvd., Gardena, CA, 90247, (310) 515-1720.

14. DIGI-KEY, PO Box 677, Thief River Falls, MN, 56701-0677, (800) 344-4539.

15. Tantalum Electrolytic Capacitor SPICE Models, Kemet Electronics, Box 5928, Greenville, SC, 29606, (803) 963-6300.

16. Eichhoff Electronics, Inc., 205 Hallene Road, Warwick, RI., 02886, (401) 738-1440.

LOW DROPOUT REFERENCES AND REGULATORS
Walt Jung

Many circuits require stable regulated voltages relatively close in potential to an unregulated source. An example would be a linear post regulator for a switching power supply, where voltage loss is critical. This low dropout type of regulator is readily implemented with a rail-rail output op amp. The wide output swing and low saturation voltage enables outputs to come within a fraction of a volt of the source for medium current (<30mA) loads, such as reference applications. For higher output currents, the rail-rail voltage swing feature allows direct drive to low saturation voltage pass devices, such as power PNPs or P-channel MOSFETs. Op amps which work from 3V up with the rail-rail features are most suitable here, providing power economy and maximum flexibility.

BASIC REFERENCES IN LOW POWER SYSTEMS

Among the many problems in making stable DC voltage references work from 5V and lower supplies are quiescent power consumption, overall efficiency, the ability to operate down to 3V, low input/output (dropout) capability, and minimum noise output. Because such low voltage supplies can't support zeners of ≅6V, these low voltage references must necessarily be bandgap based — a 1.2V potential.

One difficulty is simply to get a reference circuit which works well at 3V inputs, conditions which dictate a lower voltage reference diode. A workhorse circuit solution is the reference and appropriate low power scaling buffer shown in Figure 9.40. Here a low current 1.2V diode is used for D1, the 1.235V AD589. Resistor R1 sets the current, chosen for 50µA at the minimum supply of 2.7V. Obviously, loading on the unbuffered diode must be minimized at the V_{REF} node.

The amplifier U1 both buffers and optionally scales up the 1.2V reference, allowing higher source/sink currents. A higher op amp quiescent current is expended in doing this, and is a basic design tradeoff of this approach. This current can range from 150-300µA/channel with the OP295/495 and OP191/291/491 series for U1, 620µA/channel using an AD820/822/829 section, or in the range of 1000-2000µA/channel with the OP284 and OP279. The former two series are most useful for very light loads (<2mA), while the latter three series provide device dependent outputs up to 50mA. All devices are simply used in the circuit as shown, and their key specs are summarized in Figure 9.41.

HARDWARE DESIGN TECHNIQUES

RAIL-RAIL OUTPUT OP AMPS ALLOW GREATEST FLEXIBILITY IN LOW DROPOUT REFERENCES

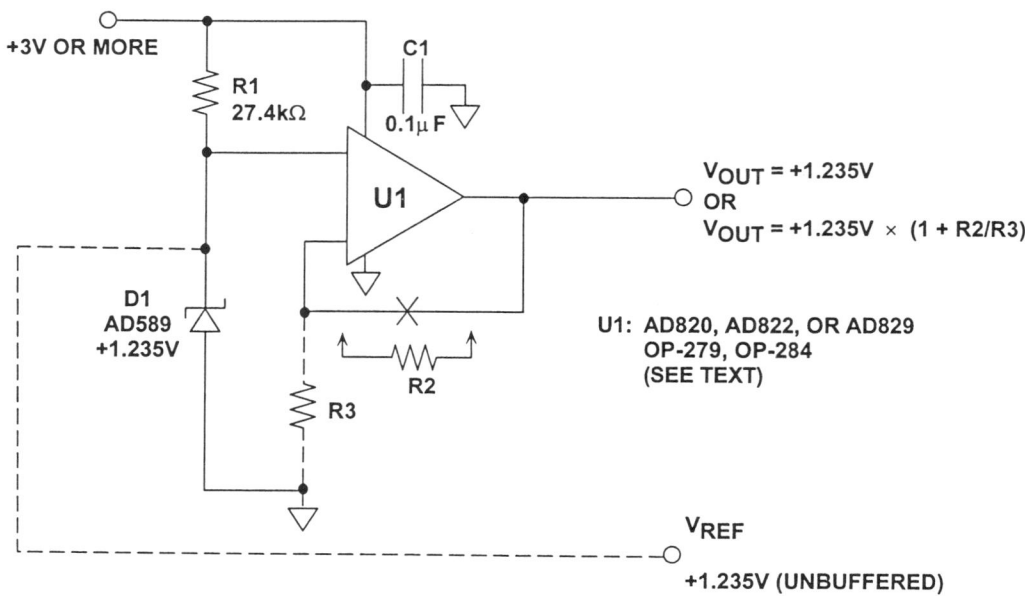

Figure 9.40

OP AMPS USEFUL IN LOW VOLTAGE RAIL-RAIL REFERENCES AND REGULATORS

Device*	Iq/channel mA	Vsat(+), V(min @ mA)	Vsat(−), V (max @ mA)	Isc, mA (min)
OP193/293/493	0.017	4.20 @ 1	0.280 @ 1 (typ)	± 8
OP295/495	0.150 (max)	4.50 @ 1	0.110 @ 1	± 11
OP191/291/491	0.300	4.80 @ 2.5	0.075 @ 2.5	± 8.75
AD820/822	0.620	4.89 @ 2	0.055 @ 2	± 15
OP284/484	1.250 (max)	4.85 @ 2.5	0.125 @ 2.5	± 7.5
OP279	2.000	4.80 @ 10	0.075 @ 10	± 45

*Typical device specifications @ Vs = +5V, T_A = 25°C, unless otherwise noted.

Figure 9.41

In Figure 9.40, without gain scaling resistors R2-R3, V_{OUT} is simply 1.235V. With the resistors, V_{OUT} can be anywhere between the rails, due to the rail-rail output swing of the op amps mentioned above. With rail-rail devices, this buffered reference is inherently "low dropout", allowing a +4.5V reference output on a +5V supply, for example. The general expression for V_{OUT} is shown in the figure, where "V_{REF}" is the reference voltage, in this case 1.235V.

Amplifier standby current can be optionally reduced to below 20µA if an amplifier from the OP193/293/493 series is used. This will be at the expense of current drive and positive rail saturation, but does provide the lowest possible quiescent current when necessary. All devices operate from voltages down to 3V (except the OP279, which operates at 5V).

Power conservation can be a critical issue with references, as can output DC precision. For such applications, simple one-package fixed voltage references which simply "drop in" with minimal external circuitry and deliver high accuracy are most attractive. Two unique features of the three terminal REF19X bandgap reference family are low power, and shutdown capability. The series allows fixed outputs from 2.048-5V to be controlled between ON and OFF, via a TTL/CMOS power control input. It provides precision reference quality for those popular voltages shown in Figure 9.42.

30mA REFERENCE FAMILY WITH OPTIONAL SHUTDOWN

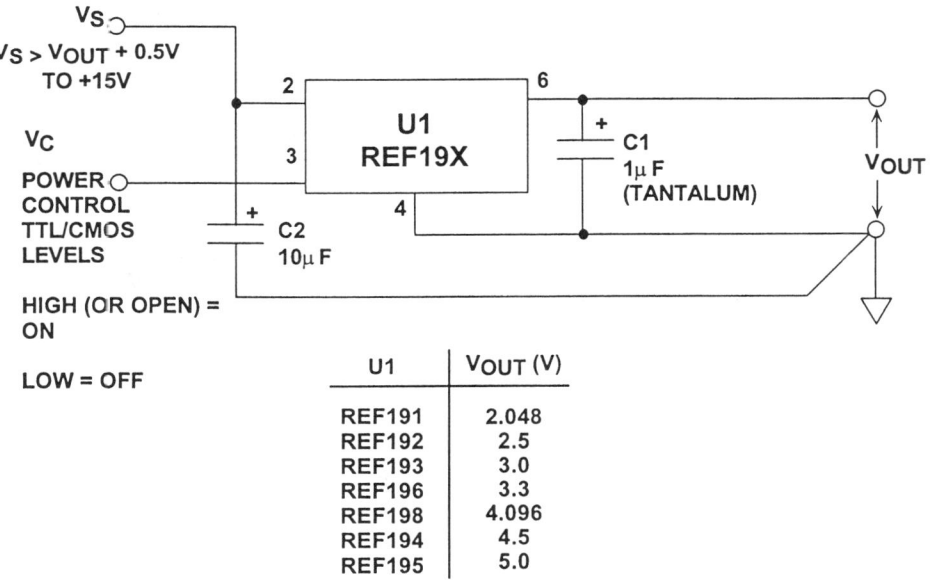

Figure 9.42

Hardware Design Techniques

The REF19X family can be used as a simple three terminal fixed reference as per the table by tying pins 2 and 3 together, or as an ON/OFF controlled device, by programming pin 3 as noted. In addition to the shutdown capacity, the distinguishing functional features are a low dropout of 0.5V at 10mA, and a low current drain for both quiescent and shutdown states, 45 and 15µA (max.), respectively. For example, working from inputs in the range of 6.3 to 15V, a REF195 used as shown drives 5V loads at up to 30mA, with grade dependent tolerances of ±2 to ± 5mV, and max TCs of 5 to 25ppm/°C. Other devices in the series provide comparable accuracy specifications, and all have low dropout features.

To maximize DC accuracy in this circuit, the output of U1 should be connected directly to the load with short heavy traces, to minimize IR drops. The common terminal (pin 4) is less critical due to lower current in this leg.

Low dropout regulators

By adding a boost transistor to the basic rail-rail output low dropout reference of Figure 9.40, output currents of 100mA or more are possible, while still retaining features of low standby current and low dropout voltage. Figure 9.43 shows a low dropout regulator with 800µA standby current, suitable for a variety of outputs at current levels of 100mA.

The 100mA output is achieved with a controlled gain bipolar power transistor for pass device Q1, an MJE170. Maximum output current control is provided by limiting base drive to Q1 with series resistor R3. This limits the base current to about 2mA, so the max H_{FE} of Q1 then allows no more than 500mA, thus limiting Q1's short circuit dissipation to safe levels.

100 mA LOW NOISE, LOW DROPOUT REGULATOR

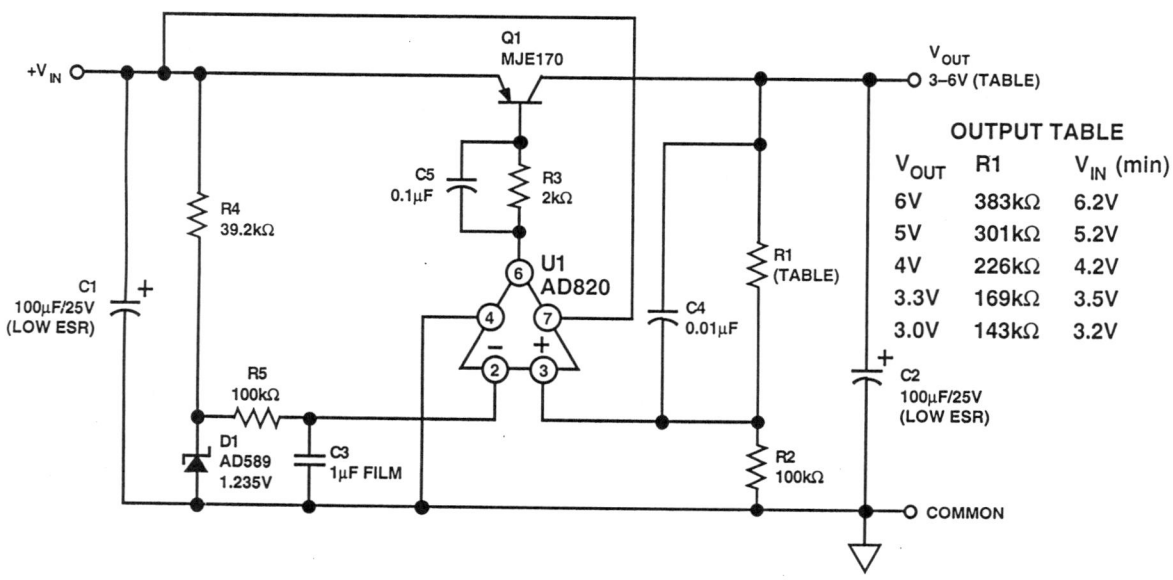

Figure 9.43

Hardware Design Techniques

Overall, the circuit operates as a follower with gain, as was true in the case of Figure 9.40, so V_{OUT} has a similar output expression. The circuit is adapted for different voltages simply by programming R1 via the table. Dropout with a 100mA load is about 200mV, thus a 5V output is maintained for inputs above 5.2V (see table), and V_{OUT} levels down to 3V are possible. Step load response of this circuit is quite good, and transient error is only a few mVp-p for a 30-100mA load change. This is achieved with low ESR switching type capacitors at C1-C2, but the circuit also works with conventional electrolytics (with higher transient errors).

If desired, lowest output noise with the AD820 is reached by including the optional reference noise filter, R5-C3. Lower current op amps can also be used for lower standby current, but with larger transient errors due to reduced bandwidth.

While the 30mA rated output current of the REF19X series is higher than most reference ICs, it can be boosted to much higher levels if desired, with the addition of a PNP transistor, as shown in Figure 9.44. This circuit uses full time current limiting for protection of pass transistor shorts.

In this circuit the supply current of reference U1 flows in R1-R2, developing a base drive for pass device Q1, whose collector provides the bulk of the output current. With a typical gain of 100 in Q1 for 100-200mA loads, U1 is never required to furnish more than a few mA, and this factor minimizes temperature related drift. Short circuit protection is provided by Q2, which clamps drive to Q1 at about 300mA of load current. With separation of control/power functions, DC stability is optimum, allowing best advantage of premium grade REF19X devices for U1. Of course, load management should still be exercised. A short, heavy, low resistance conductor should be used from U1-6 to the V_{OUT} sense point "S", where the collector of Q1 connects to the load.

150 mA BOOSTED OUTPUT REGULATOR/REFERENCE WITH CURRENT LIMITING

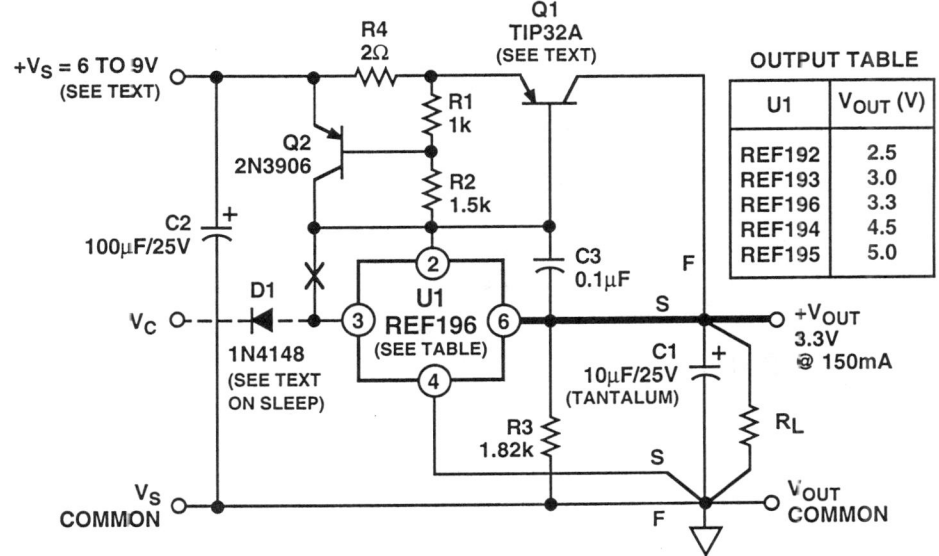

Figure 9.44

Hardware Design Techniques

Because of the current limiting, dropout voltage is raised about 1.1V over that of the REF19X devices. However, overall dropout typically is still low enough to allow operation of a 5 to 3.3V regulator/reference using the 3.3V REF-196 for U1, with a Vs of 4.5V and a load current of 150mA.

The heat sink requirements of Q1 depend upon the maximum power. With Vs = 5V and a 300mA current limit, the worst case dissipation of Q1 is 1.5W, less than the TO-220 package 2W limit. If TO-39 or TO-5 packaged devices such as the 2N4033 are used, the current limit should be reduced to keep maximum dissipation below the package rating, by raising R4. A tantalum output capacitor is used at C1 for its low ESR, and the higher value is required for stability. Capacitor C2 provides input bypassing, and can be a ordinary electrolytic.

Shutdown control of the booster stage is shown as an option, and when used, some cautions are in order. To enable shutdown control, the connection to U1-2 and U1-3 is broken at "X", and diode D1 allows a CMOS control source to drive U1-3 for ON/OFF control. Startup from shutdown is not as clean under heavy load as it is with the basic REF19X series stand-alone, and can require several milliseconds under load. Nevertheless, it is still effective, and can fully control 150mA loads. When shutdown control is used, heavy capacitive loads should be minimized.

By combining a REF19X series reference IC with a rail-rail output op amp, the best of both worlds is realized performance-wise (see Figure 9.45). The REF19X basic reference provides a stable low TC voltage source with low current drain, while the rail-rail output op amp provides high output current with both sink/source load capability.

30 mA OUTPUT CURRENT REGULATOR/REFERENCE

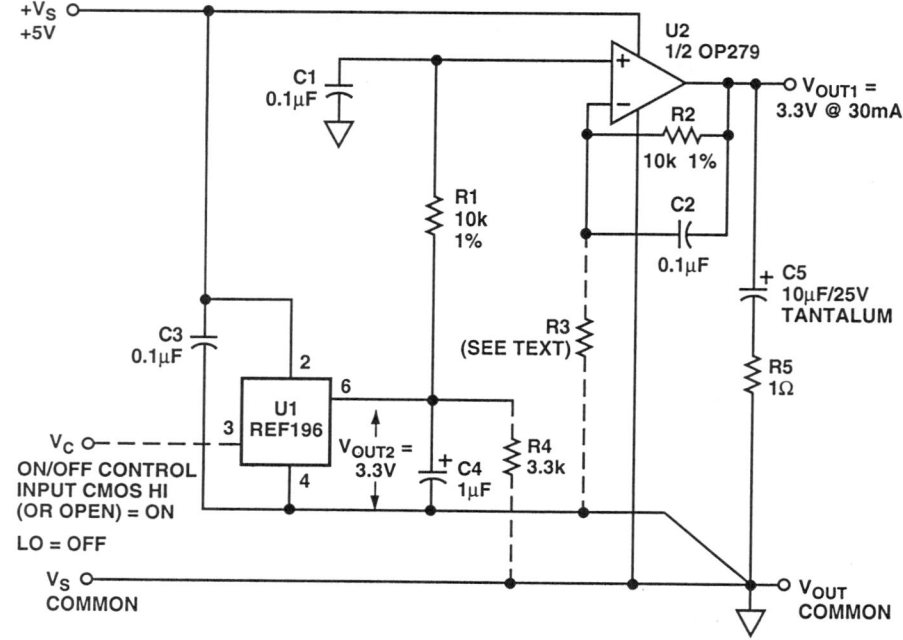

Figure 9.45

The low dropout performance of this circuit is provided by stage U2, 1/2 of an OP279 connected as a follower/buffer for the V_{OUT2} as produced by U1. The low voltage saturation characteristic of the OP279 allows up to 30 mA of load current in the illustrated use, a 5V to 3.3V converter. In this application the stable 3.3V from U1 is applied to U2 through a noise filter, R1-C1. U2 replicates the U1 voltage within a few mV, but at a higher current output at V_{OUT1}. It also has ability to both sink and source output current(s), unlike most IC references. R2 and C2 in the feedback path of U2 provide bias compensation for lowest DC error and additional noise filtering.

To scale V_{OUT2} to another (higher) output level, the optional resistor R3 (shown dotted) is added, causing the new V_{OUT1} to become:

$$V_{OUT1} = V_{OUT2}\left[1 + \frac{R2}{R3}\right]$$

As an example, for a V_{OUT1} = 4.5V, and V_{OUT2} = 2.5V from a REF192, the gain required of U2 is 1.8 times, so R3 and R2 would be chosen for a ratio of 1.25/1, or 22.5kΩ/18kΩ. Note that for the lowest V_{OUT1} DC error, R2||R3 should be maintained equal to R1 (as here), and the R2-R3 resistors should be stable, close tolerance metal film types.

Performance of the circuit is good in both AC and DC senses, with the measured DC output change for a 30mA load change under 1mV, equivalent to an output impedance of <0.03Ω. The transient performance for a step change of 0-10mA of load current is determined largely by the R5-C5 output network. With the values shown, the transient is about 10mV peak, and settles to within 2mV in 8μs, for either polarity. Further reduction in transient amplitude is possible by reducing R5 and possibly increasing C3, but this should be verified by experiment to minimize excessive ringing with some capacitor types. Load current step changes smaller than 10mA will of course show less transient error.

The circuit can be used either as shown as a 5 to 3.3V reference/regulator, or it can also be used with ON/OFF control. By driving pin 3 of U1 with a logic control signal as noted, the output is switched ON/OFF. Note that when ON/OFF control is used, resistor R4 must be used with U1, to speed ON-OFF switching.

As noted, the "low dropout" style of regulator is readily implemented with a rail-rail output op amps such as those of Figure 9.41, as their wide output swing allows easy drive to a low saturation voltage pass device. Further, it is most useful when the op amp has a rail-rail input feature, as this allows high-side current sensing, for positive rail current limiting. Typical applications are voltages developed from a 3-9V range system sources, or anywhere where low dropout performance is required for power efficiency. The 4.5V case here works from 5V nominal sources, with worst-case levels down to 4.6V or less.

Figure 9.46 shows such a regulator using an OP284 plus a low $R_{ds(on)}$ P-channel MOSFET pass device. Low dropout performance is provided by Q1, with a rating of 0.11Ω with a gate drive of 2.7V. This relatively low gate drive allows operation on supplies as low as 3V without compromise to overall performance.

Hardware Design Techniques

300mA LOW DROPOUT REGULATOR WITH HIGH-SIDE CURRENT SENSING

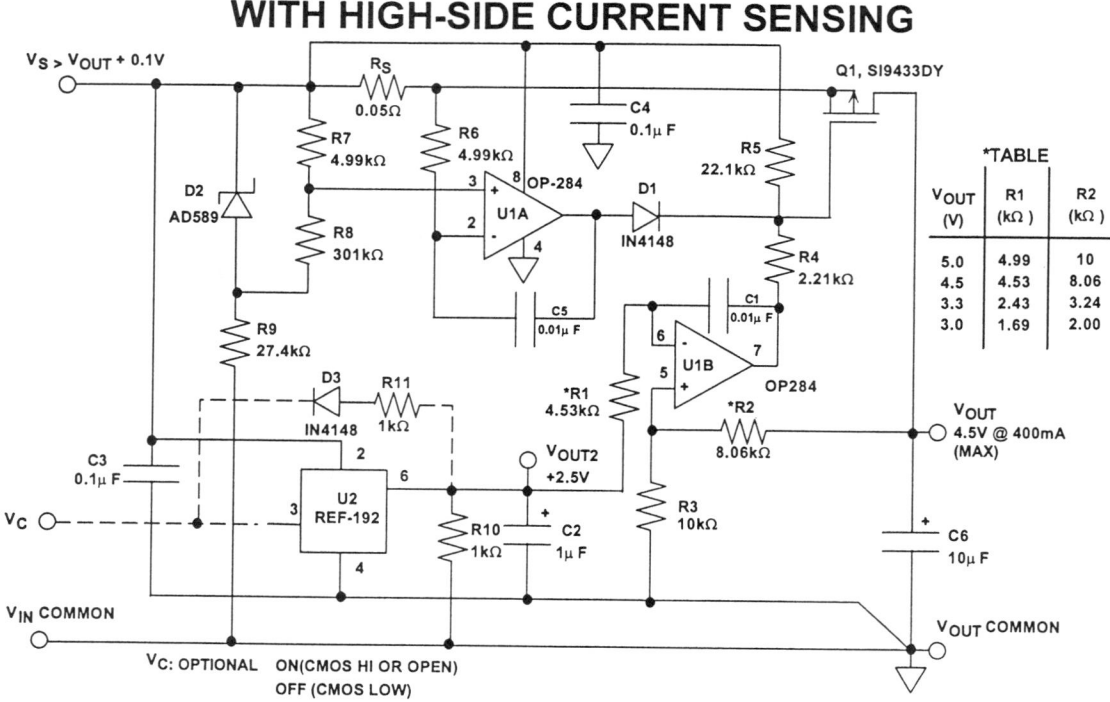

Figure 9.46

The circuit's main voltage control loop operation is provided by U1B, half of the OP284. This voltage control amplifier amplifies the 2.5V reference voltage produced by U2, a REF192. The regulated output voltage V_{OUT} is then of the same form as noted in the previous circuit.

Note that for the lowest V_{OUT} DC error, R2||R3 should be maintained equal to R1 (as here), and the R2-R3 resistors should be stable, close tolerance metal film types. The table suggests R1-R3 values for popular output voltages. In general, V_{OUT} can be anywhere between V_{OUT2} and the 12V maximum rating of Q1.

While the low voltage saturation characteristic of Q1 is part of the low dropout key, the other advantage is a low and accurate current-sense comparison.

Here, this is provided by current sense amplifier U1A, which is provided a 20mV reference from the 1.235V AD589 reference diode D2 and the R7-R8 divider. When the product of the output current and Rs match this threshold, current control is activated, and U1A drives Q1's gate via D1. Overall circuit operation is then under current mode control, with a current limit I_{limit} defined as:

$$I_{limit} = \left[\frac{V_{R(D2)}}{R_s}\right]\left[\frac{R7}{R7+R8}\right]$$

Obviously the comparison voltage should be small, since it becomes a significant portion of the overall dropout voltage. Here, the 20mV value used is higher than the typical offset of the OP284, but still reasonably low as a

Hardware Design Techniques

percentage of V_{OUT} (<0.5%). For other I_{limit} levels, sense resistor Rs should be set along with R7-R8, to maintain this threshold voltage between 20 and 50mV.

For a 4.5V output version, measured DC output change for a 225mA load change was on the order of a few µV, while the dropout voltage at this same current level was about 30mV. The current limit as shown is 400mA, allowing operation at levels up to 300mA or more. While the Q1 device can support currents of several amperes, a practical current rating takes into account the SO-8 device's 2.5W 25°C dissipation. A short-circuit current of 400mA at an input level of 5V will cause a 2W dissipation in Q1, so other input conditions should be considered carefully in terms of Q1's potential overheating. If higher powered devices are used for Q1, the circuit will support outputs of tens of amperes as well as the higher V_{OUT} levels noted above.

The circuit can be used either as shown for a standard low dropout regulator, or it can also be used with ON/OFF control. Note that when the output is OFF in this circuit, it is still active (i.e., not an open circuit). This is because the OFF state simply reduces the voltage input to R1, leaving the U1A/B amplifiers and Q1 still active.

When ON/OFF control is used, resistor R10 should be used with U1, to speed ON-OFF switching, and to allow the output of the circuit to settle to a nominal zero voltage. Components D3 and R11 also aid in speeding up the ON-OFF transition, by providing a dynamic discharge path for C2. OFF-ON transition time is less than 1ms, while the ON-OFF transition is longer, but under 10ms.

REFERENCES: LOW DROPOUT REFERENCES AND REGULATORS

1. Walt Jung, *Build an Ultra-Low-Noise Voltage Reference*, **Electronic Design Analog Applications Issue**, June 24, 1993.

2. Walt Jung, *Getting the Most from IC Voltage References*, **Analog Dialogue 28-1**, 1994.

3. Walt Jung, *The Ins and Outs of 'Green' Regulators/References*, **Electronic Design Analog Applications Issue**, June 27, 1994.

4. Walt Jung, *Very-Low-Noise 5-V Regulator*, **Electronic Design**, July 25, 1994.

ND TECHNIQUES

EMI/RFI CONSIDERATIONS
Adolfo A. Garcia

Electromagnetic interference (EMI) has become a hot topic in the last few years among circuit designers and systems engineers. Although the subject matter and prior art have been in existence for over the last 50 years or so, the advent of portable and high-frequency industrial and consumer electronics has provided a comfortable standard of living for many EMI testing engineers, consultants, and publishers. With the help of EDN Magazine and Kimmel Gerke Associates, this section will highlight general issues of EMC (electromagnetic compatibility) to familiarize the system/circuit designer with this subject and to illustrate proven techniques for protection against EMI.

A Primer on EMI Regulations

The intent of this section is to summarize the different types of electromagnetic compatibility (EMC) regulations imposed on equipment manufacturers, both voluntary and mandatory. Published EMC regulations apply at this time only to equipment and systems, and not to components. Thus, EMI *hardened* equipment does not necessarily imply that each of the components used (integrated circuits, especially) in the equipment must also be EMI *hardened*.

Commercial Equipment

The two driving forces behind commercial EMI regulations are the FCC (Federal Communications Commission) in the U. S. and the VDE (Verband Deutscher Electrotechniker) in Germany. VDE regulations are more restrictive than the FCC's with regard to emissions and radiation, but the European Community will be adding immunity to RF, electrostatic discharge, and power-line disturbances to the VDE regulations, and will require mandatory compliance in 1996. In Japan, commercial EMC regulations are covered under the VCCI (Voluntary Control Council for Interference) standards and, implied by the name, are much looser than their FCC and VDE counterparts.

All commercial EMI regulations primarily focus on *radiated* emissions, specifically to protect nearby radio and television receivers, although both FCC and VDE standards are less stringent with respect to *conducted* interference (by a factor of 10 over radiated levels). The FCC Part 15 and VDE 0871 regulations group commercial equipment into two classes: Class A, for all products intended for business environments; and Class B, for all products used in residential applications. For example, Table 9.1 illustrates the electric-field emission limits of commercial computer equipment for both FCC Part 15 and VDE 0871 compliance.

Radiated Emission Limits for Commercial Computer Equipment

Frequency (MHz)	Class A (at 3 m)	Class B (at 3 m)
30 - 88	300 µV/m	100 µV/m
88 - 216	500 µV/m	150 µV/m
216 - 1000	700 µV/m	200 µV/m

Reprinted from EDN Magazine (January 20, 1994), © CAHNERS PUBLISHING COMPANY 1995, A Division of Reed Publishing USA.

Table 9.1

In addition to the already stringent VDE emission limits, the European Community EMC standards (IEC and IEEE) will require mandatory compliance in 1996 to these additional EMI threats: Immunity to RF fields, electrostatic discharge, and power-line disturbances. All equipment/systems marketed in Europe must exhibit an immunity to RF field strengths of 1-10V/m (IEC standard 801-3), electrostatic discharge (generated by human contact or through material movement) in the range of 10-15kV (IEC standard 801-2), and power-line disturbances of 4kV EFTs (extremely fast transients, IEC standard 801-4) and 6kV lightning surges (IEEE standard C62.41).

Military Equipment

The defining EMC specification for military equipment is MIL-STD-461 which applies to radiated equipment emissions and equipment susceptibility to interference. Radiated emission limits are very typically 10 to 100 times more stringent than the levels shown in Table 9.1. Required limits on immunity to RF fields are typically 200 times more stringent (RF field strengths of 5-50mV/m) than the limits for commercial equipment.

Medical Equipment

Although not yet mandatory, EMC regulations for medical equipment are presently being defined by the FDA (Food and Drug Administration) in the USA and the European Community. The primary focus of these EMC regulations will be on immunity to RF fields, electrostatic discharge, and power-line disturbances, and may very well be more stringent than the limits spelled out in MIL-STD-461. The primary objective of the medical EMC regulations is to guarantee safety to humans.

Industrial- and Process-Control Equipment

Presently, equipment designed and marketed for industrial- and process-control applications are not required to meet pre-existing mandatory EMC regulations. In fact, manufacturers are exempt from complying to any standard in the USA. However, since industrial environments are very much electrically *hostile*, all equipment manufacturers will be required to comply with all European Community EMC regulations by 1996.

Automotive Equipment

Perhaps the most difficult and hostile environment in which electrical circuits and systems must operate is that found in the automobile. All of the key EMI threats to electrical systems exist here. In addition, operating temperature extremes, moisture, dirt, and toxic chemicals further exacerbate the problem. To complicate matters further, standard techniques (ferrite beads, feed-through capacitors, inductors, resistors, shielded cables, wires, and connectors) used in other systems are not generally used in automotive applications because of the cost of the additional components.

Presently, automotive EMC regulations, defined by the very comprehensive SAE Standards J551 and J1113, are not yet mandatory. They are, however, very rigorous. SAE standard J551 applies to vehicle-level EMC specifications, and standard J1113 (functionally similar to MIL-STD-461) applies to all automotive electronic modules. For example, the J1113 specification requires that electronic modules cannot radiate electric fields greater than 300nV/m at a distance of 3 meters. This is roughly 1000 times more stringent than the FCC Part 15 Class A specification. In many applications, automotive manufacturers are imposing J1113 RF field immunity limits on *each of the active components* used in these modules. Thus, in the very near future, automotive manufacturers will require that IC products comply with existing EMC standards and regulations.

EMC Regulations' Impact on Design

In all these applications and many more, complying with mandatory EMC regulations will require careful design of individual circuits, modules, and systems using established techniques for cable shielding, signal and power-line filtering against both small- and large-scale disturbances, and sound multi-layer PCB layouts. The key to success is to incorporate sound EMC principles early in the design phase to avoid time-consuming and expensive redesign efforts.

HARDWARE DESIGN TECHNIQUES

PASSIVE COMPONENTS: YOUR ARSENAL AGAINST EMI

Minimizing the effects of EMI requires that the circuit/system designer be completely aware of the primary arsenal in the battle against interference: *passive components*. To successfully use these components, the designer must understand their non-ideal behavior.

For example, Figure 9.47 illustrates the *real* behavior of the passive components used in circuit design. At very high frequencies, wires become transmission lines, capacitors become inductors, inductors become capacitors, and resistors behave as resonant circuits.

ALL PASSIVE COMPONENTS EXHIBIT "NON IDEAL" BEHAVIOR

Reprinted from EDN Magazine (January 20, 1994) © CAHNERS PUBLISHING COMPANY 1995, A Division of Reed Publishing USA

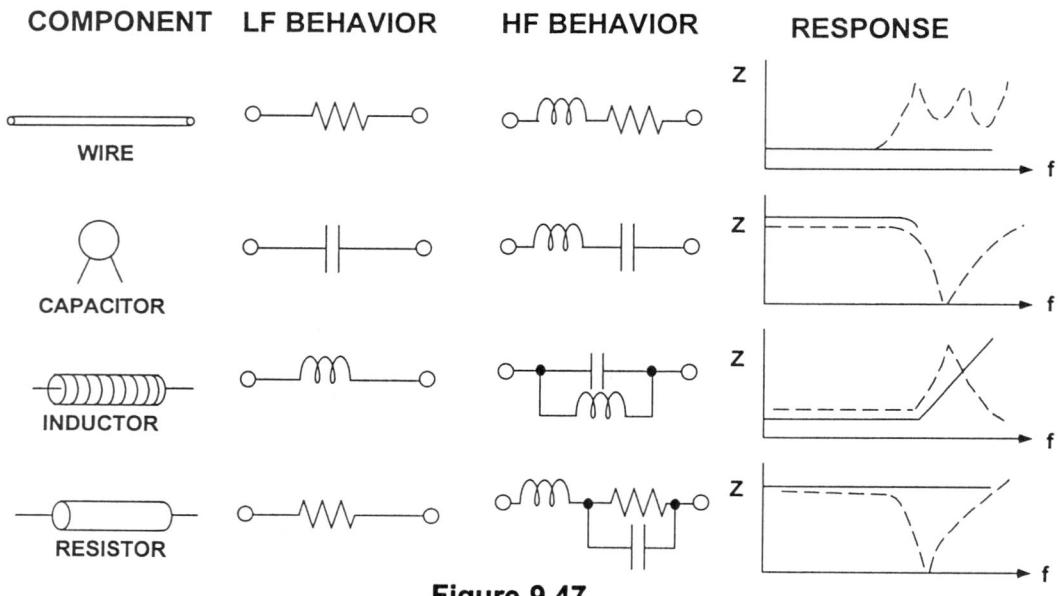

Figure 9.47

A specific case in point is the frequency response of a simple wire compared to that of a ground plane. In many circuits, wires are used as either power or signal returns, and there is no ground plane. A wire will behave as a very low resistance (less than 0.02Ω/ft for 22-gauge wire) at low frequencies, but because of its parasitic inductance of approximately 20nH/ft, it becomes inductive at frequencies above 160kHz. Furthermore, depending on size and routing of the wire and the frequencies involved, it ultimately becomes a transmission line with an uncontrolled impedance. From our knowledge of RF, unterminated transmission lines become antennas with gain, as illustrated in Figure 9.48. On the other hand, large area ground planes are much more well-behaved, and maintain a low impedance over a wide range of frequencies. With a good understanding of the behavior of *real* components, a strategy can now be developed to find solutions to most EMI problems.

HARDWARE DESIGN TECHNIQUES

IMPEDANCE COMPARISON: WIRE VS. GROUND PLANE

Reprinted from EDN Magazine (January 20, 1994) © CAHNERS PUBLISHING COMPANY 1995, A Division of Reed Publishing USA

BEHAVIOR:

LOW FREQUENCY - RESISTIVE
MEDIUM FREQUENCY - INDUCTIVE
HIGH FREQUENCY -
TRANSMISSION-LINE AND
ANTENNA EFFECTS

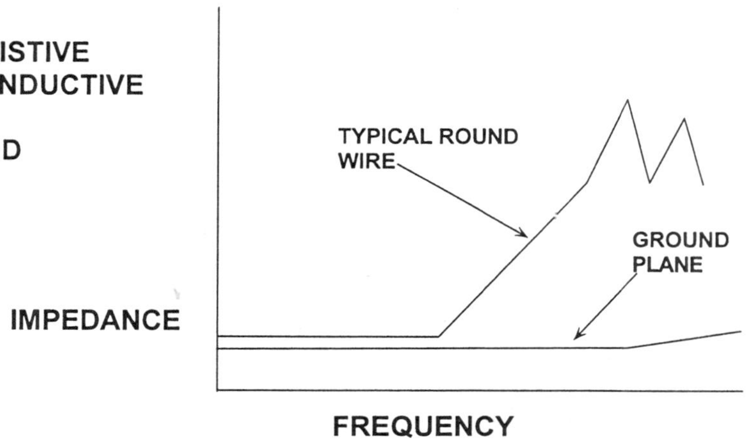

IMPEDANCE

TYPICAL ROUND WIRE

GROUND PLANE

FREQUENCY

Figure 9.48

With any problem, a strategy should be developed before any effort is expended trying to solve it. This approach is similar to the scientific method: initial circuit misbehavior is noted, theories are postulated, experiments designed to test the theories are conducted, and results are again noted. This process continues until all theories have been tested and expected results achieved and recorded. With respect to EMI, a problem solving framework has been developed. As shown in Figure 9.49, the model suggested by Kimmel-Gerke in [Reference 1] illustrates that all three elements (a *source*, a *receptor* or *victim*, and a *path* between the two) must exist in order to be considered an EMI problem. The sources of electromagnetic interference can take on many forms, and the ever-increasing number of portable instrumentation and personal communications/computation equipment only adds the number of possible sources and receptors.

Interfering signals reach the receptor by *conduction* (the circuit or system interconnections) or *radiation* (parasitic mutual inductance and/or parasitic capacitance). In general, if the frequencies of the interference are less than 30MHz, the primary means by which interference is coupled is through the *interconnects*. Between 30MHz and 300MHz, the primary coupling mechanism is *cable radiation and connector leakage*. At frequencies greater than 300MHz, the primary mechanism is *slot and board radiation*. There are many cases where the interference is broadband, and the coupling mechanisms are combinations of the above.

A DIAGNOSTIC FRAMEWORK FOR EMI

Reprinted from EDN Magazine (January 20,1994), © CAHNERS PUBLISHING COMPANY 1995, A Division of Reed Publishing USA

ANY INTERFERENCE PROBLEM CAN BE BROKEN DOWN INTO:

- The SOURCE of interference
- The RECEPTOR of interference
- The PATH coupling the source to the receptor

SOURCES	PATHS	RECEPTORS
Microcontroller ♦ Analog ♦ Digital ESD Communications Transmitters Power Disturbances Lightning	Radiated ♦ EM Fields ♦ Crosstalk 　Capacitive 　Inductive Conducted ♦ Signal ♦ Power ♦ Ground	Microcontroller ♦ Analog ♦ Digital Communications ♦ Receivers Other Electronic Systems

Figure 9.49

When all three elements exist together, a framework for solving any EMI problem can be drawn from Figure 9.50. There are three types of interference with which the circuit or system designer must contend. The first type of interference is that generated by and emitted from an instrument; this is known as circuit/system *emission* and can be either *conducted* or *radiated*. An example of this would be the personal computer. Portable and desktop computers must pass the stringent FCC Part 15 specifications prior to general use.

The second type of interference is circuit or system *immunity*. This describes the behavior of an instrument when it is exposed to large electromagnetic fields, primarily electric fields with an intensity in the range of 1 to 10V/m at a distance of 3 meters. Another term for immunity is *susceptibility*, and it describes circuit/system behavior against radiated or conducted interference.

The third type of interference is *internal*. Although not directly shown on the figure, internal interference can be high-speed digital circuitry within the equipment which affects sensitive analog (or other digital circuitry), or noisy power supplies which can contaminate both analog and digital circuits. Internal interference often occurs between digital and analog circuits, or between motors or relays and digital circuits. In mixed signal environments, the digital portion of the system often interferes with analog circuitry. In some systems, the internal interference reaches such high levels that even very high-speed digital circuitry can affect other low-speed digital circuitry as well as analog circuits.

HARDWARE DESIGN TECHNIQUES

THREE TYPES OF INTERFERENCE
EMISSIONS - IMMUNITY - INTERNAL

Reprinted from EDN Magazine (January 20, 1994) © CAHNERS PUBLISHING COMPANY 1995, A Division of Reed Publishing USA

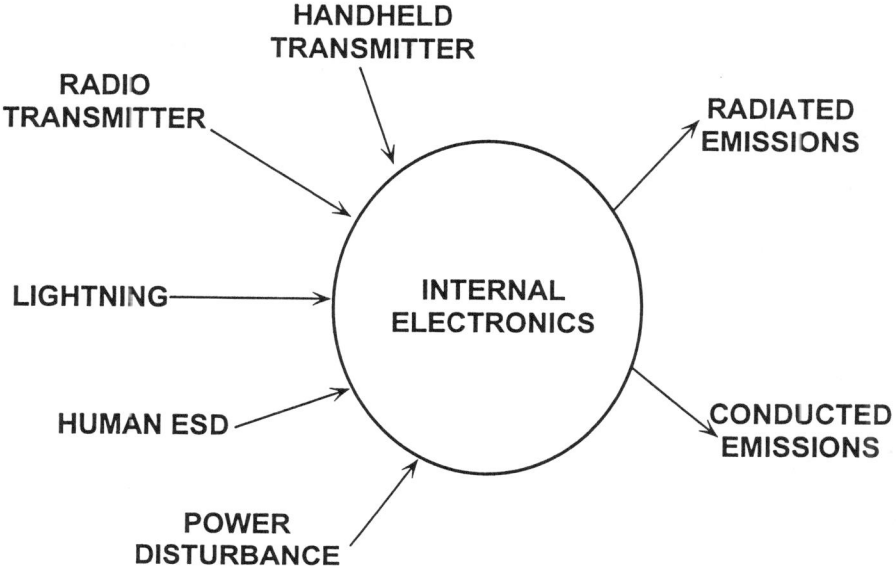

Figure 9.50

In addition to the source-path-receptor model for analyzing EMI-related problems, Kimmel Gerke Associates have also introduced the FAT-ID concept [Reference 1]. FAT-ID is an acronym that describes the five key elements inherent in any EMI problem. These five key parameters are: *frequency, amplitude, time, impedance,* and *distance.*

The *frequency* of the offending signal suggests its path. For example, the path of low-frequency interference is often the circuit conductors. As the interference frequency increases, it will take the path of least impedance, usually stray capacitance. In this case, the coupling mechanism is radiation.

Time and frequency in EMI problems are interchangeable. In fact, the physics of EMI have shows that the time response of signals contains all the necessary information to construct the spectral response of the interference. In digital systems, both the signal rise time and pulse repetition rate produce spectral components according to the following relationship:

$$f_{EMI} = \frac{1}{\pi \cdot t_{rise}} \qquad \text{Eq. 9.1}$$

For example, a pulse having a 1ns rise time is equivalent to an EMI frequency of over 300MHz. This time-frequency relationship can also be applied to high-speed analog circuits, where slew rates in excess of 1000V/µs and gain-bandwidth products greater than 500MHz are not uncommon.

When this concept is applied to instruments and systems, EMI emissions are

Hardware Design Techniques

again functions of signal rise time and pulse repetition rates. Spectrum analyzers and high speed oscilloscopes used with voltage and current probes are very useful tools in quantifying the effects of EMI on circuits and systems.

Another important parameter in the analysis of EMI problems is the physical dimensions of cables, wires, and enclosures. Cables can behave as either passive antennas (receptors) or very efficient transmitters (sources) of interference. Their physical length and their shield must be carefully examined where EMI is a concern. As previously mentioned, the behavior of simple conductors is a function of length, cross-sectional area, and frequency. Openings in equipment enclosures can behave as slot antennas, thereby allowing EMI energy to affect the internal electronics.

Radio Frequency Interference

The world is rich in radio transmitters: radio and TV stations, mobile radios, computers, electric motors, garage door openers, electric jackhammers, and countless others. All this electrical activity can affect circuit/system performance and, in extreme cases, may render it inoperable. Regardless of the location and magnitude of the interference, circuits/systems must have a minimum level of immunity to radio frequency interference (RFI). The next section will cover two general means by which RFI can disrupt normal instrument operation: the direct effects of RFI sensitive analog circuits, and the effects of RFI on shielded cables.

Two terms are typically used in describing the sensitivity of an electronic system to RF fields. In communications, radio engineers define *immunity* to be an instrument's *susceptibility to the applied RFI power density at the unit*. In more general EMI analysis, the *electric-field intensity* is used to describe RFI stimulus. For comparative purposes, Equation 9.2 can be used to convert electric-field intensity to power density and vice-versa:

$$\vec{E}\left(\frac{V}{m}\right) = 61.4 \sqrt{P_T\left(\frac{mW}{cm^2}\right)} \quad \text{Eq. 9.2}$$

where E = Electric Field Strength, in volts per meter, and
P_T = Transmitted power, in milliwatts per cm^2.

From the standpoint of the source-path-receptor model, the *strength of the electric field*, E, surrounding the receptor is a function of *transmitted power*, *antenna gain*, and *distance* from the source of the disturbance. An approximation for the electric-field intensity (for both near- and far-field sources) in these terms is given by Equation 9.3:

$$\vec{E}\left(\frac{V}{m}\right) = 5.5 \left(\frac{\sqrt{P_T \cdot G_A}}{d}\right) \quad \text{Eq. 9.3}$$

where E = Electric field intensity, in V/m;
P_T = Transmitted power, in mW/cm^2;
G_A = Antenna gain (numerical); and
d = distance from source, in meters

Hardware Design Techniques

For example, a 1W hand-held radio at a distance of 1 meter can generate an electric-field of 5.5V/m, whereas a 10kW radio transmission station located 1km away generates a field smaller than 0.6V/m.

Analog circuits are generally more sensitive to RF fields than digital circuits because analog circuits, operating at high gains, must be able to resolve signals in the microvolt/millivolt region. Digital circuits, on the other hand, are more immune to RF fields because of their larger signal swings and noise margins. As shown in Figure 9.51, RF fields can use inductive and/or capacitive coupling paths to generate noise currents and voltages which are amplified by high-impedance analog instrumentation. In many cases, out-of-band noise signals are detected and rectified by these circuits. The result of the RFI rectification is usually unexplained offset voltage shifts in the circuit or in the system.

RFI CAN CAUSE RECTIFICATION IN SENSITIVE ANALOG CIRCUITS

Reprinted from EDN Magazine (January 20, 1994) © CAHNERS PUBLISHING COMPANY 1995, A Division of Reed Publishing USA

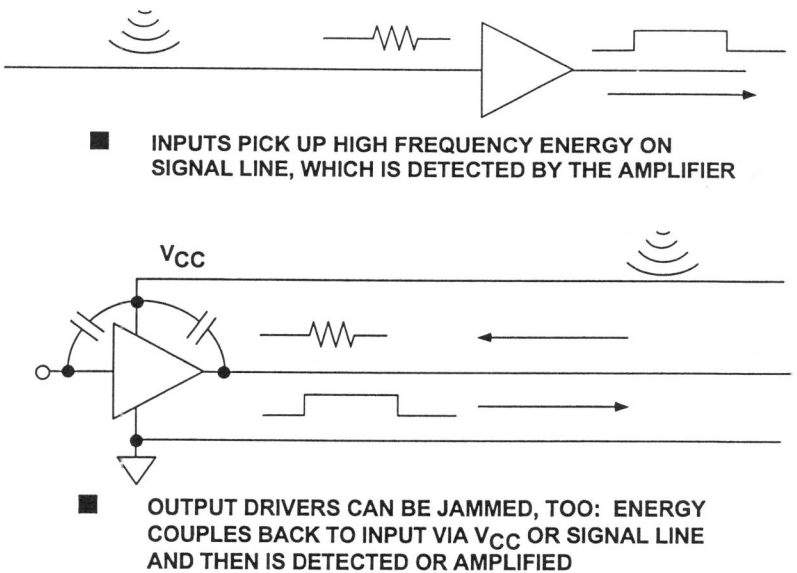

- INPUTS PICK UP HIGH FREQUENCY ENERGY ON SIGNAL LINE, WHICH IS DETECTED BY THE AMPLIFIER

- OUTPUT DRIVERS CAN BE JAMMED, TOO: ENERGY COUPLES BACK TO INPUT VIA V_{CC} OR SIGNAL LINE AND THEN IS DETECTED OR AMPLIFIED

Figure 9.51

There are techniques that can be used to protect analog circuits against interference from RF fields (see Figure 9.52). The three general points of RFI coupling are *signal inputs*, *signal outputs*, and *power supplies*. At a minimum, all power supply pin connections on analog and digital ICs should be decoupled with 0.1µF ceramic capacitors. As was shown in Reference 3, low-pass filters, whose cutoff frequencies are set no higher than 10 to 100 times the signal bandwidth, can be used at the inputs and the outputs of signal conditioning circuitry to filter noise.

KEEPING RFI AWAY FROM ANALOG CIRCUITS

Reprinted from EDN Magazine (January 20, 1994) © CAHNERS PUBLISHING COMPANY 1995, A Division of Reed Publishing USA

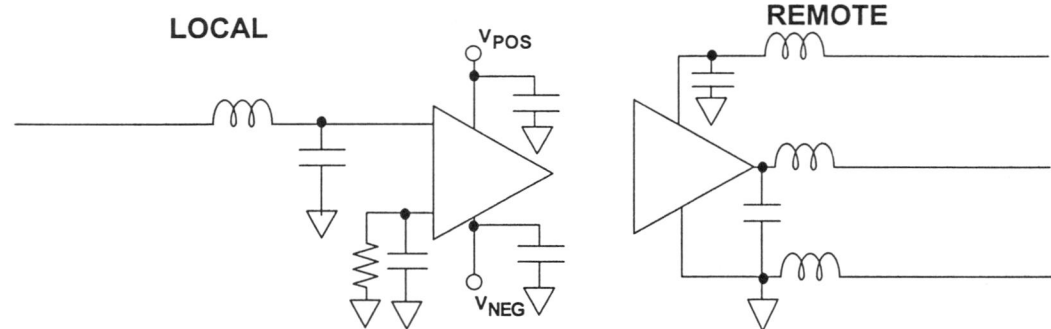

- Decouple all voltage supplies to analog chip with high-frequency capacitors
- Use high-frequency filters on all lines that leave the board
- Use high-frequency filters on the voltage reference if it is not grounded

Figure 9.52

Care must be taken to ensure that the low pass filters (LPFs) are effective at the highest RF interference frequency expected. As illustrated in Figure 9.53, real low-pass filters may exhibit *leakage* at high frequencies. Their inductors can lose their effectiveness due to parasitic capacitance, and capacitors can lose their effectiveness due to parasitic inductance. A rule of thumb is that a conventional low-pass filter (made up of a single capacitor and inductor) can begin to *leak* when the applied signal frequency is 100 to 1000 higher than the filter's cutoff frequency. For example, a 10kHz LPF would not be considered very efficient at filtering frequencies above 1MHz.

HARDWARE DESIGN TECHNIQUES

A SINGLE LOW PASS FILTER LOSES EFFECTIVENESS AT 100 - 1000 f_{3dB}

Reprinted from EDN Magazine (January 20, 1994) © CAHNERS PUBLISHING COMPANY 1995, A Division of Reed Publishing USA

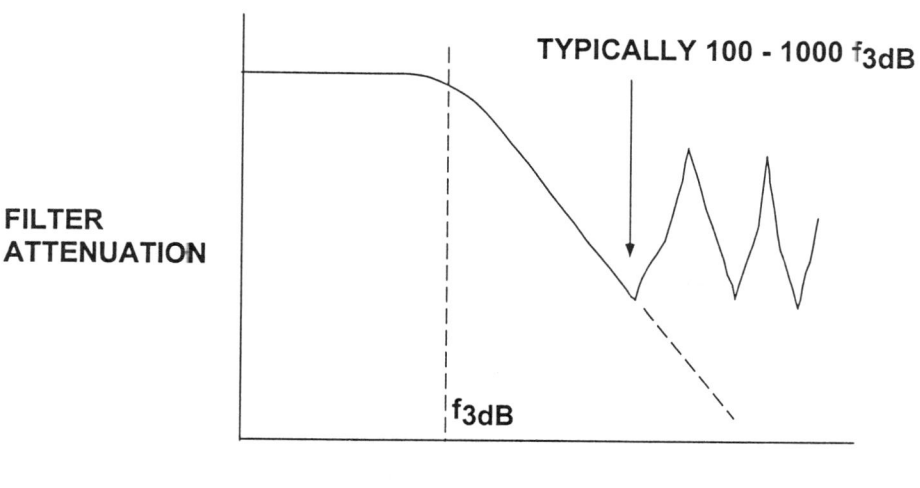

Figure 9.53

Rather than use one LPF stage, it is recommended that the interference frequency bands be separated into *low-band*, *mid-band*, and *high-band*, and then use individual filters for each band. Kimmel Gerke Associates use the stereo speaker analogy of *woofer-midrange-tweeter* for RFI low-pass filter design illustrated in Figure 9.54. In this approach, low frequencies are grouped from 10kHz to 1MHz, mid-band frequencies are grouped from 1MHz to 100MHz, and high frequencies grouped from 100MHz to 1GHz. In the case of a shielded cable input/output, the high frequency section should be located close to the shield to prevent high-frequency leakage at the shield boundary. This is commonly referred to as *feed-through* protection. For applications where shields are not required at the inputs/outputs, then the preferred method is to locate the high frequency filter section as close the analog circuit as possible. This is to prevent the possibility of pickup from other parts of the circuit.

Another cause of filter failure is illustrated in Figure 9.55. If there is any impedance in the ground connection (for example, a long wire or narrow trace connected to the ground plane), then the high-frequency noise uses this impedance path to bypass the filter completely. Filter grounds must be broadband and tied to low-impedance points or planes for optimum performance. High frequency capacitor leads should be kept as short as possible, and low-inductance surface-mounted ceramic chip capacitors are preferable.

MULTISTAGE FILTERS ARE MORE EFFECTIVE

Reprinted from EDN Magazine (January 20, 1994) © CAHNERS PUBLISHING COMPANY 1995, A Division of Reed Publishing USA

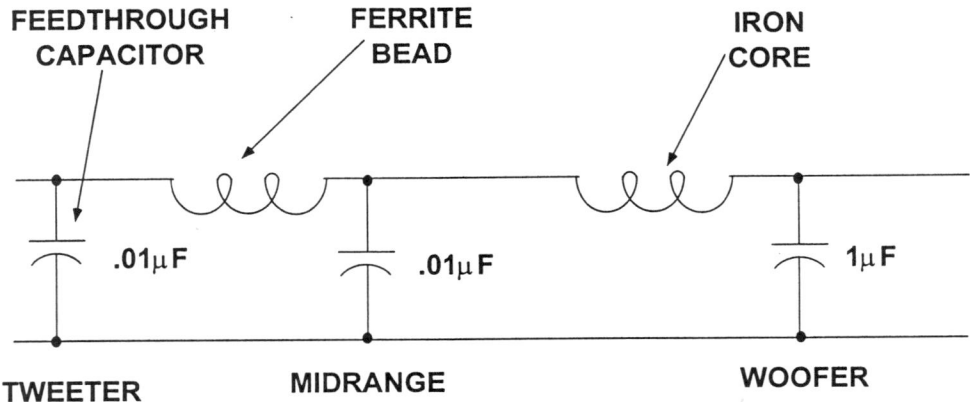

STEREO SPEAKER ANALOGY

Figure 9.54

NON-ZERO (INDUCTIVE AND/OR RESISTIVE) FILTER GROUND REDUCES EFFECTIVENESS

Reprinted from EDN Magazine (January 20, 1994) © CAHNERS PUBLISHING COMPANY 1995, A Division of Reed Publishing USA

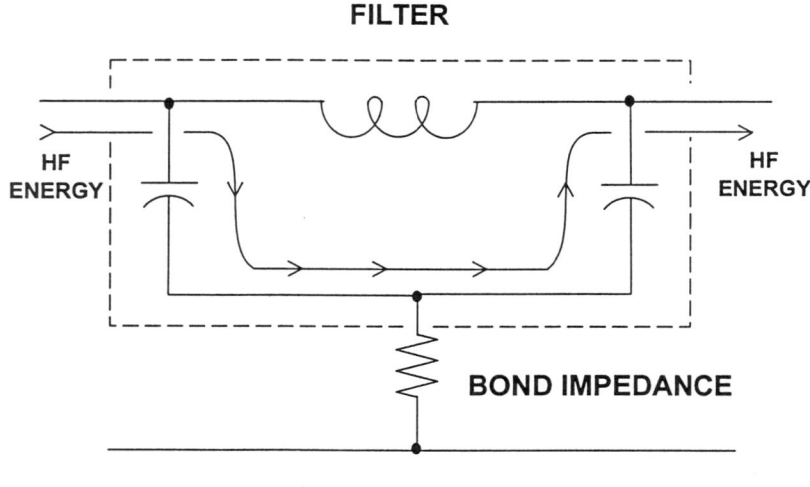

Figure 9.55

HARDWARE DESIGN TECHNIQUES

In the first part of this discussion on RF immunity, circuit level techniques were discussed. In this next section, the second strategic concept for RF immunity will be discussed: *all cables behave as antennas*. As shown in Figure 9.56, pigtail terminations on cables very often cause systems to fail radiated emissions tests because high-frequency noise has coupled into the cable shield, generally through stray capacitance. If the length of the cable is considered *electrically long* (a concept to be explained later) at the interference frequency, then it can behave as a very efficient quarter-wave antenna. The cable pigtail forms a matching network, as shown in the figure, to radiate the noise which coupled into the shield. In general, pigtails are only recommended for applications below 10kHz, such as 50/60Hz interference protection. For applications where the interference is greater than 10kHz, shielded connectors, electrically and physically connected to the chassis, should be used. In applications where shielding is not used, filters on input/output signal and power lines work well. Small ferrites and capacitors should be used to filter high frequencies, provided that: (1) the capacitors have short leads and are tied directly to the chassis ground, and (2) the filters are physically located close to the connectors to prevent noise pickup.

"SHIELDED" CABLE CAN CARRY HIGH FREQUENCY CURRENT AND BEHAVES AS AN ANTENNA

Reprinted from EDN Magazine (January 20, 1994) © CAHNERS PUBLISHING COMPANY 1995, A Division of Reed Publishing USA

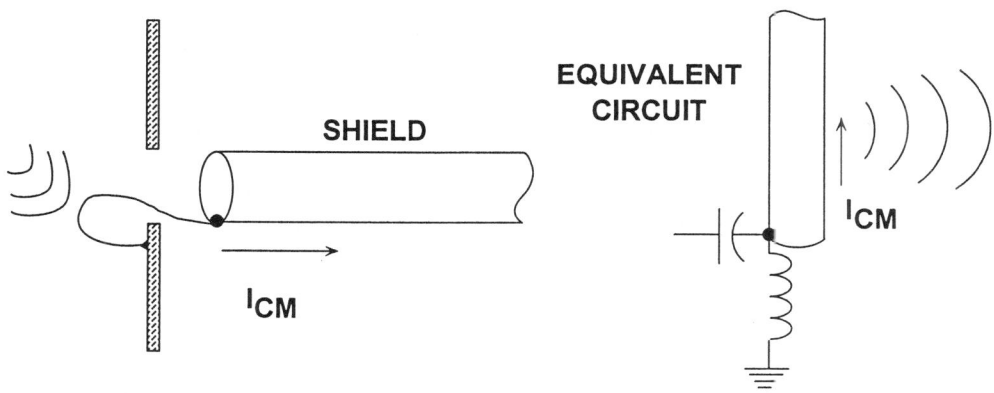

I_{CM} = COMMON-MODE CURRENT

Figure 9.56

The key issues and techniques described in this section on solving RFI related problems are summarized in Figure 9.57. Some of the issues were not discussed in detail, but are equally important. For a complete treatment of this issue, the interested reader should consult References 1 and 2. The main thrust of this section was to provide the reader with a problem-solving strategy against RFI and to illustrate solutions to commonly encountered RFI problems.

SUMMARY OF RADIO FREQUENCY INTERFERENCE AND PROTECTION TECHNIQUES

Reprinted from EDN Magazine (January 20,1994), © CAHNERS PUBLISHING COMPANY 1995, A Division of Reed Publishing USA

- Radio-Frequency Interference is a Serious Threat
  - Equipment causes interference to nearby radio and television
  - Equipment upset by nearby transmitters
- RF-Failure Modes
  - Digital circuits prime source of emissions
  - Analog circuits more vulnerable to RF than digital circuits
- Two Strategic Concepts
  - Treat all cables as antennas
  - Determine the most critical circuits
- RF Circuit Protection
  - Filters and multilayer boards
  - Multistage filters often needed
- RF Shielding
  - Slots and seams cause the most problems
- RF Cable Protection
  - High-quality shields and connectors needed for RF protection

Figure 9.57

Solutions for Power-Line Disturbances

The goal of this next section is not to describe in detail all the circuit/system failure mechanisms which can result from power-line disturbances or faults. Nor is it the intent of this section to describe methods by which power-line disturbances can be prevented. Instead, this section will describe techniques that allow circuits and systems to accommodate *transient* power-line disturbances.

Figure 9.58 is an example of a hybrid power transient protection network commonly used in many applications where lightning transients or other power-line disturbances are prevalent. These networks can be designed to provide protection against transients as high as 10kV and as fast as 10ns. Gas discharge tubes (crowbars) and large geometry zener diodes (clamps) are used to provide both differential and common-mode protection. Metal-oxide varistors (MOVs) can be substituted for the zener diodes in less critical, or in more compact designs. Chokes are used to limit the surge current until the gas discharge tubes fire.

HARDWARE DESIGN TECHNIQUES

POWER LINE DISTURBANCES CAN GENERATE EMI

Reprinted from EDN Magazine (January 20, 1994) © CAHNERS PUBLISHING COMPANY 1995, A Division of Reed Publishing USA

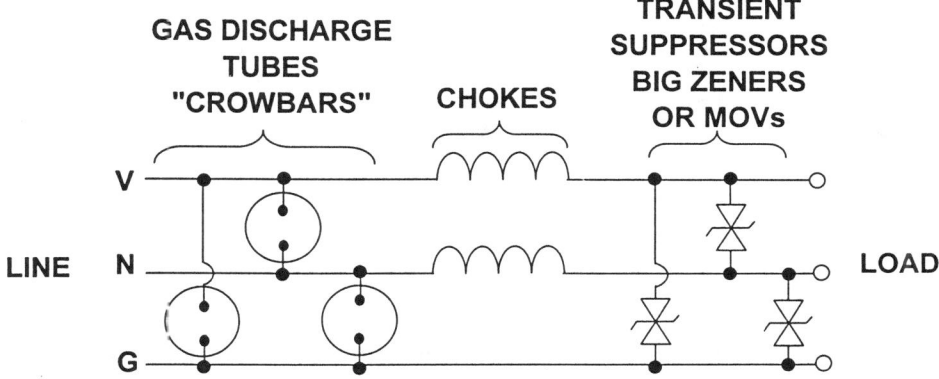

■ COMMON-MODE AND DIFFERENTIAL MODE PROTECTION

Figure 9.58

Commercial EMI filters, as illustrated in Figure 9.59, can be used to filter less catastrophic transients or high-frequency interference. These EMI filters provide both common-mode and differential mode filtering as in Figure 9.58. An optional choke in the safety ground can provide additional protection against common-mode noise. The value of this choke cannot be too large, however, because its resistance may affect power-line fault clearing.

Transformers provide the best common-mode power line isolation. They provide good protection at low frequencies (<1MHz), or for transients with rise and fall times greater than 300ns. Most motor noise and lightning transients are in this range, so isolation transformers work well for these types of disturbances. Although the isolation between input and output is galvanic, isolation transformers do not provide sufficient protection against extremely fast transients (<10ns) or those caused by high-amplitude electrostatic discharge (1 to 3ns). As illustrated in Figure 9.60, isolation transformers can be designed for various levels of differential- or common-mode protection. For differential-mode noise rejection, the Faraday shield is connected to the neutral, and for common-mode noise rejection, the shield is connected to the safety ground.

HARDWARE DESIGN TECHNIQUES

SCHEMATIC FOR A COMMERCIAL POWER LINE FILTER

Reprinted from EDN Magazine (January 20, 1994) © CAHNERS PUBLISHING COMPANY 1995, A Division of Reed Publishing USA

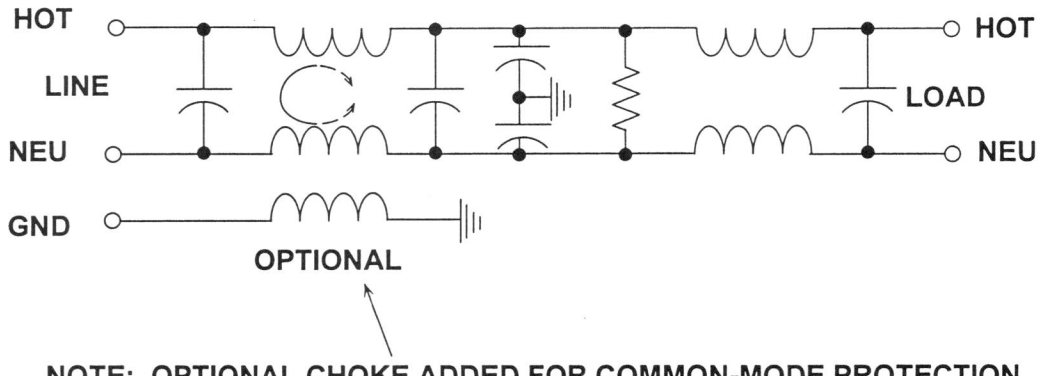

NOTE: OPTIONAL CHOKE ADDED FOR COMMON-MODE PROTECTION

Figure 9.59

FARADAY SHIELDS IN ISOLATION TRANSFORMERS PROVIDE INCREASING LEVELS OF PROTECTION

Reprinted from EDN Magazine (January 20, 1994) © CAHNERS PUBLISHING COMPANY 1995, A Division of Reed Publishing USA

- ■ **STANDARD TRANSFORMER - NO SHIELD**
  - NOTE CONNECTION FROM SECONDARY TO SAFETY GROUND TO ELIMINATE GROUND-TO-NEUTRAL VOLTAGE

- ■ **SINGLE FARADAY SHIELD**
  - CONNECT TO SAFETY GROUND FOR COMMON-MODE PROTECTION

- ■ **SINGLE FARADAY SHIELD**
  - CONNECT TO NOISY-SIDE NEUTRAL WIRE FOR DIFFERENTIAL-MODE PROTECTION

- ■ **TRIPLE FARADAY SHIELD**
  - CONNECT TO SAFETY GROUND FOR COMMON MODE
  - CONNECT TO NEUTRALS FOR DIFFERENTIAL MODE

Figure 9.60

Hardware Design Techniques

Printed Circuit Board Design for EMI Protection

This section will summarize general points regarding the most critical portion of the design phase: the printed circuit board layout. It is at this stage where the performance of the system is most often compromised. This is not only true for signal-path performance, but also for the system's susceptibility to electromagnetic interference and the amount of electromagnetic energy radiated by the system. Failure to implement sound PCB layout techniques will very likely lead to system/instrument EMC failures.

Figure 9.61 is a real-world printed circuit board layout which shows all the paths through which high-frequency noise can couple/radiate into/out of the circuit. Although the diagram shows digital circuitry, the same points are applicable to precision analog, high-speed analog, or mixed analog/digital circuits. Identifying critical circuits and paths helps in designing the PCB layout for both low emissions and susceptibility to radiated and conducted external and internal noise sources.

A key point in minimizing noise problems in a design is to *choose devices no faster than actually required by the application*. Many designers assume that faster is better: fast logic is better than slow, high bandwidth amplifiers are clearly better than low bandwidth ones, and fast DACs and ADCs are better, even if the speed is not required by the system. Unfortunately, faster is not better, but worse where EMI is concerned.

METHODS BY WHICH HIGH FREQUENCY ENERGY COUPLES AND RADIATES INTO CIRCUITRY VIA PLACEMENT

Reprinted from EDN Magazine (January 20, 1994) © CAHNERS PUBLISHING COMPANY 1995, A Division of Reed Publishing USA

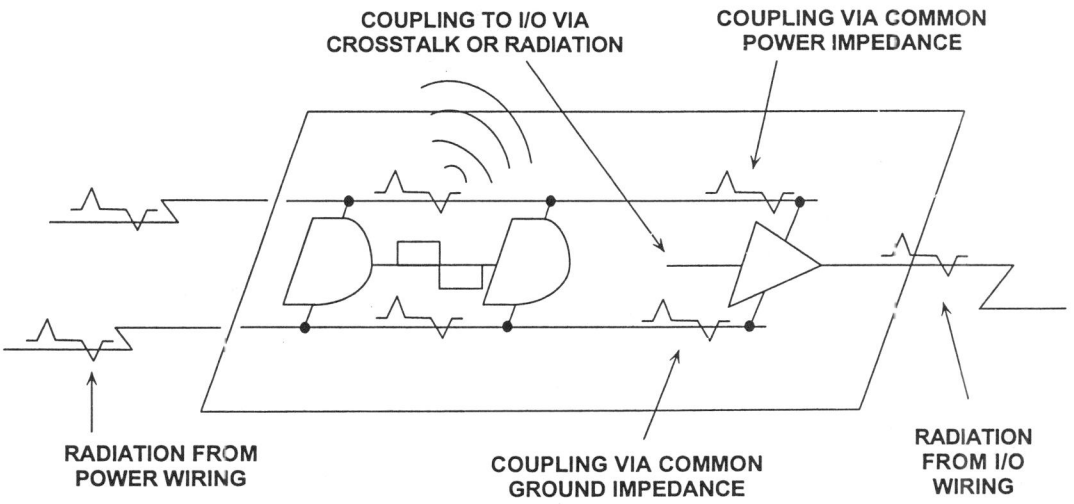

Figure 9.61

HARDWARE DESIGN TECHNIQUES

Many fast DACs and ADCs have digital inputs and outputs with rise and fall times in the nanosecond region. Because of their wide bandwidth, the sampling clock and the digital inputs and can respond to any form of high frequency noise, even glitches as narrow as 1 to 3ns. These high speed data converters and amplifiers are easy prey for the high frequency noise of microprocessors, digital signal processors, motors, switching regulators, hand-held radios, electric jackhammers, etc. With some of these high-speed devices, a small amount of input/output filtering may be required to desensitize the circuit from its EMI/RFI environment. Adding a small ferrite bead just before the decoupling capacitor as shown in Figure 9.62 is very effective in filtering high frequency noise on the supply lines. For those circuits that require bipolar supplies, this technique should be applied to both positive and negative supply lines.

To help reduce the emissions generated by extremely fast moving digital signals at DAC inputs or ADC outputs, a small resistor or ferrite bead may be required at each digital input/output.

POWER SUPPLY FILTERING AND SIGNAL LINE SNUBBING GREATLY REDUCES EMI EMISSIONS

Reprinted from EDN Magazine (January 20, 1994) © CAHNERS PUBLISHING COMPANY 1995, A Division of Reed Publishing USA

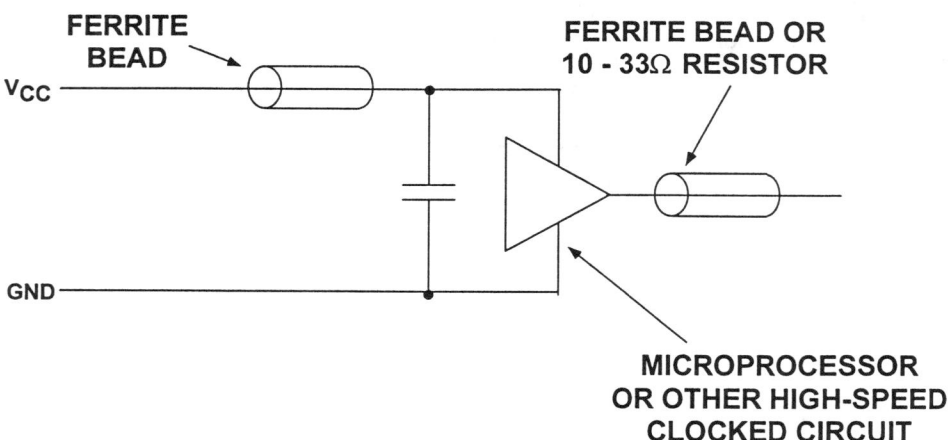

Figure 9.62

Once the system's critical paths and circuits have been identified, the next step in implementing sound PCB layout is to partition the printed circuit board according to circuit function. This involves the appropriate use of power, ground, and signal planes. Good PCB layouts also isolate critical analog paths from sources of high interference (I/O lines and connectors, for example). High frequency circuits (analog and digital) should be separated from low frequency

HARDWARE DESIGN TECHNIQUES

ones. Furthermore, automatic signal routing CAD layout software should be used with extreme caution, and critical paths routed by hand.

Properly designed multilayer printed circuit boards can reduce EMI emissions and increase immunity to RF fields by a factor of 10 or more compared to double-sided boards. A multilayer board allows a complete layer to be used for the ground plane, whereas the ground plane side of a double-sided board is often disrupted with signal crossovers, etc.

The preferred multi-layer board arrangement is to embed the signal traces between the power and ground planes, as shown in Figure 9.63. These low-impedance planes form very high-frequency *stripline* transmission lines with the signal traces. The return current path for a high frequency signal on a trace is located directly above and below the trace on the ground/power planes. The high frequency signal is thus contained inside the PCB, thereby minimizing emissions. The embedded signal trace approach has an obvious disadvantage: debugging circuit traces that are hidden from plain view is difficult.

"TO EMBED OR NOT TO EMBED" THAT IS THE QUESTION

Reprinted from EDN Magazine (January 20, 1994) © CAHNERS PUBLISHING COMPANY 1995, A Division of Reed Publishing USA

- **Advantages of Embedding**
    - Lower impedances, therefore lower emissions and crosstalk
    - Reduction in emissions and crosstalk is significant above 50MHz
    - Traces are protected
- **Disadvantages of Embedding**
    - Lower interboard capacitance, harder to decouple
    - Impedances may be too low for matching
    - Hard to prototype and troubleshoot buried traces

Figure 9.63

Much has been written about terminating printed circuit board traces in their characteristic impedance to avoid reflections. A good rule-of-thumb to determine when this is necessary is as follows: *Terminate the line in its characteristic impedance when the one-way propagation delay of the PCB track is equal to or greater than one-half the applied signal rise/fall time (whichever edge is faster).* A conservative approach is to use a 2 inch (PCB track length)/ nanosecond (rise-, fall-time) criterion. For example, PCB tracks for high-speed logic with rise/fall time of 5ns should be terminated in their characteristic impedance and if the track length is equal to or greater than 10 inches (including any meanders). The 2 inch/ nanosecond track length criterion is summarized in Figure 9.64 for a number of logic families.

LINE TERMINATION SHOULD BE USED WHEN LENGTH OF PCB TRACK EXCEEDS 2 inches / ns

Reprinted from EDN Magazine (January 20,1994), © CAHNERS PUBLISHING COMPANY 1995, A Division of Reed Publishing USA

DIGITAL IC FAMILY	t_r, t_f (ns)	PCB TRACK LENGTH (inches)	PCB TRACK LENGTH (cm)
GaAs	0.1	0.2	0.5
ECL	0.75	1.5	3.8
Schottky	3	6	15
FAST	3	6	15
AS	3	6	15
AC	4	8	20
ALS	6	12	30
LS	8	16	40
TTL	10	20	50
HC	18	36	90

t_r = rise time of signal in ns
t_f = fall time of signal in ns

■ For analog signals @ f_{max}, calculate $t_r = t_f = 0.35 / f_{max}$

Figure 9.64

This same 2 inch/nanosecond rule of thumb should be used with analog circuits in determining the need for transmission line techniques. For instance, if an amplifier must output a maximum frequency of f_{max}, then the equivalent risetime, t_r, can be calculated using the equation $t_r = 0.35/f_{max}$. The maximum PCB track length is then calculated by multiplying the risetime by 2 inch/nanosecond. For example, a maximum output frequency of 100MHz corresponds to a risetime of 3.5ns, and a track carrying this signal greater than 7 inches should be treated as a transmission line.

Equation 9.4 can be used to determine the characteristic impedance of a PCB track separated from a power/ground plane by the board's dielectric (microstrip transmission line):

$$Z_O(\Omega) = \frac{87}{\sqrt{\varepsilon_r + 1.41}} \ln\left[\frac{5.98d}{0.89w + t}\right] \quad \text{Eq. 9.4}$$

where ε_r = dielectric constant of printed circuit board material;
   d = thickness of the board between metal layers, in mils;
   w = width of metal trace, in mils; and
   t = thickness of metal trace, in mils.

The one-way transit time for a single metal trace over a power/ground plane can be determined from Eq. 9.5:

$$t_{pd}(ns/ft) = 1.017\sqrt{0.475\varepsilon_r + 0.67} \quad \text{Eq. 9.5}$$

For example, a standard 4-layer PCB board might use 8-mil wide, 1 ounce (1.4 mils) copper traces separated by 0.021" FR-4 (ε_r=4.7) dielectric material. The characteristic impedance and one-way transit time of such a signal trace would be 88Ω and 1.7ns/ft (7"/ns), respectively. Transmission lines can be effectively terminated in several ways depending on the application, as described in Section 2 of this book.

Figure 9.65 is a summary of techniques that should be applied to printed circuit board layouts to minimize the effects of electromagnetic interference, both emissions and immunity.

CIRCUIT BOARD DESIGN AND EMI

Reprinted from EDN Magazine (January 20,1994), © CAHNERS PUBLISHING COMPANY 1995, A Division of Reed Publishing USA

"ALL EMI PROBLEMS BEGIN AND END AT A CIRCUIT"

- Identify critical, sensitive circuits
- Where appropriate, choose ICs no faster than needed
- Consider and implement sound PCB design
- Spend time on the initial layout (by hand, if necessary)
- Power supply decoupling (digital and analog circuits)
- High-speed digital and high-accuracy analog don't mix
- Beware of connectors for input / output circuits
- Test, evaluate, and correct early and often

Figure 9.65

A Review of Shielding Concepts

The concepts of shielding effectiveness presented next are background material. Interested readers should consult References 1,3, and 4 cited at the end of the section for more detailed information.

Applying the concepts of shielding requires an understanding of the source of the interference, the environment surrounding the source, and the distance between the source and point of observation (the receptor or victim). If the circuit is operating close to the source (in the near-, or induction-field), then the field characteristics are determined by the source. If the circuit is remotely located (in the far-, or radiation-field), then the field characteristics are determined by the transmission medium.

A circuit operates in a near-field if its distance from the source of the interference is less than the wavelength (λ) of the interference divided by 2π, or $\lambda/2\pi$. If the distance between the circuit and the source of the interference is larger than this quantity, then the circuit operates in the far field. For instance, the interference caused by a 1ns pulse edge has an upper bandwidth of approximately 350MHz. The wavelength of a 350MHz signal is approximately 32 inches (the speed of light is approximately 12"/ns). Dividing the wavelength by 2π yields a distance of approximately 5 inches, the boundary between near- and far-field. If a circuit is within 5 inches of a 350MHz interference source, then the circuit operates in the near-field of the interference. If the distance is greater than 5 inches, the circuit operates in the far-field of the interference.

Regardless of the type of interference, there is a characteristic impedance associated with it. The characteristic, or wave impedance of a field is determined by the ratio of its electric (or E-) field to its magnetic (or H-) field. In the far field, the ratio of the electric field to the magnetic field is the characteristic (wave impedance) of free space, given by $Z_0 = 377\Omega$. In the near field, the wave-impedance is determined by the nature of the interference and its distance from the source. If the interference source is high-current and low-voltage (for example, a loop antenna or a power-line transformer), the field is predominately magnetic and exhibits a wave impedance which is less than 377Ω. If the source is low-current and high-voltage (for example, a rod antenna or a high-speed digital switching circuit), then the field is predominately electric and exhibits a wave impedance which is greater than 377Ω.

Conductive enclosures can be used to shield sensitive circuits from the effects of these external fields. These materials present an impedance mismatch to the incident interference because the impedance of the shield is lower than the wave impedance of the incident field. The effectiveness of the conductive shield depends on two things: First is the loss due to the *reflection* of the incident wave off the shielding material. Second is the loss due to the *absorption* of the transmitted wave *within* the shielding material. Both concepts are illustrated in Figure 9.66. The amount of reflection loss depends upon the type of interference and its wave impedance. The amount of absorption loss, however, is independent of the type of interference. It is the same for near- and far-field radiation, as well as for electric or magnetic fields.

REFLECTION AND ABSORPTION ARE THE TWO PRINCIPAL SHIELDING MECHANISMS

Reprinted from EDN Magazine (January 20, 1994) © CAHNERS PUBLISHING COMPANY 1995, A Division of Reed Publishing USA

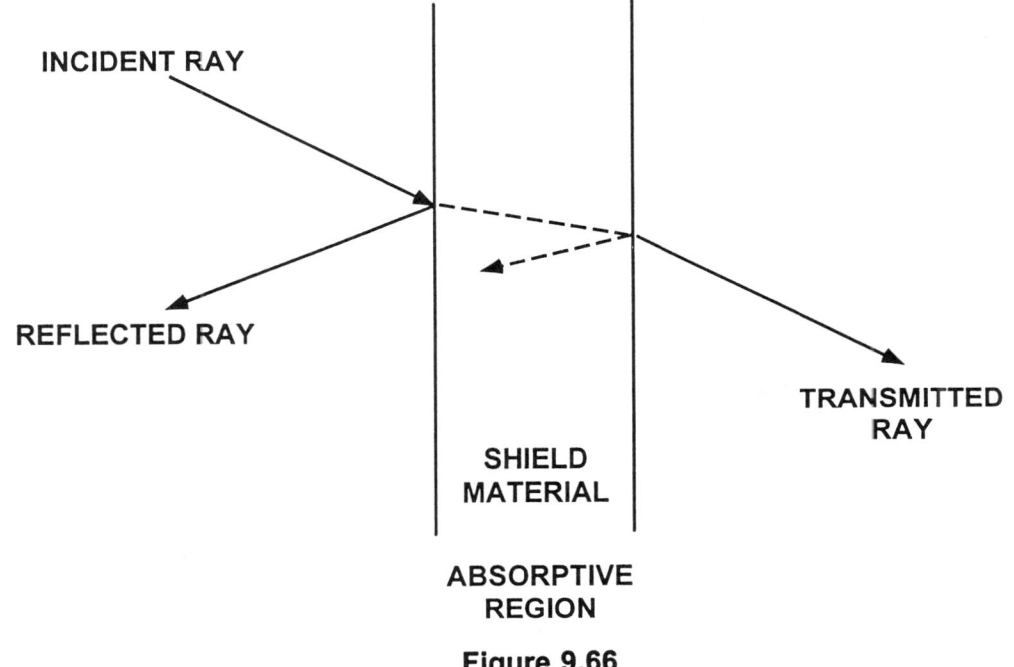

Figure 9.66

Reflection loss at the interface between two media depends on the difference in the characteristic impedances of the two media. For electric fields, reflection loss depends on the frequency of the interference and the shielding material. This loss can be expressed in dB, and is given by:

$$R_e(dB) = 322 + 10\log_{10}\left[\frac{\sigma_r}{\mu_r f^3 r^2}\right] \quad \text{Eq. 9.6}$$

where σ_r = relative conductivity of the shielding material, in Siemens per meter;
μ_r = relative permeability of the shielding material, in Henries per meter;
f = frequency of the interference, and
r = distance from source of the interference, in meters

For magnetic fields, the loss depends also on the shielding material and the frequency of the interference. Reflection loss for magnetic fields is given by:

$$R_m(dB) = 14.6 + 10\log_{10}\left[\frac{fr^2\sigma_r}{\mu_r}\right] \quad \text{Eq. 9.7}$$

and, for plane waves ($r > \lambda/2\pi$), the reflection loss is given by:

$$R_{pw}(dB) = 168 + 10\log_{10}\left[\frac{\sigma_r}{\mu_r f}\right] \quad \text{Eq. 9.8}$$

Absorption is the second loss mechanism in shielding materials. Wave attenuation due to absorption is given by:

$$A(dB) = 3.34\, t\sqrt{\sigma_r \mu_r f} \quad \text{Eq. 9.9}$$

where t = thickness of the shield material, in inches. This expression is valid for plane waves, electric and magnetic fields. Since the intensity of a transmitted field decreases exponentially relative to the thickness of the shielding material, the absorption loss in a shield one skin-depth (δ) thick is 9dB. Since absorption loss is proportional to thickness and inversely proportional to skin depth, increasing the thickness of the shielding material improves shielding effectiveness at high frequencies.

Reflection loss for plane waves in the far field decreases with increasing frequency because the shield impedance, Z_s, increases with frequency. Absorption loss, on the other hand, increases with frequency because skin depth decreases. For electric fields and plane waves, the primary shielding mechanism is reflection loss, and at high frequencies, the mechanism is absorption loss. For these types of interference, high conductivity materials, such as copper or aluminum, provide adequate shielding. At low frequencies, both reflection and absorption loss to magnetic fields is low; thus, it is very difficult to shield circuits from low-frequency magnetic fields. In these applications, high-permeability materials that exhibit low-reluctance provide the best protection. These low-reluctance materials provide a magnetic shunt path that diverts the magnetic field away from the protected circuit. Some characteristics of metallic materials commonly used for shielded enclosures are shown in Table 9.2.

Impedance and Skin Depths for Various Shielding Materials

Material	Conductivity σ_r	Permeability μ_r	Shield Impedance $|Z_s|$	Skin depth δ (inch)
Cu	1	1	$3.68\text{E-}7 \cdot \sqrt{f}$	$\dfrac{2.6}{\sqrt{f}}$
Al	1	0.61	$4.71\text{E-}7 \cdot \sqrt{f}$	$\dfrac{3.3}{\sqrt{f}}$
Steel	0.1	1000	$3.68\text{E-}5 \cdot \sqrt{f}$	$\dfrac{0.26}{\sqrt{f}}$
µ Metal	0.03	20,000	$3\text{E-}4 \cdot \sqrt{f}$	$\dfrac{0.11}{\sqrt{f}}$

Table 9.2

where $\sigma_o = 5.82 \times 10^7$ S/m
$\mu_o = 4\pi \times 10^{-7}$ H/m
$\varepsilon_o = 8.85 \times 10^{-12}$ F/m

A properly shielded enclosure is very effective at preventing external interference from disrupting its contents as well as confining any internally-generated interference. However, in the real world, openings in the shield are often required to accommodate adjustment knobs, switches, connectors, or to provide ventilation (see Figure 9.67). Unfortunately, these openings may compromise shielding effectiveness by providing paths for high-frequency interference to enter the instrument.

ANY OPENING IN AN ENCLOSURE CAN ACT AS AN EMI WAVEGUIDE BY COMPROMISING SHIELDING EFFECTIVENESS

Reprinted from EDN Magazine (January 20, 1994) © CAHNERS PUBLISHING COMPANY 1995, A Division of Reed Publishing USA

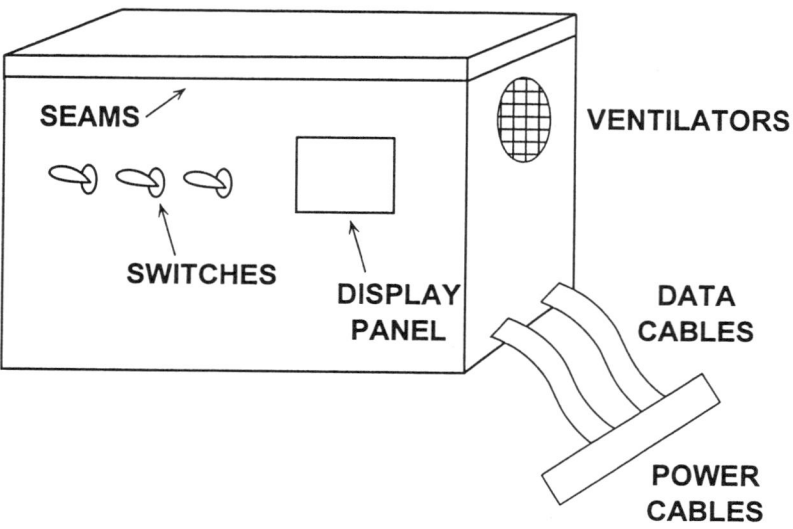

Figure 9.67

The longest dimension (not the total area) of an opening is used to evaluate the ability of external fields to enter the enclosure, because the openings behave as slot antennas. Equation 9.10 can be used to calculate the shielding effectiveness, or the susceptibility to EMI leakage or penetration, of an opening in an enclosure:

$$\text{Shielding Effectiveness (dB)} = 20 \log_{10}\left(\frac{\lambda}{2 \cdot L}\right) \quad \text{Eq. 9.10}$$

where λ = wavelength of the interference and
L = maximum dimension of the opening

Maximum radiation of EMI through an opening occurs when the longest dimension of the opening is equal to one half-wavelength of the interference frequency (0dB shielding effectiveness). A rule-of-thumb is to keep the longest dimension less than 1/20 wavelength of the interference signal, as this provides 20dB shielding effectiveness. Furthermore, a few small openings on each side of an enclosure is preferred over many openings on one side. This is because the openings on different sides radiate energy in different directions, and as a result, shielding effectiveness is not compromised. If openings and seams cannot be avoided, then conductive gaskets, screens, and paints alone or in combination should be used judiciously to limit the longest dimension of any opening to less than 1/20 wavelength. Any cables, wires, connectors, indicators, or control shafts penetrating the enclosure should have circumferential metallic shields physically bonded to the enclosure at the point of entry. In those applications where unshielded cables/wires are used, then filters are recommended at the point of shield entry.

Hardware Design Techniques

General Points on Cables and Shields

Although covered in more detail later, the improper use of cables and their shields is a significant contributor to both radiated and conducted interference. Rather than developing an entire treatise on these issues, the interested reader should consult References 1, 2, 4, and 5. As illustrated in Figure 9.68, effective cable and enclosure shielding confines sensitive circuitry and signals within the entire shield without compromising shielding effectiveness.

Depending on the type of interference (pickup/radiated, low/high frequency), proper cable shielding is implemented differently and is very dependent on the length of the cable. The first step is to determine whether the length of the cable is *electrically short* or *electrically long* at the frequency of concern. A cable is considered *electrically short* if the length of the cable is less than 1/20 wavelength of the highest frequency of the interference, otherwise it is *electrically long*. For example, at 50/60Hz, an *electrically short* cable is any cable length less than 150 miles, where the primary coupling mechanism for these low frequency electric fields is capacitive. As such, for any cable length less than 150 miles, the amplitude of the interference will be the same over the entire length of the cable. To protect circuits against low-frequency electric-field pickup, only one end of the shield should be returned to a low-impedance point. A generalized example of this mechanism is illustrated in Figure 9.69.

In this example, the shield is grounded at the receiver. An exception to this approach (which will be highlighted again later) is the case where line-level (>1Vrms) audio signals are transmitted over long distances using twisted pair,

LENGTH OF SHIELDED CABLES DETERMINES AN "ELECTRICALLY LONG" OR "ELECTRICALLY SHORT" APPLICATION

Reprinted from EDN Magazine (January 20, 1994) © CAHNERS PUBLISHING COMPANY 1995, A Division of Reed Publishing USA

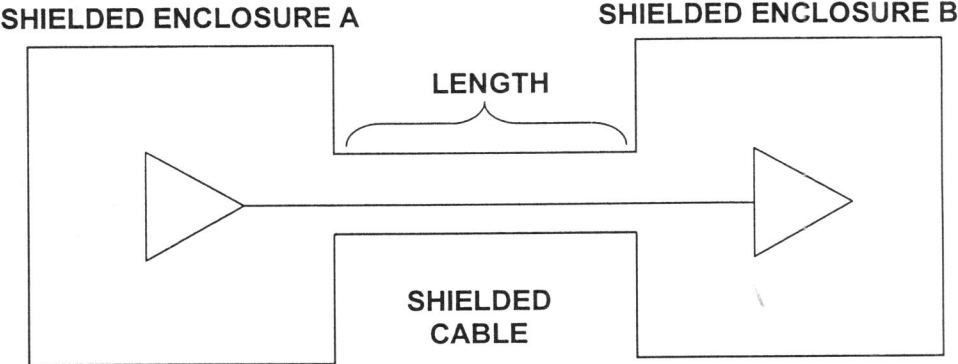

FULLY SHIELDED ENCLOSURES CONNECTED BY FULLY SHIELDED CABLE KEEP ALL INTERNAL CIRCUITS AND SIGNAL LINES INSIDE THE SHIELD.
- TRANSITION REGION: 1/20 WAVELENGTH

Figure 9.68

CONNECT THE SHIELD AT ONE POINT AT THE LOAD TO PROTECT AGAINST LOW FREQUENCY (50/60Hz) THREATS

Reprinted from EDN Magazine (January 20, 1994) © CAHNERS PUBLISHING COMPANY 1995, A Division of Reed Publishing USA

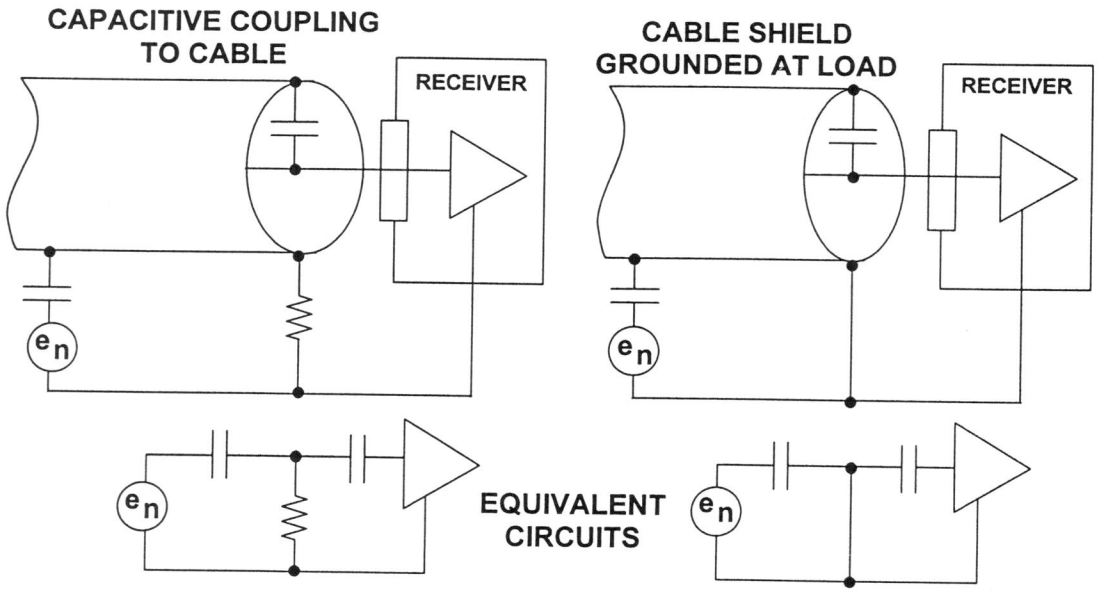

Figure 9.69

shielded cables. In these applications, the shield again offers protection against low-frequency interference, and an accepted approach is to ground the shield at the driver end (LF and HF ground) and ground it at the receiver with a capacitor (HF ground only).

In those applications where the length of the cable is *electrically long*, or protection against high-frequency interference is required, then the preferred method is to connect the cable shield to low-impedance points at both ends (direct connection at the driving end, and capacitive connection at the receiver). Otherwise, unterminated transmission lines effects can cause reflections and standing waves along the cable. At frequencies of 10MHz and above, circumferential (360°) shield bonds and metal connectors are required to main low-impedance connections to ground.

In summary, for protection against low-frequency (<1MHz), electric-field interference, grounding the shield at one end is acceptable. For high-frequency interference (>1MHz), the preferred method is grounding the shield at both ends, using 360° circumferential bonds between the shield and the connector, and maintaining metal-to-metal continuity between the connectors and the enclosure. Low-frequency ground loops can be eliminated by replacing one of the DC shield connections to ground with a low inductance 0.01µF capacitor. This capacitor prevents low frequency ground loops and shunts high frequency interference to ground.

EMI Trouble Shooting Philosophy

System EMI problems often occur after the equipment has been designed and is operating in the field. More often than not, the original designer of the instrument has retired and is living in Tahiti, so the responsibility of repairing it belongs to someone else who may not be familiar with the product. Figure 9.70 summarizes the EMI problem solving techniques discussed in this section and should be useful in these situations.

EMI TROUBLESHOOTING PHILOSOPHY
Reprinted from EDN Magazine (January 20, 1994), © CAHNERS PUBLISHING COMPANY 1995, A Division of Reed Publishing USA

- Diagnose before you fix
- Ask yourself:
    - What are the symptoms?
    - What are the causes?
    - What are the constraints?
    - How will you know you have fixed it?
- Use available models for EMI to identify source - path - victim
- Start at low frequency and work up to high frequency
- EMI doctor's bag of tricks:
    - Aluminum foil
    - Conductive tape
    - Bulk ferrites
    - Power line filters
    - Signal filters
    - Resistors, capacitors, inductors, ferrites
    - Physical separation

Figure 9.70

REFERENCES ON EMI/RFI

1. *EDN's Designer's Guide to Electromagnetic Compatibility*, **EDN**, January, 20, 1994, material reprinted by permission of Cahners Publishing Company, 1995.

2. *Designing for EMC (Workshop Notes)*, Kimmel Gerke Associates, Ltd., 1994.

3. **Systems Application Guide**, Chapter 1, pg. 21-55, Analog Devices, Incorporated, Norwood, MA, 1993.

4. Henry Ott, **Noise Reduction Techniques In Electronic Systems, Second Edition**, New York, John Wiley & Sons, 1988.

5. Ralph Morrison, **Grounding And Shielding Techniques In Instrumentation, Third Edition**, New York, John Wiley & Sons, 1986.

6. **Amplifier Applications Guide**, Chapter XI, pg. 61, Analog Devices, Incorporated, Norwood, MA, 1992.

7. B.Slattery and J.Wynne, *Design and Layout of a Video Graphics System for Reduced EMI*, **Analog Devices Application Note AN-333**.

8. Paul Brokaw, *An IC Amplifier User Guide To Decoupling, Grounding, And Making Things Go Right For A Change*, **Analog Devices Application Note**, Order Number E1393-5-590.

9. A. Rich, *Understanding Interference-Type Noise*, **Analog Dialogue**, 16-3, 1982, pp. 16-19.

10. A. Rich, *Shielding and Guarding*, **Analog Dialogue**, 17-1, 1983, pp. 8-13.

11. **EMC Test & Design**, Cardiff Publishing Company, Englewood, CO. An excellent, general purpose trade journal on issues of EMI and EMC.

SENSORS AND CABLE SHIELDING
John McDonald

The environments in which analog systems operate are often rich in sources of EMI. Common EMI noise sources include power lines, logic signals, switching power supplies, radio stations, electric lighting, and motors. Noise from these sources can easily couple into long analog signal paths, such as cables, which act as efficient antennas. Shielded cables protect signal conductors from electric field (E-field) interference by providing low impedance paths to ground at the offending frequencies. Aluminum foil, copper, and braided stainless steel are materials very commonly used for cable shields due to their low impedance properties.

Simply increasing the separation between the noise source and the cable will yield significant additional attenuation due to reduced coupling, but shielding is still required in most applications involving remote sensors.

There are two paths from an EMI source to a susceptible cable: capacitive (or E-field) and magnetic (or H-field) coupling (see Figure 9.73). Capacitive coupling occurs when parasitic capacitance exists between a noise source and the cable. The amount of parasitic capacitance is determined by the separation, shape, orientation, and the medium between the source and the cable.

Magnetic coupling occurs through parasitic mutual inductance when a magnetic field is coupled from one conductor to another as shown in Figure 9.73. Parasitic mutual inductance depends on the shape and relative orientation of the circuits in question, the magnetic properties of the medium, and is directly proportional to conductor loop area. Minimizing conductor loop area reduces magnetic coupling proportionally.

Shielded *twisted pair* cables offer further noise immunity to magnetic fields. Twisting the conductors together reduces the net loop area, which has the effect of canceling any magnetic field pickup, because the sum of positive and negative incremental loop areas is ideally equal to zero.

Hardware Design Techniques

PRECISION SENSORS AND CABLE SHIELDING

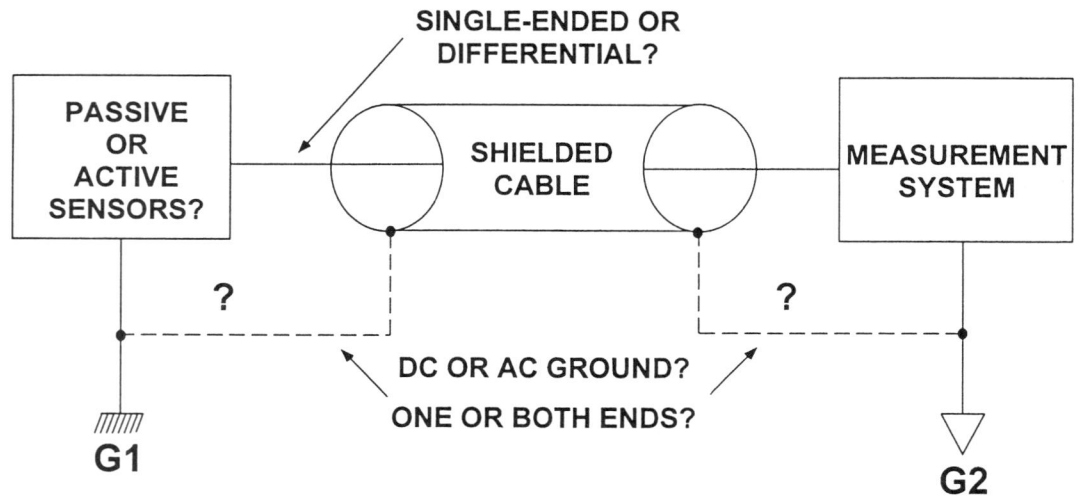

Figure 9.71

WHY SHIELD CABLES?

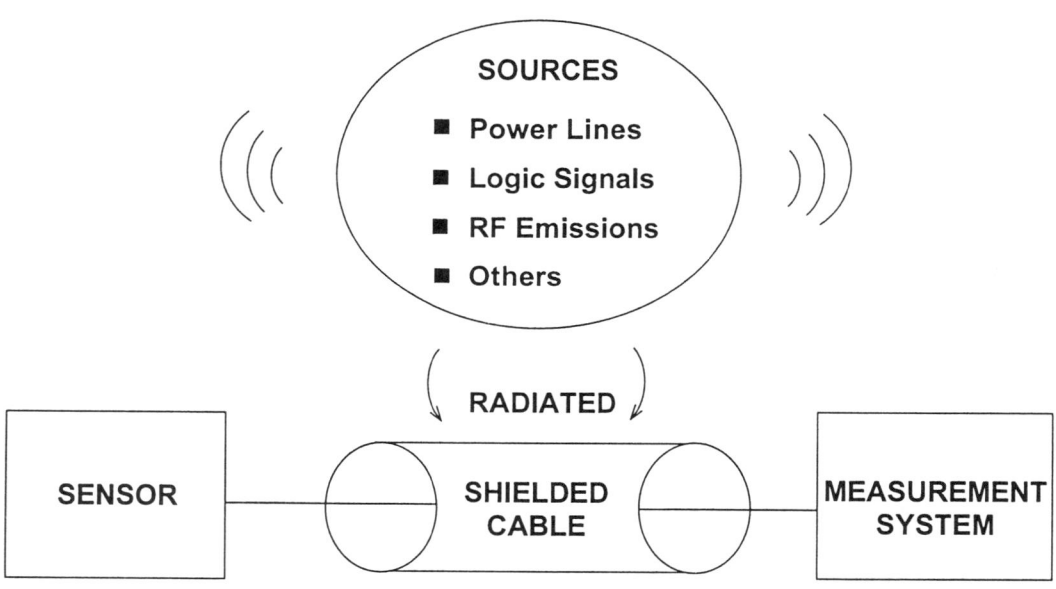

Figure 9.72

HOW DOES INTERFERENCE ENTER THE SYSTEM?

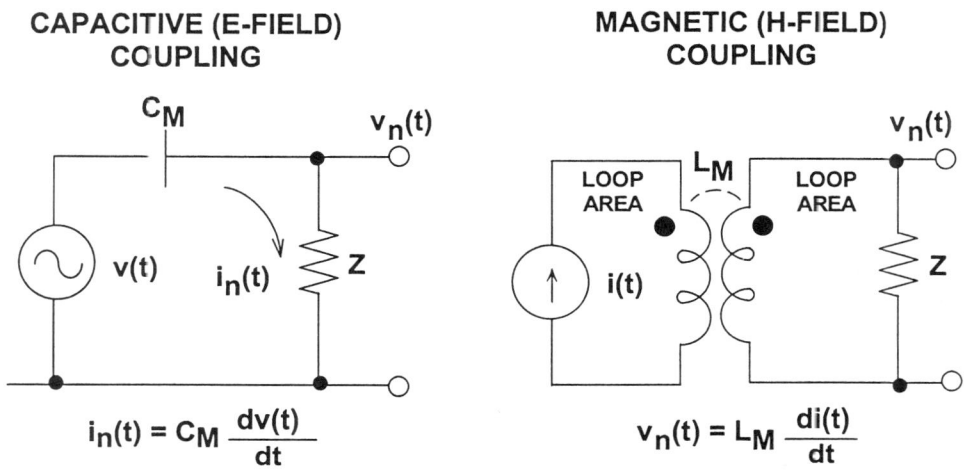

Figure 9.73

To study the shielding problem, a precision *RTD (Resistance Temperature Detector)* amplifier circuit was used as the basis for a series of experiments. A remote 100Ω RTD was connected to the bridge, bridge driver, and the bridge amplifier circuit (Figure 9.74) using 10 feet of a shielded twisted pair cable. The RTD is one element of a 4-element bridge (the three other resistor elements are located in the bridge and bridge driver circuit). The gain of the instrumentation amplifier was adjusted so that the sensitivity at the output was 10mV/°C, with a 5V full scale. Measurements were made at the output of the instrumentation amplifier with the shield grounded in various ways. The experiments were conducted in lab standard environment where a considerable amount of electronic equipment was in operation.

The first experiment was conducted with the shield ungrounded. As shown in Figure 9.74, shields left floating are not useful and offer no attenuation to EMI-induced noise, in fact, they act as antennas. Capacitive coupling is unaffected, because the floating shield provides a coupling path to the signal conductors. Most cables exhibit parasitic capacitances between 10-30pF/ft. Likewise, HF magnetically coupled noise is not attenuated because the floating cable shield does not alter either the geometry or the magnetic properties of the cable conductors. LF magnetic noise is not attenuated significantly, because most shield materials absorb very little magnetic energy.

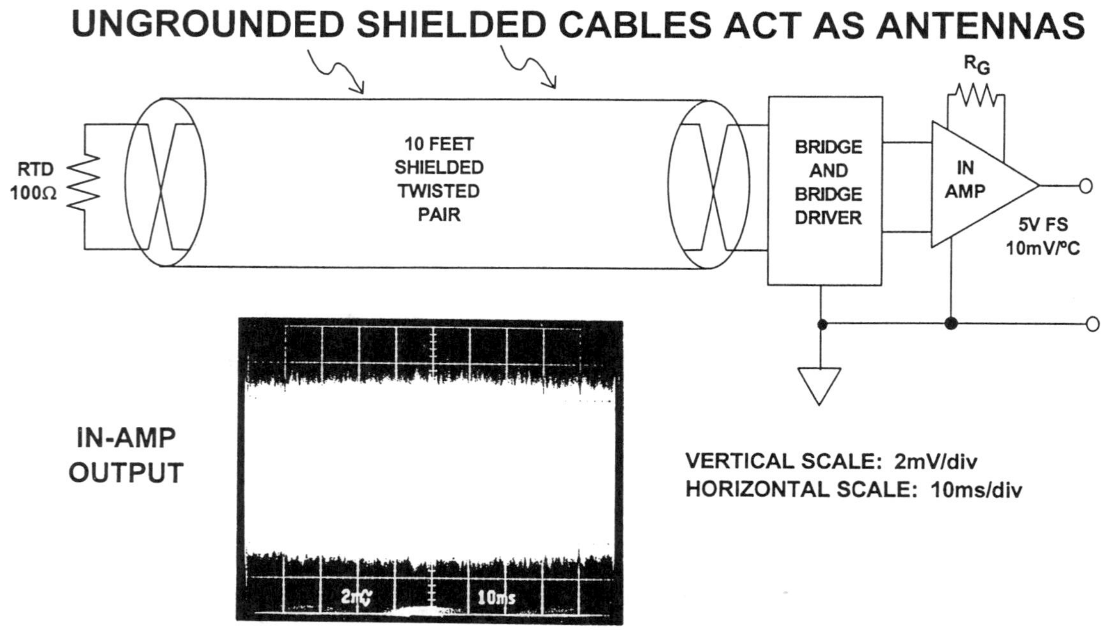

Hardware Design Techniques

UNGROUNDED SHIELDED CABLES ACT AS ANTENNAS

[Diagram: RTD 100Ω connected through 10 feet shielded twisted pair to Bridge and Bridge Driver, feeding an In-Amp with R_G, output 5V FS, 10mV/°C]

IN-AMP OUTPUT

VERTICAL SCALE: 2mV/div
HORIZONTAL SCALE: 10ms/div

Figure 9.74

To implement effective EMI/RFI shielding, the shield must be grounded. A grounded shield reduces the value of the impedance of the shield to ground (Z in Figure 9.73) to small values. Implementing this change will reduce the amplitude of the E-Field noise substantially.

Designers often ground both ends of a shield in an attempt to reduce shield impedance and gain further E-Field attenuation. Unfortunately, this approach can create a new set of potential problems. The AC and DC ground potentials are generally different at each end of the shield. Figure 9.75 illustrates how low-frequency ground loop current is created when both ends of a shield are grounded. This low frequency current flows through the large loop area of the shield and couples into the center conductors through the parasitic mutual inductance. If the twisted pairs are precisely balanced, the induced voltage will appear as a common-mode rather than a differential voltage. Unfortunately, the conductors may not be perfectly balanced, the sensor and excitation circuit may not be fully balanced, and the common mode rejection at the receiver may not be sufficient. There will therefore be some differential noise voltage developed between the conductors at the output end, which is amplified and appears at the final output of the instrumentation amplifier. With the shields of the experimental circuit grounded at both ends, the results are shown in Figure 9.76.

Hardware Design Techniques

SIGNAL GROUND AND EARTH GROUND HAVE DIFFERENT POTENTIALS WHICH MAY INDUCE GROUND LOOP CURRENT

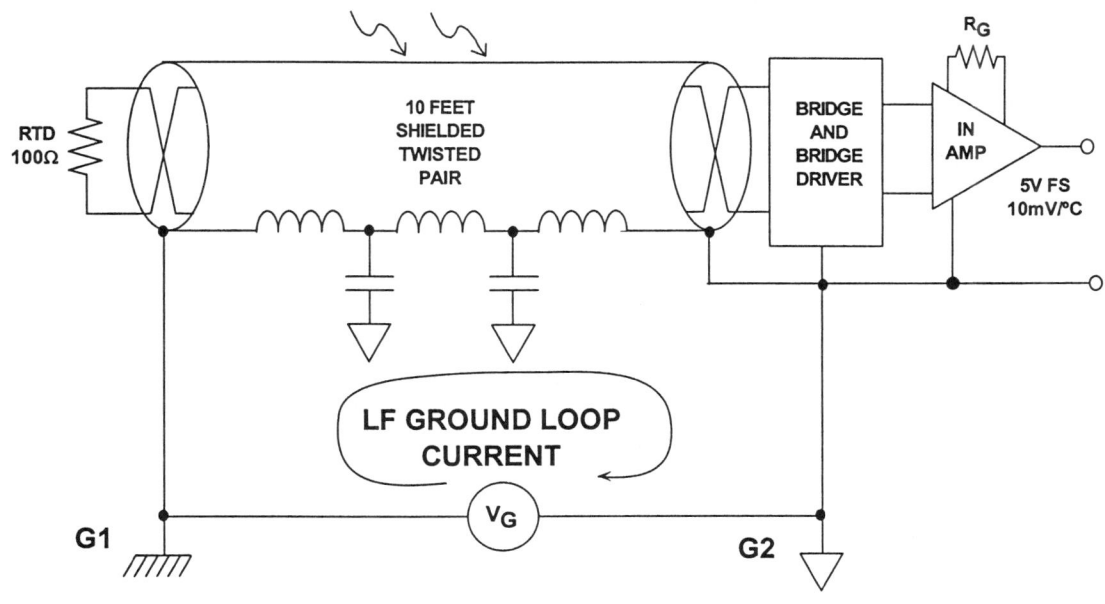

Figure 9.75

GROUNDING BOTH ENDS OF A SHIELD PRODUCES LOW FREQUENCY GROUND LOOPS

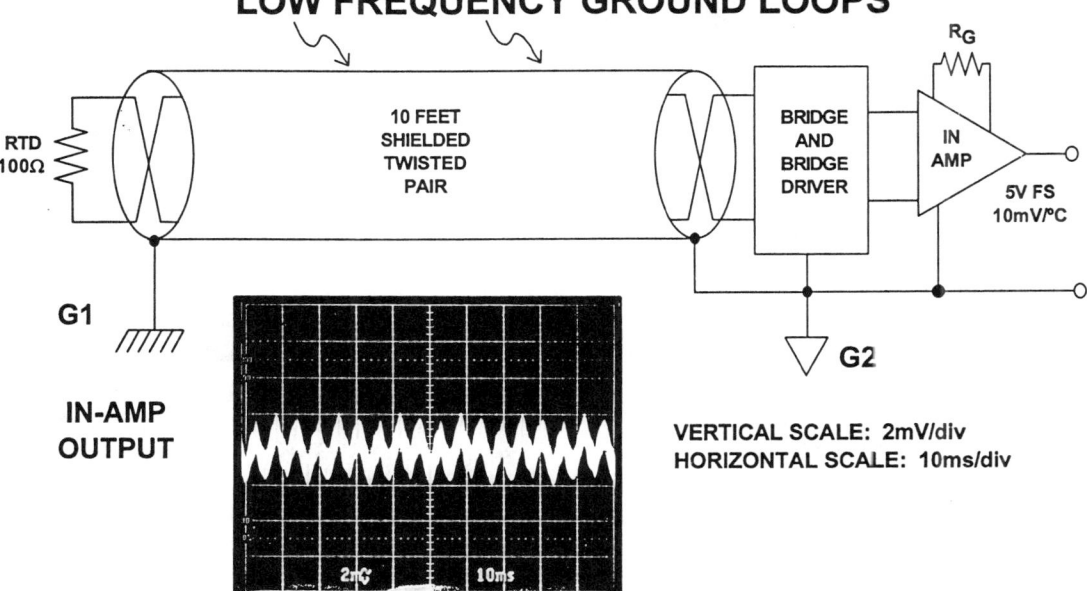

Figure 9.76

HARDWARE DESIGN TECHNIQUES

Figure 9.77 illustrates a properly grounded system with good electric field shielding. Notice that the ground loop has been eliminated. The shield has a single point ground, located at the signal conditioning circuitry, and noise coupled into the shield is effectively shunted into the receiver ground and does not appear at the output of the instrumentation amplifier.

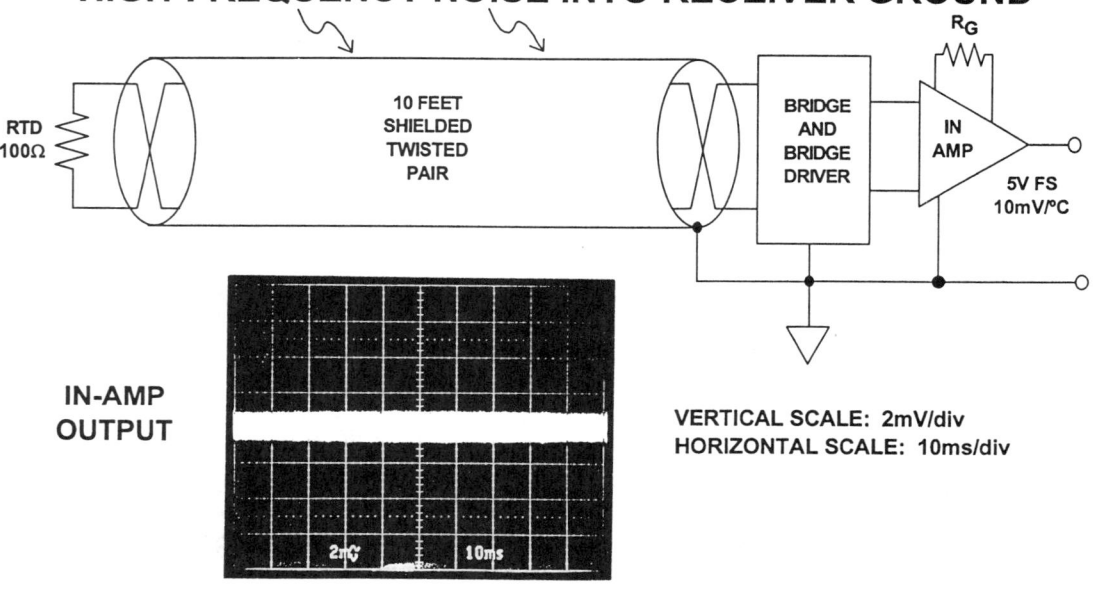

Figure 9.77

Figure 9.78 shows an example of a remotely located, ungrounded, passive sensor (EEG electrodes) which is connected to a high-gain, low power AD620 instrumentation amplifier through a shielded twisted pair cable. Note that the shield is properly grounded at the signal conditioning circuitry. The AD620 gain is 1000×, and the amplifier is operated on ±3V supplies. Notice the absence of 60Hz interference in the amplifier output.

HARDWARE DESIGN TECHNIQUES

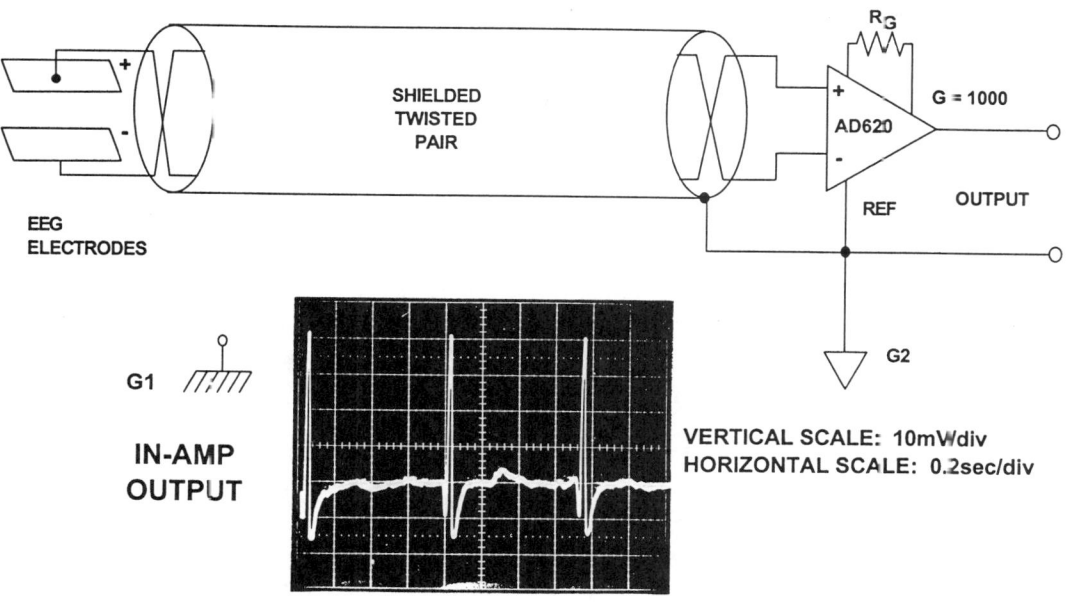

FOR UNGROUNDED PASSIVE SENSORS, GROUND SHIELD AT THE RECEIVING END

Figure 9.78

Most high impedance sensors generate low-level current or voltage outputs, such as a photodiode responding to incident light. These low-level signals are especially susceptible to EMI, and often are of the same order of magnitude as the parasitic parameters of the cable and input amplifier.

Even properly shielded cables can degrade the signals by introducing parasitic capacitance that limits bandwidth, and leakage currents that limit sensitivity. An example is shown in Figure 9.79, where a high-impedance photodiode is connected to a preamp through a long shielded twisted pair cable. Not only will the cable capacitance limit bandwidth, but cable leakage current limits sensitivity. A preamplifier, located close to the high-impedance sensor, is recommended to amplify the signal and to minimize the effect of cable parasitics.

Figure 9.80 is an example of a high-impedance photodiode detector and preamplifier, driving a shielded twisted pair cable. Both the amplifier and the shield are grounded at a remote location. The shield is connected to the cable driver common, G1, ensuring that the signal and the shield at the driving end are both referenced to the same point. The capacitor on the receiving side of the cable shunts high frequency noise on the shield into ground G2 without introducing a low-frequency ground loop. This popular grounding scheme is known as *hybrid* grounding.

9 - 91

Hardware Design Techniques

SHIELDS ARE NOT EFFECTIVE WITH HIGH IMPEDANCE REMOTE SENSORS

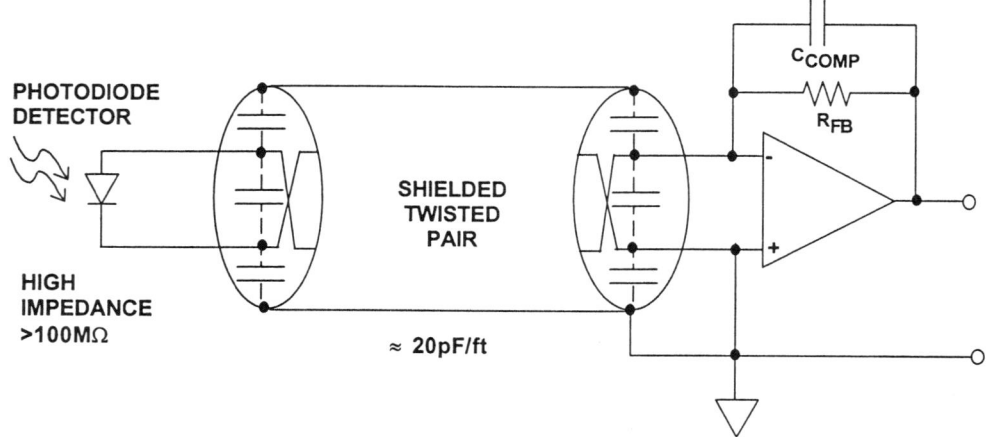

- CABLE CAPACITANCE LIMITS BANDWIDTH
- CABLE LEAKAGE CURRENT LIMITS SENSITIVITY

Figure 9.79

REMOTELY LOCATED HIGH IMPEDANCE SENSOR WITH PREAMP

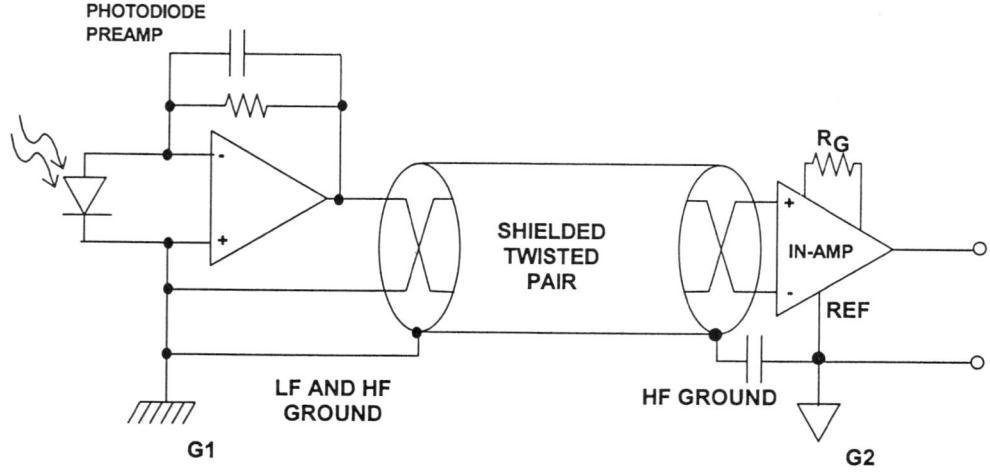

Figure 9.80

Hardware Design Techniques

Figure 9.81 illustrates a balanced active line driver with a hybrid shield ground implementation. When a system's operation calls for a wide frequency range, the hybrid grounding technique often provides the best choice (Reference 8). The capacitor at the receiving end shunts high-frequency noise on the shield into G2 without introducing a low-frequency ground loop. At the receiver, a common-mode choke can be used to help prevent RF pickup entering the receiver, and subsequent RFI rectification (see References 9 and 10). Care should be taken that the shields are grounded to the chassis entry points to prevent contamination of the signal ground (Reference 11).

HYBRID (LF AND HF) GROUNDING WITH ACTIVE DRIVER

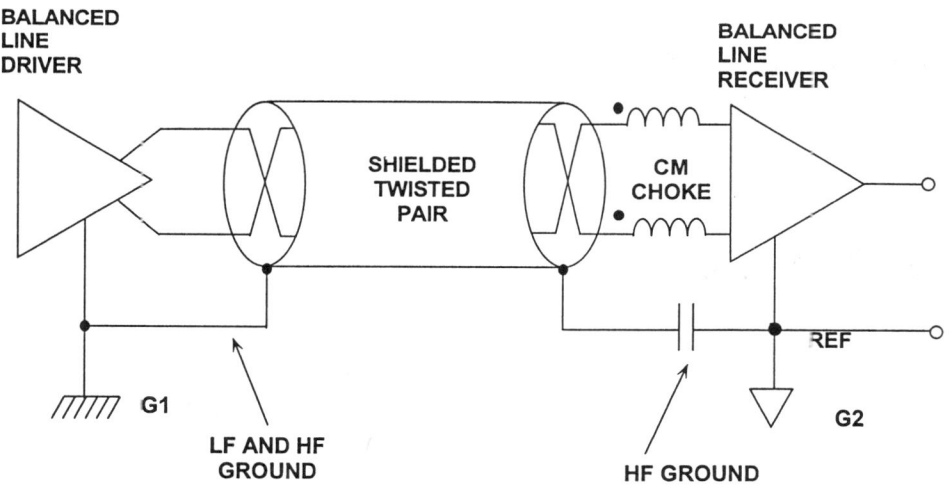

Figure 9.81

To summarize this discussion, shield grounding techniques must take into account the type and the configuration of the sensor as well as the nature of the interference. When a low-impedance passive sensor is used, grounding the shield to the receiving end is the best choice. Active sensor shields should generally be grounded at the source (direct connection to source ground) and at the receiver (connect to receiver ground using a capacitor). This hybrid approach minimizes high-frequency interference and prevents low-frequency ground loops. Shielded twisted conductors offer additional protection against shield noise because the coupled noise occurs as a common-mode, and not a differential signal.

The best shield can be compromised by poor connection techniques. Shields often use "pig-tail" connections to make the connection to ground. A "pig-tail" connection is a single wire connection from shield to either chassis or circuit ground. This type of connection is inexpensive, but at high frequency, it does not provide low impedance. Quality shields do not leave large gaps in the cable/instrument shielding system. Shield gaps provide paths for high frequency EMI to enter the system. The cable shielding system should include the cable end connectors. Ideally, cable shield connectors should make 360° contact with the chassis ground.

This section has highlighted the more common techniques used in cable shielding. There are other techniques which involve the use of driven shields, twin-shields, common-mode chokes, etc. References 1, 2, 5, 6, 7, and 8 provide an exhaustive study of the entire topic of noise reduction techniques including cable shielding.

SUMMARY OF CABLE SHIELDING TECHNIQUES

- Do not let the shield "float"
- Do not connect both ends of directly to ground
- No LF current should flow in the shield
- Use the *hybrid* approach for LF and HF electric field interference
- The shield includes the connector, therefore avoid using *pigtails* to connect shields to ground. Use *chassis ground* to prevent *signal ground* contamination
- Use Common-Mode chokes at receiver to enhance RF rejection
- Other techniques exist:
    Driven Shields
    Twin-Shields

Figure 9.82

REFERENCES: SENSORS AND CABLE SHIELDING

1. H.W. Ott, **Noise Reduction Techniques in Electronic Systems, Second Edition**, John Wiley & Sons, Inc., New York, 1988.

2. Ralph Morrison, **Grounding and Shielding Techniques in Instrumentation, Third Edition**, John Wiley & Sons, Inc., New York, 1988.

3. **Systems Application Guide**, Section 1, Analog Devices, Inc., Norwood, MA, 1993.

4. AD620 Instrumentation Amplifier, Data Sheet, Analog Devices, Inc.

5. A. Rich, *Understanding Interference-Type Noise*, **Analog Dialogue**, 16-3, 1982, pp. 16-19.

6. A. Rich, *Shielding and Guarding*, **Analog Dialogue**, 17-1, 1983, pp. 8-13.

7. *EDN's Designer's Guide to Electromagnetic Compatibility*, **EDN**, January, 20, 1994, material reprinted by permission of Cahners Publishing Company, 1995.

8. *Designing for EMC (Workshop Notes)*, Kimmel Gerke Associates, Ltd., 1994.

9. James Bryant and Herman Gelbach, *High Frequency Signal Contamination*, **Analog Dialogue**, Vol. 27-2, 1993.

10. Walt Jung, *System RF Interference Prevention*, **Analog Dialogue**, Vol. 28-2, 1994.

11. Neil Muncy, *Noise Susceptibility in Analog and Digital Signal Processing Systems*, presented at 97th **Audio Engineering Society Convention**, Nov. 1994.

GENERAL REFERENCES: HARDWARE DESIGN TECHNIQUES

1. **Linear Design Seminar**, Section 11, Analog Devices, Inc., 1995.

2. **E.S.D. Prevention Manual**
   Available free from Analog Devices, Inc.

3. B.I. & B. Bleaney, **Electricity & Magnetism**, OUP 1957, pp 23,24, & 52.

4. Paul Brokaw, *An I.C. Amplifier User's Guide to Decoupling, Grounding and Making Things Go Right for a Change*, **Analog Devices Application Note**, Available free of charge from Analog Devices, Inc.

5. Jeff Barrow, *Avoiding Ground Problems in High Speed Circuits*, **R.F. Design**, July 1989.

   AND

   Paul Brokaw & Jeff Barrow, *Grounding for Low- and High-Frequency Circuits*, **Analog Dialogue**, 23-3 1989.
   Free from Analog Devices.

6. International EMI Emission Regulations
Canada	CSA C108.8-M1983	FDR	VDE 0871/VDE 0875
Japan	CISPR (VCCI)/PUB 22	USA	FCC-15 Part J

7. Bill Slattery & John Wynne, *Design & Layout of a Video Graphics System for Reduced EMI*, Analog Devices Application Note (E1309-15-10/89)
   Free from Analog Devices.

8. William R. Blood, Jr., **MECL System Design Handbook**
   (HB205, Rev. 1), Motorola Semiconductor Products, Inc., 1988.

9. Wainwright Instruments Inc., 7770 Regents Rd., #113, Suite 371, San Diego, CA 92122, Tel. 619-558-1057, Fax. 619-558-1019.

   Wainwright Instruments GmbH, Widdersberger Strasse 14, DW-8138 Andechs-Frieding, Germany. Tel: +49-8152-3162, Fax: +49-8152-40525.

10. Ralph Morrison, **Grounding and Shielding Techniques in Instrumentation, Third Edition**, John Wiley, Inc., 1986.

11. Henry W. Ott, **Noise Reduction Techniques in Electronic Systems, Second Edition**, John Wiley, Inc., 1988.

12. Robert A. Pease, **Troubleshooting Analog Circuits**, Butterworth-Heinemann, 1991.

13. Jim Williams, Editor, **Analog Circuit Design: Art, Science, and Personalities**, Butterworth-Heinemann, 1991.

14. Doug Grant and Scott Wurcer, *Avoiding Passive Component Pitfalls*, **The Best of Analog Dialogue**, pp. 143-148, Analog Devices, Inc., 1991.

15. Walt Jung and Richard Marsh, *Picking Capacitors, Part I.*, **Audio**, February, 1980.

16. Walt Jung and Richard Marsh, *Picking Capacitors, Part II.*, **Audio**, March, 1980.

17. Daryl Gerke and Bill Kimmel, *The Designer's Guide to Electromagnetic Compatibility*, **EDN Supplement**, January 20, 1994.

18. Walt Kester, *Basic Characteristics Distinguish Sampling A/D Converters*, **EDN**, September 3, 1992, pp.135-144.

19. Walt Kester, *Peripheral Circuits Can Make or Break Sampling ADC System*, **EDN**, October 1, 1992, pp. 97-105.

20. Walt Kester, *Layout, Grounding, and Filtering Complete Sampling ADC System*, **EDN**, October 15, 1992, pp. 127-134.

Hardware Design Techniques

INDEX

■ SUBJECT INDEX

■ ANALOG DEVICES PARTS INDEX

Index

Subject and Author Index

A

Absolute value generator, 4.42
ACCEL Technologies, Inc., 9.19
AD626:
    CMRR, 1.26
    equivalent circuit, 1.26-27
    high-side current monitor interface, 1.28
    sensing capability, 1.28
AD797:
    noise, calculation, 8.4-5
    THD, 8.4
AD811, capacitive load, 2.5
AD812:
    characteristics, 4.28
    drive circuit for 10-bit ADC, 4.28
    as inverting level shifter, 4.29
    key specifications, 4.25
AD817, capacitive loads, internal compensation, 2.8
AD830:
    feedback amplifier, 2.42
        CMR, 2.43
AD876:
    analog input transients, 4.23
    block diagram, 4.21
    characteristics, 4.28
    drive circuit, 4.28
    key specifications, 4.21
    switched capacitor input:
        circuit, 4.23
        operation, 4.22, 4.24
AD771X family:
    24-bit devices, 3.20
    applications, 3.16
    background calibration, 3.35
    calibration, 3.35-36
    differences, summary, 3.15
    effective resolution of bits, 3.20
    ENOB, 3.20, 3.23-25
    external capacitor, and PGA, 3.27
    external filter, 3.25
    gain, 3.36
    high oversampling, 3.25
    microprocessor interfaces, 3.39-40
    noise optimization, 3.29-30, 3.32-33
    reference voltage, 3.31
    noise, 3.32-33
    SNR, 3.20
AD7710, AD7711, AD7712, AD7713, AD7714 Data Sheets, 3.43
AD7710:
    applications, 3.37-38
    digital filter frequency response, 6.15
    input overvoltage, protection, 3.30
    input transient loading, 3.28
    key features, 3.14
    weigh scale application, 3.38
AD7713:
    applications, 3.38
    isolated 4-wire interface, 3.39
AD7714:
    ADC evaluation board, 9.14-16
        software control, 9.15
    applications, 3.41-42
    block diagram, 3.16
    conversion time, 3.19
    digital filter, 3.18
        frequency response, 3.18
    external filter, restrictions, 3.27
    input structure:
        buffered, 3.28
        unbuffered, 3.26
    low-power data converters, 3.41
    RMS noise variation, 3.22
        ENOB, 3.23
    smart transmitters, 3.42
        operation, 3.42
    total supply current, 3.41
AD7715:
    RMS noise variation, 3.22
        ENOB, 3.23
AD7716, key features, 6.17
AD7890, specifications, 6.12
AD8002, frequency response, 2.39
AD8011, key specifications, 4.31
AD8036:
    block diagram, 4.35
    distortion, 4.38
    equivalent circuit, 4.35
    key specifications, 4.38
    overdrive recovery, 4.38

performance, 4.36
summary of specifications, 4.39
AD8037:
   amplitude modulator, 4.45
   distortion, 4.38
   equivalent circuit, 4.35
   inverting amplifier, gain of unity, 4.43
   key specifications, 4.38
   overdrive recovery, 4.38
   performance, 4.36
   summary of specifications, 4.39
AD8300:
   block diagram, 6.37
   diagram, 6.37
   key features, 6.38
   remote multichannel data distribution, 6.39
AD8522:
   analog section schematic, 6.34
   diagram, 6.32
   key features, 6.33
   on-chip bandgap reference, 6.33
   rail-to-rail output, 6.35
   remote multichannel data distribution, 6.39
AD8582, key features, 6.36
AD8842:
   key features, 6.30-31
   TrimDAC, diagram, 6.30
AD9002, dynamic performance, 5.21
AD9022:
   architecture, 4.12
   effective input noise, 4.50
   equivalent input noise, 4.47
   key specifications, 4.13
   output histogram, 4.47
   S/(N+D), 4.14-15
   SFDR, 4.14-15
AD9026:
   evaluation board, 9.15, 9.17-18
      block diagram, 9.16
AD9042:
   block diagram, 5.12
   characteristics, 5.11
   input structure, 5.13
   key specifications, 5.13
   performance, 5.13
AD9050:
   ac coupling into input, 4.20
   analog input circuit, 4.18-19
   block diagram, 4.18
   input circuit, 4.18
   key specifications, 4.19
   single-supply ADC, 4.18-20
      driving, 4.18-20
AD9100:
   ultrafast SHA, 5.15
      closed-loop architecture, 5.16-17
      FFT output, 5.20
      hold-mode SFDR, 5.17-19
      measuring SFDR, test configuration and timing, 5.18
      SFDR effects, 5.19
      SFDR performance, 5.19
AD9101:
   applications, 5.21
   block diagram, 5.21
   circuit, 5.20
   ultrafast SHA, 5.15
      closed-loop architecture, 5.16-17
      performance, 5.19
AD9104, ultraperformance low distortion ADC, 5.15
AD9622:
   third order IMD intercept:
      calculation of IMD, 8.7
      and frequency, plot, 8.7
AD9631:
   decoupling, harmonic distortion, 2.12
   op amp:
      key specifications, 4.13
      THD, 4.14-15
      THD + N, 4.14-15
AD9632:
   noise, calculation, 4.49
   op amp, key specifications, 4.13
AD620 Instrumentation Amplifier, 9.95
ADC:
   3-bit, unipolar, transfer characteristic, 3.4
   8-bit, flash, with external SHA, 5.14-15
   10-bit:
      CMOS, 4.20-24
      drive circuit, 4.28
      single-supply, 4.18-20, 4.29
         block diagram, 4.21

drive circuit, 4.30
key specifications, 4.19, 4.21
12-bit:
    data acquisition, 6.12-13
    equivalent input noise, 4.47
    evaluation board, 9.15-18
    FFT windowing using Hanning window, 8.18-19
    low power data acquisition, 6.12-13
    noise, calculation, 4.49
    noise floor, FFT, 5.25
    sampling, 4.12
        SFDR versus input frequency, 5.11-12
applications, digital spectral analysis, 5.24-25
bandpass sampling, 5.1-2
convert command, 6.4
and DAC, data acquisition system, 6.1
design issues, 3.1
drive:
    functions, 4.10-11
    gain setting and level shifting, 4.27-31
    input transient settling time, model, 4.25
dynamic performance, 4.2-10
    aperture jitter, 4.7-8
    characteristics, 4.3
    distortion, 4.10
        transfer characteristics, 4.11
    and drive selection criteria, 4.26
    effective aperture delay time, 4.9
    ENOB, 4.3-4, 4.7
    full power bandwidth, 4.6
    gain, 4.7
        signal bandwidth, 4.5-6
    Gaussian noise, 4.3
    linearity, 4.10
    noise and distortion, 4.6
    quantization noise, 4.3
    S/(N+D), 4.4, 4.6-7, 4.12
    sampling clock jitter, 4.7-8
        SNR, 4.8
    SFDR, 4.4
    slew rate, 4.6
    THD, 4.4
dynamic range, increased using dither signals, 5.22-23
dynamic range increase, using dither signals, 5.22-23
fast, and EMI/RFI, 9.72
FFT analysis, 8.13
filter:
    bandpass, 4.50
    lowpass, 4.50
    passive antialiasing, 4.50
flash:
    8-bit, 4.39-40
    block diagram, 4.16
    driving, 4.15-17
    with external SHA, 5.14-15
    input model, 4.17
    nonlinear input impedance, 4.15
    THD, 4.16-17
grounding and decoupling, 4.53-54
harmonic sampling, 5.1-2, 5.6-7
high resolution, 3.1-43
    low-frequency measurement, 3.12-42
high speed:
    decimation, 8.11
    differential driver, 2.35
    distortion measurement, FFTs, 8.11-12
    input protection and clamping, 4.34-40
    low voltage reference, summary, 4.32
    noise, sources, 4.51
    PC-based test system, diagram, 8.12
    power supply decoupling, 4.51-52
    sampling, 4.1-62
        clamping, applications, 4.41-46
        distortion, 4.2
        drive amplifier selection, 4.10-15
        external reference voltage, 4.1, 4.31-33
        flash, driving, 4.15-17
        gain setting and level shifting, 4.27-31
        input protection and clamping, 4.34-40
        key specifications, 4.2

modern trends, 4.2
noise, 4.2, 4.47-61
single-supply, driving, 4.18-20
switched-capacitor input,
   driving, 4.20-26
ideal, versus internal noise, 3.21
input buffer, using differential driver,
2.34
input overvoltage protection circuits,
4.35
linearity and resolution:
   effect on SFDR and noise, 5.24-25
      applications, 5.24-25
lower resolution, quantization noise
and input signal, 5.25
measurement, 3.12-42
multiplexed:
   data acquisition systems, 6.10
   SAR, 6.4-5
   SHA, 6.6
noise, 3.22-24
   internal, for dc input, 3.21
nonsampling, 6.18
Nyquist bandwidth, 5.7-8
oversampling:
   decimation, 3.7
   digital filtering, 3.7
   dither added to input, 3.5
   ENOB, 3.7
   SNR, 3.7
quantization error signal, 5.22
quantization noise, randomization,
5.23
sampling, 6.6, 6.19
   basics, 5.1
   external SHA, 5.14
   FFT simulation, 5.22
   high speed, 4.1-62
   with integral SHA, 4.9
   nonlinearities, 4.4
   quantization noise, 3.6
   SFDR and ENOB, 5.14
   signal ranges, 5.10-11
sampling clock generation, using
crystal oscillator, 4.56
settling time, 6.10
SHA:
   external, 5.14-21
      SFDR and ENOB increase,
         5.14-21
   input processing, 6.6
sigma-delta, 3.2-12
   22-bit, 9.14
      quad, 6.16-17
         key features, 6.17
   architecture and theory, 3.2
   bandpass, 3.11
      advantages, 5.28
      applications, 5.28-29
      architecture, 5.26-27
      digital filter, 5.29
      experimental, 5.29
      future 16-bit resolution, 5.29
      intermediate frequency
         sampling, 5.30
      noise shaping, 5.28
   buffer, 3.17
   characteristics, 3.3
   circuitry, 3.2
   decimation, 3.2-3
   description, 3.2
   ENOB, 3.6
   filter:
      bandpass, 3.10
         applications, 3.10
      digital, 3.2-3
         low pass, 3.6
      low pass, 3.10
      programmable, 3.12
      setting, 3.19
   high accuracy systems, 3.41
   high integration level, 3.14
   key concepts, 3.3
   low-frequency measurement,
      applications, 3.13
   manufacture, 3.2
   minimum charge time, 3.26
   modulator, 3.8
      first-order, block diagram, 3.8
      PGA function, 3.17, 3.18
      quantization noise, 3.9
      second-order, block diagram,
         3.9
      SNR versus oversampling
         ratio, 3.10
   multiplexing, 6.14-17
   multiplexing inputs, 6.14-17
   noise:

Gaussian, 3.23
    distribution, 3.23
  optimization, 3.32-33
  quantization, 3.4
  shaping, 3.2-3
oversampling, 3.2-3
PGA, 3.13
quantization:
  error, 3.4
  uncertainty, 3.4
reference:
  decoupling, 3.34
  noise calculation, 3.33
  voltage, noise, 3.32-33
sensor interfacing, 3.12-13
signal conditioning, transducer input, 3.12-13
single-supply measurement, 3.16
  PGA, 3.17
SNR, 3.11
summary, 3.12
synchronized, applications, 6.16
traditional, noise shaping, 5.28
transducers, 3.29
signal conditioning, 3.1-43
single-supply, 2.24-30
  driving, 4.18-20
  input common-mode range, 2.24
  optimum input voltage, 2.24
SNR, 3.20
super-Nyquist sampling, 5.1-2
switched capacitor inputs, driving, 4.20-26
transient response, 6.9
trends, 3.1
undersampling, 5.1-2
  applications, 5.14
    selection procedures, 5.11
  broadband digital receivers, 5.10
  intermediate frequency digital receiver:
    architecture, 5.26
    using PGA, 5.26
  Nyquist limit, 5.26
  sampling clock jitter, effects, 5.30-32
  trends, 5.26-30
voltage feedback, third order intercept value, 8.6-7

Aliasing:
  definition, 5.2
  frequency domain effects, 5.3-4
  intermediate frequency signal, 5.6, 5.9
  non-overlapping, minimum sampling rate, 5.4-5
  time domain effects, 5.3
AMP04:
  inverting-mode output, 1.23-24
  low frequency filtering, 1.26
Amplifier:
  clamped, see: Clamped amplifiers
  input bias current, and CMR, 1.6
  input stage:
    configuration, 7.4
    current-voltage characteristics, 7.3
    overvoltage, 7.1-5
      curve tracer, 7.3-4
  output phase reversal, prevention, 7.6
  output voltage phase reversal, 7.5-10
  overvoltage:
    protection:
      offset and noise performance, 7.10
      using Schottky diodes, 7.8-10
      using series resistance, 7.8, 7.10
  pulsed-bridge transducer-driver, 1.24-25
  single-supply, 1.1-31
  two op amp in amp:
    applications, 1.11
    configuration, 1.10-14
    limitation, 1.14
    voltage nodal analysis, 1.11-14
  two op amp FET-input in amp, 1.15-16
    performance summary, 1.16
    programmable, 1.15
  three op amp in amp:
    advantages, 1.16
    applications, 1.17
    topology, 1.16-21
    voltage nodal analysis, 1.17-20
  in amps, 1.21-30, 1.26

INDEX

composite, 1.21-23
   application, 1.24
   performance summary, 1.23
   transient response, 1.22
input-stage rectification, 1.29-30
signal conditioning, 1.23-28
characteristics, 1.1
input stage, 1.3
limitations, 1.2
rail-to-rail:
   two op amp in amp topology, 1.10-14
   two op amp FET-input in amps, 1.15-16
   characteristics, 1.3-4
   complementary design processes, 1.2
   FET-input, 1.5-6
   ground reference, 1.2
   input stage topology, 1.6
      transistor pairs, 1.7
   op amp input stages, 1.4-7
   output, 1.8-21
      considerations, 1.9-10
      limitations, 1.9
      voltage swings, 1.10
   three op amp in amp topology, 1.16-21
Amplifier Applications Guide (1992), 2.44, 4.62, 5.34, 7.24, 9.84
Amplitude modulator:
   circuit operation, 4.46
   waveform, 4.45
Analog circuit:
   breadboarding, 9.3-10
   high performance, breadboarding, 9.3-10
   prototyping and simulating, 9.1-10
   rectification, 9.63
   RFI protection, 9.64
   RFI sensitivity, 9.63
   SPICE modeling, disadvantages, 9.1-2
Analog integrated circuit:
   amplifier input stage overvoltage, 7.1-5
   amplifier output voltage phase reversal, 7.5-10
   overvoltage effects, 7.1-24
      electrostatic discharge protection, 7.11-23
Analog multiplexer, diagram, 6.3
Analog signal:
   aliasing, 5.2
   intermediate, aliasing, 5.6, 5.9
   sampling, 5.1
      effects, 5.3-5
Asta, Dan, 5.33-34
Audio applications:
   bandstop filter, 8.2
   THD+N measurement, 8.3

B

Bandgap reference, optional shutdown, 9.47
Bandpass sampling, applications, digital cellular radio, 5.9
Barrow, Jeff, 9.96
Baseband sampling, 5.7
Baum, J., 1.31
Bennett, W.R., 8.23
Blackman, R.B., 8.23
Bleaney, B., 9.96
Bleaney, B.I., 9.96
Blood, William R. Jr., 2.44, 9.96
Board radiation, 9.59
Bode diagram, 2.4
Boltzmann's Constant, 3.30
Breadboard, characteristics, 9.2-3
Breadboarding:
   prototyping, 9.3-10
      bird's nest, 9.3-5
      CAD techniques for layout, 9.8
      cage jacks, 9.5
      deadbug, 9.3-5, 9.8
      IC sockets, disadvantages, 9.5
      larger circuits, 9.8
      milled, 9.9
      multilayer, problems, 9.9
      pin sockets, 9.5
      solder-mount, 9.7
         advantages, 9.7
         components, 9.6
      summary, 9.10
Broadband digital receiver:
   ADC input, 5.10

INDEX

applications, 5.9-11
  SFDR, 5.10
Brokaw, Paul, 9.84, 9.96
Bryant, James M., 3.1, 5.34, 9.1, 9.95
Buffer latch, grounding and decoupling, 4.55
Buxton, Joe, 2.44, 3.1

C

Cable:
  antenna-like behavior, 9.67
  capacitance, 2.15
  coaxial, attenuation versus frequency, 2.16
  electrical length, 9.67
  EMI source, 9.62
    capacitive coupling, 9.85
    magnetic coupling, 9.85
  filters, ferrites and capacitors, 9.67
  shield grounding, 9.88, 9.90
  shielded, 9.86
    current-carrying, 9.67
    grounding at load, 9.82
    parasitic capacitance, 9.91-92
    summary of techniques, 9.94
    ungrounded, as antennas, 9.88
  source-terminated, disadvantages, 2.19
  twisted-pair, noise immunity to magnetic fields, 9.85
Cable driving, 2.14-23
Cable radiation, 9.59
CAD techniques, breadboard prototyping, 9.8
Cage jacks, 4.59
Calibration, considerations, 3.35-37
Capacitance, parasitic, 9.30, 9.91-92
Capacitive coupling, 9.85
Capacitive loading:
  active compensation, 2.7
  compensation, 2.2-5
  consideration, 2.2
  drawbacks, reducing speed, 2.10
  forcing high noise gain, 2.4
  in-the-loop compensation, 2.7
  internal compensation, 2.8
    caveats, 2.9
  overcompensation, 2.2
  passive compensation, 2.4
Capacitor:
  ceramic, 9.27, 9.29
    chip caps, 9.29
  dielectric types, 9.27
  electrolytic, 9.27-30
    aluminum:
      general purpose, 9.28
      OS-CON, 9.28, 9.30
      switching, 9.27-28
    ESR degradation by temperature, 9.29
    impedance versus frequency, 9.30-31
    tantalum, 9.27-28
      SPICE model, 9.31
    types, 9.28
  ESR, 9.29
  ferrites, 9.32
    beads, 9.33
    characteristics, 9.32
    choices, 9.34
    impedance, 9.33
    leaded beads, 9.33
    PSpice models, 9.33
  film, 9.27-28
    frequencies, 9.29
    values, 9.28
  selection criteria, 9.27
  stacked-film, 9.28-29
  switching power supply, noise reduction, 9.26-27
Circuit, and interference, 9.73
Clamped amplifier, 4.34-40
  in absolute value generators, 4.42
  as amplitude modulator, 4.44-45
  applications, 4.41-46
  distortion levels, 4.37-38
  and flash converter, 4.39-40
  in full wave rectifiers, 4.42
  input and output clamping, comparison, 4.37
  key specifications, 4.38
  overdrive recovery, 4.38
  specifications, 4.39
Clelland, Ian, 9.43
Code resolution, noise-free, estimating, 3.24
Coherent sampling, requirements, 8.14

Coleman, Brendan, 8.23
Colotti, James J., 8.23
Common-mode rejection, see: CMR
Computers, radiated emission limits, 9.56
Conductive shield:
   absorption, 9.76
      loss, 9.78
   loss mechanisms, 9.77-78
   reflection, 9.76
      loss, 9.77-78
Connector leakage, 9.59
Conversion time, and filter setting, 3.19
Counts, Lew, 1.31
Coupling:
   E-field, 9.87
   H-field, 9.87
Crosstalk, 6.2
Current:
   sinking, 1.11-13
   sourcing, 1.11-13

D

DAC:
   12-bit, 6.36-39
      SHA error, 6.34
   and ADC, data acquisition system, 6.1
   complete, 6.34
   demultiplexed, 6.22
      advantages, 6.23-25
      cost advantage, 6.27
      data update rates, 6.25
      operation, 6.23-24
      refreshing, 6.24-25
   dual, 12-bit, single-supply, 6.35
   fast, and EMI/RFI, 9.72
   grounding and decoupling, 4.53-54
   multiple, 6.21
      8-bit:
         octal, 6.26
            TrimDAC, 6.29-30
         quad, 6.26
      12-bit:
         octal, 6.26
         quad, 6.26
         world's smallest, 6.32
      16-channel, 8-bit, 6.26-27
      advantages, 6.25-26
      applications, 6.20
      data distribution, 6.20-39
      in data distribution, 6.20-39
      data readback, 6.26
      serial data interface, 6.26
      SHAs, 6.20
      single-chip, advantages, 6.27
      versus demultiplexed,
         comparisons, 6.28-30
   multiplexed single, 6.21
   rail-to-rail operation, 6.34
   trends, 3.1
Data acquisition system:
   8-channel:
      12-bit, 6.11
         specifications, 6.12
   complete, on chip, 6.10-13
   definition, 6.1
   filtering, 6.7-9
   monolithic, 6.11
   multiplexed:
      ADC/DAC, 6.2
      diagram, 6.4-5
      SHA, diagram, 6.7
   simultaneously sampled:
      using nonsampling ADC, 6.18
      using sampling ADC, 6.19
Data distribution system:
   options, 6.21
   using demultiplexed DACs, 6.22-23
1dB compression points, 8.6
Decimation, 3.6
   sigma-delta ADC, 3.2-3
Decoupling, 4.52-56
Designing for EMC, 9.84, 9.95
Differential driver, 2.34-39
   cross-coupled, 2.37-39
      balanced outputs, 2.37
      benefits, 2.37-38
      diagram, 2.37
      frequency response, 2.39
      high CMR, 2.37
      high gain-bandwidths, 2.38
   inverter-follower, 2.35-37
      circuit gain, 2.36
      design, 2.36
      diagram, 2.35
      net input impedance, 2.36
      trimmed phase matching, 2.36
Differential line receiver, see: Differential

receivers
Differential receiver, 2.39-43
    active feedback, 2.41-43
        advantages, 2.41
        video loop-through connection, diagram, 2.42
    CMR, 2.39
     4 resistor, 2.40-41
        CMR, 2.40-41
    receiver output, 2.39
    video:
        diagram, 2.40
        drawback, 2.41
        gain/phase performance, 2.41
DIGI-KEY, 9.43
Digital filtering, sigma-delta ADC, 3.2-3
Digital receiver:
    baseband sampling, 5.7
    intermediate frequency sampling, 5.8
Digital spectral analysis, FFT, 5.24
Distortion:
    ADC, 4.10
        measuring using FFTs, 8.11-12
    audio applications, bandstop filter, 8.2
    FFT testing, 8.12-19
        analyzing, 8.21-22
        troubleshooting, 8.19-20
    harmonic:
        components, 8.1-2
        definition, 8.1-2
        measurement, 8.1
    high frequency:
        spectrum analyzers, 8.10-11
        two-tone generation, 8.8-9
    high speed sampling ADCs, 4.2
    intermodulation products, 8.5
    measurements, 8.1-24
    op amp, high speed, 8.1-7
Doernberg, Joey, 8.23
Dostal, J., 2.44
Droop, 6.23, 6.34
DSP, interface, 3.40

E

Edge connections, 4.58
EDN's DEsigner's Guide to Electromagnetic Compatibility, 9.84, 9.95
Effective aperture delay time, 4.9
Effective number of bits, see ENOB
Effective resolution of bits:
    calculation, 3.20
    determination, 3.21
EIAJ ED-4701 Test Method C-111,
Electrostatic Discharges, 7.24
Eichhoff Electronics, Inc., 9.44
Electric field, strength, 9.62
Electromagnetic compatibility, equipment regulations, 9.55-57
Electromagnetic interference, see: EMI
Electrostatic discharge, see: ESD
EMC Design Workshop Notes, 9.43
EMC Test & Design, 9.84
EMI:
    amplitude, 9.61
    capacitive coupling, 9.87
    conducted interference, 9.55
    container openings as waveguides, 9.80
    diagnostic framework, 9.60
    distance, 9.61
    FAT-ID concept, 9.61
    filters:
        common-mode, 9.69-70
        differential-mode, 9.69-70
    frequency, 9.61
    generation, power-line disturbances, 9.69
    impedance, 9.61
    improper cable grounding, 9.89
    magnetic coupling, 9.87
    physical dimensions of materials, 9.62
    radiated emissions, 9.55
    regulating bodies, 9.55
    regulations:
        European Community standards, 9.56-57
        FCC, 9.55
        FDA standards, 9.56
        impact on design, 9.57
        MIL-STD-461, 9.56
        SAE standards, 9.57
        types of disturbance, 9.56
        VDE, 9.55
    sources, 9.59
    time, 9.61-62
    trouble shooting, philosophy, 9.83
EMI/RFI considerations, 9.55-84

ENOB, ADC performance, 4.4
Equivalent circuit, 4.29
Equivalent series inductance, see: ESL
Equivalent series resistance, see: ESR
ESD:
    damage:
        characteristics, 7.18-19
        elimination, 7.11
        prevention, Analog Devices'
        committment, 7.22
    definitions, 7.11
    failure:
        mechanisms, 7.18-19
        prevention, 7.18-19
        threshold level, 7.11
    generation, examples, 7.12
    models, 7.12-15
        applicable to ICs, 7.13
        Charged Device Model (CDM), 7.13-15
            standard, 7.15
        correlation, 7.13
        Human Body Model (HBM), 7.12, 7.14-15
            standard, 7.12
        Machine Model (MM), 7.13-15
            standard, 7.15
            testing guidelines, 7.14
        schematic representation, 7.15
        significant features, 7.14-15
        standards for ICs, 7.17
        waveforms, 7.16-17
    origins, 7.12
    protection:
        between Analog Devices and customer, 7.23
        for ICs, 7.11-23
    sensitive devices:
        handling, 7.20
        recognition, 7.20
        safe handling workstation, 7.21
    triboelectric effect, 7.12
ESD Association Draft Standard DS5.3 for Electrostatic Discharge (ESD) Sensitivity Testing-CDM, 7.24
ESD Association Standard S5.2 for Electrostatic Discharge (ESD) Sensitivity Testing-MM, 7.24
E.S.D. Prevention Manual, 9.96

ESD Protection Manual, 7.22-24
Evaluation boards, advantages, 9.10-11
Extended frequency complementary bipolar high speed devices, 2.1

F
Fair-Rite Linear Ferrites Catalog, 9.43
Faraday shield, 4.56, 9.69-70
Fast Fourier transform, see: FFT
Ferguson, Paul F. Jr., 3.43, 5.34
Ferrites:
    beads, for EMI protection, 9.72
    characteristics, 9.32-34
    power line filters, 9.41
    switching power supply, noise reduction, 9.26
FFT, 5.6
    and aliasing, 8.21
    bin width, 8.12
        average noise versus broadband RMS quantization noise, 8.13
    coherent sampling, 8.13, 8.16
        requirements, 8.14
    endpoint discontinuity, 8.16
    Mathcad output plot, 8.18-19
    as narrowband filter, 5.24
    output analysis, 8.21-22
        ENOB, 8.21-22
        harmonic distortion, 8.22
        S/(N+D), 8.21-22
        SFDR, 8.21-22
        SNR, 8.22
        THD, 8.21-22
        two-tone IMD, 8.21-22
    output troubleshooting, 8.19-20
        excess harmonic distortion, 8.20
        excess noise floor, 8.20
    PC-based analysis, 8.12
    sinewave:
        input, spectral purity, 8.19-20
        integral number of cycles, 8.14-16
        non-integral number of cycles, 8.15-16
    software, verification, 8.20
    testing, 8.12-19
    weighting, 8.16
    windowing, 8.16
        frequency-domain characteristics,

8.16-17
    time-domain characteristics, 8.16-17
Filter:
    antialiasing, to reduce op amp noise, 4.50
    bandstop, audio applications, 8.2
    capacitors, 9.27
    card entry, 9.35-38
    in data acquisition system, diagram, 6.8
    data acquisition systems, 6.7-9
    failure, 9.65-66
    loss:
        ESL, 9.27
        ESR, 9.27
    low-pass:
        effectiveness, 9.65
        leakage, 9.64
        RFI frequency effectiveness, 9.64
    multistage, effectiveness, 9.66
    noise, 6.8
    non-zero, ground, 9.66
    notch, in spectrum analyzer, 8.11
    setting:
        and conversion rate, 6.15
        settling time, 6.8, 6.15
    single-pole, settling times, 6.9
    switching supply outputs, 4.60-61
Flash converter:
    8-bit, with external SHA, 5.14-15
    block diagram, 4.16
    driving, 4.15-17
    nonlinear input impedance, 4.15
Fleming, Tarlton, 5.33
Frederiksen, Thomas M., 2.44
Freeman, Wes, 6.1, 6.20, 7.1, 7.11
Full wave rectifier:
    applications, 4.44
    input/output scope waveform, 4.44
    schematic, 4.42
    waveforms, 4.43

G

Garcia, Adolfo A., 1.1, 6.40, 7.1, 9.1, 9.55
Gelbach, Herman, 9.95
Gerke, Daryl, 9.97
Ghausi, M.S., 8.24
Gilbert, Barrie, 5.34
Gold, Bernard, 8.24
Grant, Doug, 9.97
Gratzek, Tom, 5.33
Groshong, Richard, 5.33
Ground, analog and digital separated, 4.53
Ground loop current:
    induced by improper grounding, 9.89
    shield grounding, 9.89
Ground plane, 4.56
Grounding, 4.52-56
    electric field shielding, 9.90

H

Hageman, Steve, 9.43
Hanning window, 8.16-18
    time and frequency representation, 8.17
Hardware:
    design, 9.1-97
        basic references in low power systems, 9.45-48
        breadboarding techniques, 9.3-10
        EMI regulations, 9.55-57
            automotive equipment, 9.57
            commercial equipment, 9.55-56
            impact on design, 9.57
            industrial and process equipment, 9.57
            medical equipment, 9.56
            military equipment, 9.56
        EMI/RFI considerations, 9.55-84
            EMI trouble-shooting, 9.83
            passive components, 9.58-75
                power-line disturbances, 9.68-70
                printed circuit board design, 9.71-75
                RFI, 9.62-68
            shielding, 9.76-82
                cables, 9.81-82
        evaluation boards, 9.10-19
        low dropout references and regulators, 9.45-54
        low dropout regulators, 9.48-53
        prototyping and simulating analog

   circuits, 9.1-10
   sensors and cable shielding,
   9.85-97
   switching power supplies,
   noise reduction and filtering,
   9.21-44
Harris, Frederick J., 8.23
Henry, Tim, 2.45
Higgins, Richard J., 8.24
Hilton, Howard, 5.33
Hodges, David A., 8.23
HP Journal, 8.23
HP Product Note 5180A-2, 8.23

I
ICs:
  classifying for ESD, 7.17
  protection from ESD, 7.21
IEEE Standard 746-1984, 2.45
IEEE Trial-Use Standard for Digitizing Waveform Recorders, 8.24
IMD, 8.5
  from PC-based analysis of FFT, 8.12
  plot, 4.5
  second and third order, 8.5-6
Impedance, wire versus ground plane, 9.59
In amp:
  input-stage:
   emitter-coupled, 1.29
   filter, 1.29-30
    common-mode, 1.30
    differential-mode, 1.30
   rectification, 1.29-30
   source-coupled, 1.29
Inductors, switching power supply, noise reduction, 9.26
Input overvoltage, 7.2
  protection, 3.30
Instrumentation amplifier, see: In amp
Integrated circuit:
  ease of use, 2.1
  overvoltage effects, 7.1-24
Integrator, 3.10
Intercept points, second and third order, IMD, 8.5-6
Interference:
  characteristics, 9.25
  conduction, 9.59-60
  distance, 9.76
  emissions, 9.60-61
  environment, 9.76
  immunity, 9.60-61
  impedance, 9.76
  interconnects, 9.59
  internal, 9.60-61
  paths, 9.59-60
  radiation, 9.59-60
  receptors, 9.59-60
  shielding, 9.76-82
  sources, 9.59-60, 9.76
Intermodulation distortion, see: IMD
International EMI Emission Regulations, 9.96
Irons, Fred H., 5.33-34

J
Jacobsen, E., 1.31
Jantzi, S.A., 3.43, 5.34
Johnson noise, 4.48
  equation, 3.30
Jung, Walt, 1.31, 2.1, 2.44, 3.43, 9.1, 9.21, 9.43, 9.45, 9.54, 9.95, 9.97
Jung, Walter G., 2.44

K
Kaufman, M., 2.44
Kester, W.A., 2.44
Kester, Walt, 2.1, 2.44, 3.1, 4.1, 4.62, 5.1, 6.1, 8.1, 8.23, 9.1, 9.97
Kimmel Gerke Associates, 9.84, 9.95
Kimmel, Bill, 9.97
Kirsten, Jeff, 5.33
Kitchin, C., 1.31

L
Laker, K.R., 8.24
Lee, Hae-Seung, 8.23
Line driver, 2.20-23
  performance criteria, 2.20
  single-ended, high performance, 2.21
  single-supply, 2.27-30
   ac coupled, 2.27
   headroom, 2.28
   waveform duty cycle, 2.28
  supply bypassing, 2.21

Linear Design Seminar (1994), 7.24
Linear Design Seminar (1995), 1.31, 3.43, 4.62, 6.40, 9.96
Linear post regulation, switching power supply, noise reduction, 9.26
Low dropout regulator, 9.47-53
    high-side current sensing, 9.52
    low noise, diagram, 9.48
    on/off control, 9.53
    rail-rail output, 9.51-53
Low power system:
    basic references, 9.45-48
    scaling buffer, 9.45-46
LPKF CAD/CAM Systems, Inc., 9.19
Lyne, Niall, 7.24
Lyons, Richard G., 5.33

M
McDonald, John, 9.1, 9.21, 9.85
Magnetic coupling, 9.85
Mahoney, Matthew, 8.24
Marsh, Richard, 9.43, 9.97
Mathcad software, 8.12, 8.24
MathSoft, Inc., 8.24
Meehan, Pat, 8.23
Microprocessor, AD771X family interfaces, 3.39
MIL-STD-883 Method 3015, Electrostatic Discharge Sensitivity Classification, 7.24
Minimum 4-Term Blackman-Harris window, 8.16, 8.18
    frequency response, 8.18
Mixed Signal Design Seminar (1991), 3.43
Mixed signal systems, signal routing, 4.57
Morrison, Ralph, 9.84, 9.95, 9.96
Multichannel application, 6.1-40
    data acquisition systems, 6.1-2
        on chip, 6.10-13
    data distribution, multiple DACs, 6.20-39
    filtering, 6.7-9
    multiplexing, 6.2-7
        into sigma-delta ADCs, 6.14-17
        SHA and ADC settling time, 6.9-10
    simultaneous sampling systems, 6.18-19
Multiple ground pins, 4.59
Multiplexer:
    analog, diagram, 6.3
    key specifications, 6.3
    new features, 6.4
    on-resistance, 6.4
    PGA and SAR ADC, 6.4-5
    trench isolation, 6.4
Multiplexing:
    crosstalk, 6.2
    in data acquisition system, 6.2-7
    off-channel isolation, 6.2
    on-resistance, 6.2
        modulation, 6.2
    SHA and ADC settling time, 6.9-10
    switching time, 6.2
Muncy, Neil, 9.95
Murden, Frank, 5.33

N
Noise:
    broadband RMS:
        applications, 5.23
        quantization noise level, 5.24
    code resolution, estimating, 3.24
    Gaussian:
        ADC, 4.3
        broadband RMS, 5.23
    and harmonic distortion, 8.2
    high speed sampling ADCs, 4.2, 4.47-61
    peak-to-peak, 3.23
    performance, optimizing, 3.24
    power supply, 4.59
    quantization, 5.22
        ADC, 4.3
        and FFT record length, 8.15
        randomization, 8.15
        SNR conversion, 4.3
    RMS, 3.23
    sampling clock, characteristics, 5.32
    shaping, sigma-delta ADC, 3.2-3
    types, 3.23
    wideband, input filter, 3.25
Norton equivalent circuit, 4.29
NTSC video:

INDEX

differential gain/phase performance, 2.21
line drivers, 2.20
performance, 2.43
Nyquist bandwidth, 3.5, 5.22, 6.7, 8.16
Nyquist's Criterion, 3.6, 5.1-2
   definition, 5.1

O

Ohmtek (firm), 2.44
OP275, THD+N, 8.3
OPX91 family:
   input stage overvoltage:
      diagram, 7.7
      protection, 7.8
   output phase reversal, protection, clamp diodes, 7.9
Op amp:
   capacitive load:
      characteristics, 2.3
      internal compensation, and signal levels, 2.9
      noise gain, 2.3
   CMOS FET, rail-to-rail, output, 1.9
   complementary devices:
      common-emitter output, 2.25
      push-pull drive, 1.8
   current feedback:
      noise, calculation, 4.49
      open-loop series resistance, 2.5
      in video line driver, 2.22
         very high performance, 2.22-23
      in video PGA, 2.33
   current-feedback, 4.29-30
   current-limiting resistances, fabrication, 7.3
   drive, selection criteria, based on ADC
   dynamic performance, 4.26
   dual precision:
      single output IC:
         universal evaluation board, 9.11-13
            optimum values, 9.14
   dual supply, 7.5
   dynamic range:
      defined, 8.1
      harmonic distortion, 8.1
      THD, 8.1-2
      THD + N, 8.1-2
   high speed, 2.1-45
      application circuits, 2.30-43
         2:1 video multiplexer, 2.31
         3:1 video multiplexer, 2.32-33
         differential drivers, 2.34
            cross-coupled, 2.37-39
            inverter-follower, 2.35-37
         differential receivers, 2.39-43
            4 resistor, 2.40-41
            active feedback, 2.41-43
         video, 2.30
         video programmable gain amplifier, 2.33-34
      cable driving, 2.14-23
         line drivers, 2.20-23
         load-terminated, pulse response, 2.17
         lumped capacitance, pulse response, 2.19-20
         source and load-terminated, pulse response, 2.18
         source-terminated, pulse response, 2.19
      decoupling:
         harmonic distortion, 2.11-12
         pulse response, 2.13
      distortion, 8.1-7
      driving capacitive loads, 2.2-14
      PC board layout, 2.11
      RF rectification, 2.12
      single-ended driver and differential receiver, diagram, 2.14-23
      single-supply, 2.24-30
         direct coupling, 2.26-27
         line drivers, 2.27-30
      stray inverting input capacitance, 2.13
      stray parasitic capacitance, 2.12
   IC output, earliest, 1.8
   level shifting circuits, 4.27
   low distortion, 4.48
      output voltage range, 2.25
   low power/low distortion, 4.29-31
   noise:
      calculations, 4.49-50
      generalized model, 4.48

output signal:
   gain compression, 8.6
   IMD, 8.6
   intercept points, 8.6
output voltage:
   phase reversal, devices, 7.5
   range, 2.24-25
rail-to-rail, 7.5-6
   input, 1.7
   output, characteristics, 9.46
selection criteria, 4.26
signal gain, 2.4
single-supply, 7.5-6
   ac coupled:
      headroom, 2.29
      waveform duty cycle, 2.29
   direct coupling, considerations, 2.26-27
   input overvoltage, external protection, 7.9
   input voltage, 2.25
   key specifications, 4.25, 4.33
   low voltage specification, 2.25
   output phase reversal, external protection, 7.9
   video applications, 2.29-30
      rail-rail output, advantages, 2.30
third order IMD intercept, 8.6-7
voltage feedback:
   noise, 4.48
      calculation, 4.49
   in video line driver, 2.21
wide bandwidth:
   capacitive loading, 2.6-8
   transmission line interconnections, 2.14
OS-CON Aluminum Electrolytic Capacitor 93/94 Technical Book, 9.43
Ott, Henry W., 9.43, 9.84, 9.96
Ott, H.W., 9.95
Oversampling, sigma-delta ADC, 3.2-4, 3.6
Overvoltage:
   effects, 7.1-24, 7.5
   input stage, 7.2

P

PADS Software, Inc., 9.19
Parasitic capacitance, 9.91-92
Passive components:
   EMI effects, 9.58
   nonideal behavior, 9.58
PC board:
   acceptable signal routing, 4.57
   edge connections, 4.58
   ground planes, 4.52
   pin sockets, 2.11
   single-point ground, 4.52
   sockets, 4.58-59
   star ground, 4.52-53
Pease, Robert A., 9.3, 9.19, 9.97
Personal computers, FFT analysis using Mathcad, 8.12
PGA, 2.33-34
   video, 2.33-34
      diagram, 2.34
Power line:
   disturbances, transient, 9.68
   filter:
      commercial, 9.39-40
         diagram, 9.40
      common-mode chokes, 9.41
      EMI protection, 9.40, 9.72
      placement, 9.42
      schematic, 9.70
   noise modes, 9.41
Power supply:
   grounding and decoupling, 4.55-56
   switching-mode, noise, 4.59-60
Printed circuit board:
   coupling, 9.71
   crosstalk, 9.71
   design and EMI, summary, 9.75
   EMI protection, 9.71-75
      design problems, 9.71
   layout, for EMI protection, 9.72
   line termination, 9.74
   multi-layer design, for EMI protection, 9.73
   radiation, 9.71
   signal trace embedding, 9.73-74
   track impedance, equations, 9.74-75

Programmable gain amplifier, see: PGA
Programmable level pulse generator, 4.41

Q

Quantization error, 3.4-5
Quantization noise, 3.4-5
Quantization uncertainty, 3.4

R

Rabiner, Lawrence, 8.24
Radiofrequency area, IMD, 8.5-6
Radiofrequency interference, see: RFI
Ramierez, Robert W., 8.23
Rebold, T.A., 5.33-34
Reference:
    decoupling, 3.34
    low voltage, summary, 3.34
    noise, calculation, 3.33
Regulator/reference:
    boosted output, current limiting, 9.49-50
    output current, 9.50
Reidy, John, 8.23
Resistance, current-limiting, characteristics, 7.3-4
Resistor:
    noise, 4.48
        switching power supply, 9.26
Resonator, 3.11
RFI:
    coupling:
        power supplies, 9.63
        signal inputs, 9.63
        signal outputs, 9.63
    electric-field intensity, 9.62
    frequency:
        feed-through protection, 9.65
        filters, 9.65
    summary and protection techniques, 9.68
    system immunity:
        antenna behavior of cables, 9.67
        circuits, 9.62
    system sensitivity, 9.62
Rich, A., 9.84, 9.95
Rose, John F., 5.33
Ruscak, Stephen, 5.33

S

S/(N+D), from PC-based analysis of FFT, 8.12
Sample-and-hold amplifier, see: SHA
Sampling, simultaneous, 6.18-19
Sampling clock:
    circuitry, 5.32
    isolation, 5.32
    jitter, 4.7-8, 4.56
        ADC, 5.30-32
        effects on SNR and ENOB, 5.30
    noise, characteristics, 5.32
    oscillators, 5.31
Sampling rate:
    function of bandwidth, 5.1
    minimum, 5.5
        non-overlapping, 5.4-5
Schottky diodes, 3.30-31, 4.34, 4.39, 4.52, 7.6, 7.8-10, 9.74
Schreier, R., 5.34
A Selection Guide for Serial DACs, 6.40
Sensors:
    and cable shielding, 9.85-97
    passive, shield grounding, 9.91
    remote:
        high impedance:
            preamp, 9.92
            and shielding, 9.92
Settling time, and filter setting, 3.19
SFDR, from PC-based analysis of FFT, 8.12
SHA:
    8-channel, 6.21-23
    acquisition time, 6.9
    classic open-loop architecture, 5.16
    closed-loop architecture, low distortion and high speed, 5.17
    droop rate, 6.23, 6.34
    external:
        properties, 5.14
        to increase ADC SFDR and ENOB, 5.14-21
    multichannel, droop rate, 6.34
    in multiplexed data acquisition systems, 6.10
    performance, optimizing circuit timing, 5.17
Shannon's information theorem, 5.1-2, 5.6

Sheingold, Dan, 6.40, 8.23
Shield grounding, techniques, 9.94
Shielding:
- cables, electrical length, 9.81-82
- conductive enclosures, 9.76
- effectiveness, compromised by openings, 9.80
- experiments, 9.87-88
- hybrid, 9.93
- materials, impedance and skin depths, 9.79
- mechanisms, 9.77
- review, 9.76-82

Sigma-delta ADC, see: ADC, sigma-delta
Signal conditioning, low-side and high-side, 1.23-28
Signal generation:
- IMD measurement, 8.8-9
- oscillator, 8.8-9
- single tone, diagram, 8.9
- two tones, diagram, 8.8-9

Signal line, snubbing, for EMI protection, 9.72
Signal routing, in mixed signal systems, 4.57
Signal-to-noise plus distortion, see: S/(N+D)
Signal-to-noise ratio, see: SNR
Single-supply:
- ac coupling, 2.24
- implications, 2.24

Slattery, B., 9.84
Slattery, Bill, 9.96
Slot radiation, 9.59
Snelgrove, W. Martin, 3.43, 5.34
SNR:
- from PC-based analysis of FFT, 8.12
- from sampling clock jitter, 5.31

Sockets, 4.58-59
Spectrum analyzer:
- amplifier distortion:
  - measurement, 8.10-11
  - notch filter, 8.11
- high frequency low distortion measurements, 8.10-11

SPICE modeling, analog circuits, disadvantages, 9.1-2
Spurious free dynamic range, see: SFDR

Star ground system, 4.53
Steyskal, Hans, 5.33
Stout, D., 2.44
Switcher, see: Switching power supply
Switching power supply:
- analog ready, 9.22
- architecture, 9.24
- band filters, 9.35
- card entry filter, 9.35-38
  - attenuation, 9.37
  - decoupling, 9.38
  - diagram, 9.36
  - disadvantages, 9.36
  - response test setup, 9.37
  - SPICE modeling, 9.37
- characteristics, 9.21-22
- description, 9.22-23
- drawbacks, 9.21
- EMI considerations, 9.25
- filter:
  - design, 9.26
  - layout and construction guidelines, 9.38-39
  - noise, 9.21-45
  - power line, 9.39-40
- interference, 9.25
- linear post regulators, PSRR, 9.34
- noise, 4.59-60, 9.21-44
  - reduction and filtering, 9.21-44
  - reduction tools, 9.26
- peak amplitudes, 9.21
- series resistors, 9.34
- topology, 9.24
- unfiltered, output, 9.23

System Applications Guide (1993), 1.31, 3.43, 4.62, 5.33, 6.40, 7.24, 8.23-24, 9.84, 9.95

T

T-Tech, Inc., 9.19
Tantalum Electrolytic Capacitor SPICE Models, 9.44
Tantalum Electrolytic and Ceramic Capacitor Families, 9.43
THD, from PC-based analysis of FFT, 8.12
Thenevin equivalent circuit, 4.29
- output voltage, 4.40

Total harmonic distortion, see: THD
Transducer, circuitry, 3.30
Transformer:
    common-mode power line filters, 9.69
    isolation, Faraday shields, 9.70
Transmission lines:
    capacitance, 2.15
    skin effect, 2.15-16
Triboelectric effect, 7.12, 7.14
TrimDAC:
    8-bit, octal, 6.29-30
    advantages, 6.31
Tukey, J.W., 8.23
Type 5MC Metallized Polycarbonate Capacitor, 9.43
Type 5250 and 6000-101 Chokes, 9.43
Type EXC L Leadless Ferrite Bead, 9.43
Type EXCEL Leaded Ferrite Bead EMI Filter, 9.43
Type HFQ Aluminum Electrolytic Capacitor and Type V Stacked Polyester Film Capacitor, 9.43

U
Undersampling:
    applications, 5.1-34
        digital receivers, 5.7
    fundamentals, 5.1-13
UTP-5, differential driver, 2.34

V
Video line driver, 2.20-23
    supply bypassing, 2.23
Video line receiver, diagram, 2.40
Video multiplexer:
    2:1, 2.31
    3:1, 2.32
        switching characteristics, 2.33
    circuit diagram, 2.31-32
VLSI mixed-signal processing, integrated circuits, 6.10
Voltage:
    nodal analysis:
        two op amp in amp, 1.11-14
        three op amp in amp, 1.17-20
    regulation, 9.45

W
Wainwright Instruments GmbH, 9.19, 9.96
Wainwright Instruments, Inc., 9.19, 9.96
Webb, Richard C., 5.33
Weeks, Pat, 8.23
Whitney, Dave, 2.44
Williams, Jim, 9.97
Witte, Robert A., 8.23
Wong, James, 1.31
Wurcer, Scott, 2.44, 9.97
Wynne, J., 9.84
Wynne, John, 9.96

X
XFCB high speed devices, 2.1
XFCB single-supply op amp, 2.29

ANALOG DEVICES PARTS LIST

A
AD574-series, 6.5
AD589, 2.26, 3.33-34, 4.32, 9.45-46, 9.48, 9.52
AD620, 1.21-22, 9.90-91
AD626, 1.21, 1.23, 1.26-28
AD680, 4.28-29, 4.32
AD780, 3.31-34, 3.37-38, 4.32, 4.39-40, 9.14
AD797, 3.25, 8.4
AD810, 2.22-23, 2.30-31
AD811, 2.4-5, 2.22, 4.41-42, 9.37
AD812, 2.3, 2.22, 2.26-27, 2.35, 4.23-25, 4.28-29
AD813, 2.22, 2.27, 2.30, 2.32-34, 2.36
AD817, 2.8-10, 2.21, 2.27-28
AD818, 2.10, 2.21, 2.23, 2.27, 2.40
AD820, 1.4-5, 4.30, 7.5, 9.45-46, 9.48-49
AD822, 1.5, 1.15, 1.21-22, 4.32-33, 7.5, 9.45-46
AD823, 1.15
AD824, 1.5, 7.5
AD826, 2.9, 2.36
AD827, 2.9
AD828, 2.21, 2.36, 2.41
AD829, 2.2, 9.45-46
AD830, 2.41-43
AD845, 2.7
AD846, 2.1
AD847, 2.1, 2.9-10
AD876, 4.20-24, 4.28-29, 4.32-33
AD771X family, 3.12-13, 3.18-19, 3.23-25, 3.28-39, 4.31, 6.14, 6.16
AD785X series, 6.12-13
AD789X series, 6.13
AD1674, 6.6
AD1879, 3.10
AD7225, 6.26
AD7568, 6.26
AD7710, 3.10, 3.12, 3.15-16, 3.25-27, 3.30-34, 3.36-37, 6.14-15
AD7711, 3.10, 3.12, 3.15-16, 3.26-27, 3.31-32, 3.34, 3.36
AD7712, 3.10, 3.12, 3.15-16, 3.26-27, 3.31-32, 3.34, 3.36
AD7713, 3.10, 3.12-13, 3.15-16, 3.26-27, 3.31, 3.34, 3.36, 3.38-39
AD7714, 3.10, 3.12-13, 3.15-19, 3.25-28, 3.31, 3.33-36, 3.39-42, 9.14, 9.16
AD7715, 3.13, 3.15, 3.22
AD7716, 6.16-17
AD7853, 6.13
AD7854, 6.13
AD7858, 6.13
AD7859, 6.13
AD7890, 6.11-13
AD7891, 6.12-13
AD7892, 6.12-13
AD7893, 6.12-13
AD7896, 6.12-13
AD8001, 2.1, 2.6, 2.12-13, 2.16-19, 2.20, 2.23, 9.11-14
AD8002, 2.1, 2.37-39
AD8011, 4.29-31
AD8036, 2.1, 4.34-39, 4.41-42
AD8037, 2.1, 4.34-45
AD8041, 2.29-30
AD8300, 6.36-39
AD8522, 6.32-35, 6.39
AD8582, 6.32, 6.35-36
AD8600, 6.26-27, 6.29
AD8842, 6.29-30, 6.32
AD9002, 4.39-40, 5.19, 5.21
AD9022, 4.12-15, 4.47-50, 5.11, 8.18-19
AD9023, 4.12, 5.11
AD9026, 5.11, 5.19-20, 9.15-17
AD9027, 5.11, 5.19
AD9042, 5.11-13, 5.19
AD9050, 4.18-20, 4.29-30
AD9100, 5.15-19
AD9101, 5.15-17, 5.19-21
AD9104, 5.15, 5.17-19
AD9562, 9.8-9
AD9618, 5.18
AD9622, 8.6-7
AD9631, 2.1, 2.11-12, 4.12, 4.14-15
AD9632, 2.1, 4.12, 4.14, 4.48
AD9698, 9.15-17
AD9713, 9.15-17
ADG411, 6.4

INDEX

ADG412, 6.4
ADG413, 6.4
ADG508F, 6.4
ADG509F, 6.4
ADG511, 6.4
ADG512, 6.4
ADG513, 6.4
ADG528F, 6.4
ADSP-2103, 3.40
ADSP-2105, 3.40
AMP04, 1.21, 1.23-26

B
BUF04, 7.5

D
DAC8228, 6.22-23
DAC8300, 6.27-28
DAC8412, 6.26
DAC8413, 6.26
DAC8420, 6.26
DAC8800, 6.26

O
OP07 family, 1.2, 1.6
OP90, 1.5
OP97, 1.6
OP113, 7.5
OP176, 7.5
OP183, 7.5
OP191, 7.5, 7.5-6, 9.45-46
OP193, 7.5, 9.47
OP213, 1.21, 3.25, 7.5
OP275, 8.3
OP279, 1.7, 7.5, 9.45-47, 9.50-51
OP282, 1.4-5, 7.5
OP283, 7.5
OP284, 1.7, 1.9, 1.21, 9.45-46, 9.51-52
OP285, 7.5
OP291, 1.21, 7.5-6, 9.45-46
OP292, 7.5
OP293, 7.5, 9.47
OP295, 7.5, 9.45-46
OP393, 7.5
OP413, 7.5
OP467, 7.5
OP482, 1.4-5, 7.5
OP484, 1.7, 1.9

OP491, 7.5-6, 9.45-46
OP492, 7.5
OP493, 9.47
OP495, 7.5, 9.45-46
OPX13 family, 1.2
OPX91 family, 1.7, 7.5-7
OPX93, 1.4-5

R
REF02, 6.22
REF43, 3.34, 4.32
REF19X family, 9.47-50
REF191, 4.32, 9.47
REF192, 3.33-34, 4.32, 6.27, 9.47, 9.49, 9.51-52
REF193, 4.32, 9.47, 9.49
REF194, 4.32, 9.47, 9.49
REF195, 1.25, 9.47-49
REF196, 4.32, 9.47, 9.49-50
REF198, 4.32-33, 9.47

S
SMP08, 6.21-23